El-Ghazali Talbi   Pierre Liardet
Pierre Collet   Evelyne Lutton
Marc Schoenauer (Eds.)

Artificial Evolution

7th International Conference, Evolution Artificielle, EA 2005
Lille, France, October 26-28, 2005
Revised Selected Papers

 Springer

Volume Editors

El-Ghazali Talbi
Université des Sciences et Technologies de Lille
Laboratoire d'Informatique Fondamentale de Lille (LIFL)
UMR CNRS 8022, Bâtiment M3, 59655 Villeneuve d'Ascq Cedex, France
E-mail: talbi@lifl.fr

Pierre Liardet
Université de Provence
Centre de Mathématiques et Informtique, Laboratoire LATP
UMR-CNRS 6632, 39 rue Frédéric Joliot-Curie, 13453 Marseille cedex 13, France
E-mail: liardet@cmi.univ-mrs.fr

Pierre Collet
Université du Littoral Côte d'Opale
Laboratoire d'Informatique du Littoral
BP 719, 62228 Calais cedex, France
E-mail: Pierre.Collet@univ-littoral.fr

Evelyne Lutton
INRIA Rocquencourt
Domaine de Voluceau, BP 105, 78153 Le Chesnay cedex, France
E-mail: Evelyne.Lutton@inria.fr

Marc Schoenauer
Université Paris Sud
Equipe TAO, INRIA Futurs, LRI
Bâtiment 490, 91405 Orsay cedex, France
E-mail: Marc.Schoenauer@inria.fr

Library of Congress Control Number: 2006923564

CR Subject Classification (1998): F.1, F.2.2, I.2.6, I.5.1, G.1.6, J.3

LNCS Sublibrary: SL 1 – Theoretical Computer Science and General Issues

ISSN   0302-9743
ISBN-10   3-540-33589-7 Springer Berlin Heidelberg New York
ISBN-13   978-3-540-33589-4 Springer Berlin Heidelberg New York

Springer is a part of Springer Science+Business Media

springer.com

© Springer-Verlag Berlin Heidelberg 2006
Printed in Germany

Typesetting: Camera-ready by author, data conversion by Scientific Publishing Services, Chennai, India
Printed on acid-free paper   SPIN: 11740698   06/3142   5 4 3 2 1 0

Preface

This book is based on the best papers presented at the 7th Conference on Artificial Evolution, EA 2005, held in Lille (France). Previous EA meetings took place in Marseille (2003), Le Creusot (2001), Dunkerque (1999), Nimes (1997), Brest (1995), and Toulouse (1994).

There were 78 submitted papers, of which 27 were selected for presentation. They cover all aspects of artificial evolution: genetic programming, machine learning, combinatorial optimization, co-evolution, self-assembling, artificial life and bioinformatics.

In addition, the program included an invited talk by David Corne on "Evolutionary Computation in Bioinformatics: How to Save Lives and Make Scientific Breakthrough."

Thanks to the Organizing Committee and the Steering Committee for their hard work.

All the submissions were reviewed by at least three members of the Program Committee. I am very grateful to the members for their conscientious work.

We take this opportunity to thank the different partners whose financial and material support contributed to the success of the conference: Polytech'Lille, Université des Sciences et Technologies de Lille, INRIA, AFIA, CNRS, Région Nord-Pas-De-Calais, ROADEF, and EA association.

Finally, I wish to thank all the authors who submitted papers, and the authors of accepted papers for sending their final versions on time.

January 2006   El-Ghazali Talbi

Organization

Steering Committee

El-Ghazali Talbi   Université de Lille 1, France
Pierre Collet   Université du Littoral, France
Cyril Fonlupt   Université du Littoral, France
Evelyne Lutton   INRIA Rocquencourt, France
Marc Schoenauer   INRIA Futurs, France

Organizing Committee

El-Ghazali Talbi   Université des Sciences et Technologies de Lille, France
Clarisse Dhaenens   Université des Sciences et Technologies de Lille, France
Laetitia Jourdan   Université des Sciences et Technologies de Lille, France
Nouredine Melab   Université des Sciences et Technologies de Lille, France
Franck Seynhaeve   Université des Sciences et Technologies de Lille, France

Referees

E. Alba
A. Auger
P. Bessière
C. Blum
J. Branke
B. Braunschweig
E.K. Burke
A. Caminada
E. Cantu-Paz
M. Capcarrère
U.K. Chakraborty
A. Channon
C.C. Coello
P. Collard
P. Collet
K. Deb
D. Delahaye
C. Dhaenens
M. Dorigo
N. Durand
M. Ebner
D. Fogel

C. Fonlupt
J. Gottlieb
M. Grana
S. Gustafson
N. Hansen
J-K. Hao
W. Hart
A. Jaszkiewicz
L. Jourdan
M. Keijzer
J. Knowles
N. Krasnogor
W. Langdon
Y. Landrin-Schweitzer
C. Lattaud
A. Leier
P. Liardet
E. Lutton
V. Masero
N. Melab
J.J. Merelo
Z. Michalewicz

N. Monmarché
M. Pelikan
P. Preux
G. Raidl
J-P. Rennard
C.C. Ribeiro
D. Robilliard
G. Rudolph
M. Schoenauer
M. Sebag
F. Seynhaeve
P. Siarry
T. Soule
T. Stuetzle
E.-G. Talbi
J. Timmis
S. Tsutsui
G. Venturini
S. Yang
E. Zitzler
A. Zomaya

Sponsoring Institutions

Polytech'Lille (Ecole d'Ingénieurs)
USTL (Université des Sciences et Technologies de Lille) INRIA Futurs (Institut
National de Recherche en Informatique et Automatique)
AFIA (Association Française pour l'Intelligence Artificielle)
CNRS (Centre National de La Recherche Scientifique)
Région Nord Pas-de-Calais
ROADEF (Société française de recherche opérationnelle et d'aide à la décision)

Table of Contents

Genetic Programming

Machine Learning

Applications

Combinatorial Optimization

Co-evolution

Self-assembling

Artificial Life and Bioinformatics

Advances

Santa Fe Trail Hazards

Denis Robilliard, Sébastien Mahler, Dominique Verhaghe, and Cyril Fonlupt

Laboratoire d'Informatique du Littoral,
Université du Littoral-Côte d'Opale,
BP719, Calais Cedex 62228, France

Abstract. This paper focuses on methodological problems associated to the famous Santa Fe Trail (SFT) problem, a very common benchmark for evaluating Genetic Programming (GP) algorithms, introduced by Koza in its first book on GP. We put in evidence the difficulty to ensure fair comparisons especially with new genotype representations as found in works on grammar-based automatic programming, such as Grammatical Evolution, and Bayesian Automatic Programming. We extend a work by Langdon et al. by measuring the effort to solve SFT by random search with different time steps limits and a reduced but semantically equivalent function set.

1   Introduction

The Santa Fe Trail (SFT) problem was inspired by Jefferson *et al.* Genesys–Tracker system [1] and was first formally described and used as a Genetic Programming (GP) benchmark by Koza in its seminal book [2] and also in [3]. This problem can be briefly stated as finding a command program for a robotic ant such that the ant retrieves a maximum number of food pellets forming a trail with gaps and turns on a toroidal grid. This problem has become quite popular as a benchmark in the GP field and is still repeatedly used, despite (or perhaps because) it has been shown by Langdon and Poli that GP does not improve much on pure random search [4, 5].

We focus on several recent works exploring two grammar-based automatic programming paradigms, Grammatical Evolution (GE) and Bayesian Automatic Programming (BAP). In recent papers introducing these techniques, their efficiency against GP was measured on the SFT benchmark among other tests. We show that the setup of these SFT experiments includes small changes in the benchmark definition, having great consequences in solving the problem, up to the point that comparisons with GP are called into question. However some legitimate changes should be acknowledged when using the Santa Fe Trail benchmark with new automatic programming paradigm. We also compute the effort to solve SFT with various time steps limits and a reduced function set.

The rest of the paper is organized as follows: after recalling the initial definition of the SFT in Section 2, we briefly present GE, BAP and their implementations of the SFT in Section 3. In Section 4 we emphasize the differences in experimental conditions introduced by these works, and how it relates to the difficulty of the problem. Conclusion are drawn in Section 5.

E. Talbi et al. (Eds.): EA 2005, LNCS 3871, pp. 1–12, 2006.

2   Santa Fe Trail and Genetic Programming

2.1   Presentation of the Problem

In this section we recall the basics of the SFT benchmark as it appeared in [2, 3]. It consists in generating a control program for a virtual ant robot to find the maximum number of food pellets along an irregular trail on a toroidal grid. The ant has only a limited perception of its environment. This is modeled by a binary if-food-ahead conditional operator, that executes its first argument only if the ant senses food on the neighboring cell in the direction it is facing, or else executes its second argument. The ant can also move in its environment by doing a 90 degrees rotation to the left or to the right, or by moving one grid cell forward. Each move operation and each turn operation consumes one time step. During the simulation of the ant foraging behavior, this control program is iterated until a fixed number of time steps is exhausted. In a few words, the less the number of time steps, the more "clever" the program should be to retrieve the maximum amount of food.

To solve this problem with GP, Koza used his well-known Lisp-like parse tree representation for programs, incorporating the set of functions:

- In [2]: { if-food-ahead, progn2, progn3 }
- In [3]: { if-food-ahead, progn }

where progn is the sequence operator that simply executes its arguments in order, from left sibling to right sibling, and progn2, progn3 do the same with a limitation of respectively 2 and 3 siblings.

The set of terminals, to implement the basic motion of the ant, was:

{ left, right, move }

Koza said he arbitrarily fixed to 400 the amount of available time steps. Mutation was not used, fitness proportionate copy and crossover were used with respective ratio of 10% and 90%. Initial solutions were limited to depth 4, and crossover was also limited to produce at most depth 15 individuals. Different population sizes were tried such as 500, 1000, 2000 and 4000, and these were refined for 50 generations (not counting generation 0). We will call this setting SFT. Notice that Langdon and Poli in [4] set the maximum time limit to 600 steps, assuming a possible mistype in the original Koza's paper.

3   Two Context Free Grammars GP Variants

In this section we briefly present two variants of GP, Grammatical Evolution and Bayesian Automatic Programming. A detailed description of these two techniques is out of the scope of this paper, and we only sum-up their basic principles, in order to introduce their genotype representation based on integer codons string and their mapping process for translation of genotype codons string to phenotype programs through derivation rules in a Backus Naur Form (BNF) grammar.

3.1   Grammatical Evolution Paradigm

One GP variant that has been tested many times on the SFT benchmark is the Grammatical Evolution (GE) system from O'Neil and Ryan [6, 7]. Several studies have been published dealing with this paradigm, notably [8, 9, 10], and the SFT experiments with GP comparisons appear in [6, 11, 7, 12].

In GE, a genotype is a string of integers (called codons), and before an individual can be evaluated it must be translated into a program. Codons are parsed from left to right and each one is used to make a choice between available derivation rules in a BNF grammar, beginning with the grammar start symbol. The choice of the derivation for a given symbol is done by taking the integer codon modulo the number of available rules (obviously, this introduces some bias depending whether the codon range is divisible by the number of rules and this has been addressed in [8]). The process is continued until every grammar variable has been derived in terminal symbols. The resulting string of terminal symbols is the phenotype program, and it is evaluated in the usual GP way. If there are unused codons when derivation is complete, these are simply ignored, and in case there are not enough codons, the translation process wraps over to the beginning of the genotype. If the derivation is not complete after 10 wrapping operations, then the individual is considered invalid and gets a very low fitness.

Table 1. An overview of the GE decoding scheme

The SFT-GE grammar (the start symbol is *code*, alternative derivations are numbered in the right column) :	

code	::= *line*	(0)
	| *code line*	(1)
line	::= *if-statement*	(0)
	| *op*	(1)
if-statement	::= **if-food-ahead** { *line* } **else** { *line* }	(0)
op	::= **left**	(0)
	| **right**	(1)
	| **move**	(2)

Suppose we have the following genotype to decode :

| 101 | 40 | 50 | 93 | 91 | 36 | 1 | 246 | 17 | 49 | 12 | 104 |

Beginning with the start symbol *code*, two alternative derivations, denoted (0) and (1), are available. The first codon of the chromosome is 101, and as 101 mod 2 = 1, rule (1) is selected and thus *code* is transformed into *code line*. It is easy to verify that the chromosome is further translated to:

...	*code line*
40 mod 2 = 0	*line line*
50 mod 2 = 0	*if-statement line*
93 mod 1 = 0	**if-food-ahead** { *line* } **else** { *line* } *line*
91 mod 2 = 1	**if-food-ahead** { *op* } **else** { *line* } *line*
36 mod 3 = 0	**if-food-ahead** { **left** } **else** { *line* } *line*
1 mod 2 = 1	**if-food-ahead** { **left** } **else** { *op* } *line*
246 mod 3 = 0	**if-food-ahead** { **left** } **else** { **left** } *line*
17 mod 2 = 1	**if-food-ahead** { **left** } **else** { **left** } *op*
49 mod 3 = 1	**if-food-ahead** { **left** } **else** { **left** } **right**

This wrapping mechanism can be avoided, notably with a refined strategy for initializing the population as suggested in [12].

Table 1 gives the grammar used in [12] to implement the SFT, which will be denoted SFT-GE. In this table, we also give an example derivation from genotype to phenotype.

3.2 Bayesian Automatic Programming Paradigm

Bayesian Automatic Programming (BAP) has been recently introduced in [13], where it is evaluated on a regression problem and on the Santa Fe Trail. BAP proposes to tackle the task of automatic programming via the use of an Estimation of Distribution Algorithm (EDA), namely a Bayesian network. Working with integer codons strings, BNF grammars and derivation trees in the same way as Grammatical Evolution (see details in section 3.1), BAP trains a Bayesian network to learn the statistical correlations between codons in promising solutions. A Conditional Probability Table (CPT) is built using the standard K2 algorithm from a fitness-biased selected subset of the population, and then this CPT table is used to generate a new population of candidate solutions, this whole generation process being iterated as in GP. Notice that this is not the first attempt at using some sort of EDA for automatic programming, see also [14] for example.

So BAP uses the same representation as GE and needs a grammar to implement the SFT. This grammar is given in Table 2, and will be denoted SFT-BAP. The BAP paper also refers to a GP function set for SFT, which is reported as: {`if-food-ahead`, `left`, `right`, `move`} without detailing the sequence operator.

Table 2. The SFT-BAP grammar used in the BAP versus GP experiment

The SFT-BAP grammar (start symbol is *expr*) :	
expr ::= *line*	(0)
| *expr line*	(1)
line ::= `if-food-ahead` { *expr* } `else` { *expr* }	(0)
| *op*	(1)
op ::= `left`	(0)
| `right`	(1)
| `move`	(2)

4   Analysis and Discussion of SFT Variants

4.1   Representation Bias Versus Program Semantic

As seen above there is a small disagreement between the two 1992 Koza's publications, whether one should use `progn` or rather `progn2` and `progn3`. We argue that this difference in representation does not bear on the ant control possibilities:

- it could affect the chances to evolve successful solutions, as the space of program trees explored is not the same: this is a representational bias;
- it does not affect the ant control, since any `progn` subtree can be exactly translated using `progn2` and `progn3` operators and reciprocally, preserving the semantic of the program. More formally, if we call respectively A and B the search spaces associated to these two versions of the benchmark, we can find a surjective mapping preserving program semantic both from A to B and from B to A.

Thus we can say that these two versions of the Santa Fe trail have different biases in searching two spaces of program trees with equivalent semantic.

4.2   Expressiveness of SFT-GE and SFT-BAP Grammars

A close observation of Table 1 shows that the expressiveness of the SFT-GE grammar differs from what can be achieved by combining the terminal and function set from SFT. It is not possible to have a sequence of several instructions embedded in an `if-food-ahead` statement like for example:

```
if-food-ahead { move right } else { right move move}
```

Thus there is no surjective mapping preserving semantic from the SFT-GE to the SFT search spaces: if all SFT-GE programs have a translation in the SFT framework, no all SFT programs have a translation in the SFT-GE grammar.

On the opposite, the SFT-BAP grammar allows sequences of instructions everywhere an *op* instruction can appear, notably inside an `if-food-ahead` statement. Any SFT program can be translated into an semantically equivalent SFT-BAP derivation, and the converse is also true. We conclude that the SFT-BAP and SFT search spaces are semantically equivalent, while the SFT-GE grammar defines a related but different benchmark.

4.3   Assessing the Difficulty of SFT-GE Benchmark

Some clues about the importance of changes in the SFT-GE search space can be obtained by comparing how GE solves the problem with the SFT-GE and SFT-BAP grammars. This was done with the maximum allowed time steps limits to retrieve the food ranging from 400 to 700 by increments of 50 time steps. We used the cumulative success frequency (or cumulative probabiliy of success), introduced in [2], computed on 100 runs with other parameters identical to [7]. The precison of this measure has been questionned notably by [15], but it will be enough here to indicate the general trend. Results are displayed on Figure 1.

Notice that when using grammar SFT-GE no perfect solution to this problem have been found up to and including 600 time steps in our experiments, and almost all GE publications do indeed refer to a maximum allowed of 615 time steps that differs from both Koza or Langdon's settings (the GE limit is reported as 600 time steps only in [7], a possible mistype).

To assess more precisely the difficulty of this new SFT-GE benchmark, we tried to solve it by generating random strings of codons:

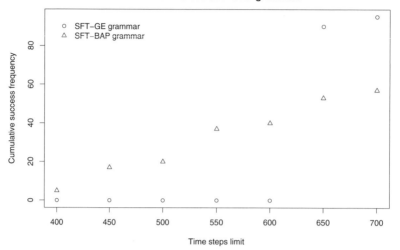

Fig. 1. Comparison of the GE system performance using 2 grammars SFT-GE and SFT-BAP with several time steps limits to retrieve the food

- as we are not supposed to know in advance the size of a typical perfect solution, we generate strings in a range of length from 3 to 102 codons;
- for every size in the interval, we generate 250 random solutions;
- we test if we obtain a perfect solution with, and without the GE wrapping mechanism, using 615 time steps;

We also try a small variation of SFT-GE: in the second possible derivation of the first rule, we swap the *line* and *code* variables. This first rule then becomes: $code ::= line \mid line\ code$, and we call SFT-GEBIS this new grammar. This will allow us to test the sensitivity of grammar-based systems.

Within these settings, we can compute a kind of cumulated success frequency (CSF), as if we were doing a GP run of 100 pseudo-generations: a pseudo-generation is a random draw of 250 individuals with a given solution size, and the next pseudo-generation increases the size by 1 codon. If a perfect solution is found, the run is stopped and a success is counted for that pseudo-generation size. The whole run is repeated 100 times to obtain the usual CSF plot in Fig. 2.

There are two main observations on this plot:

- the problem is rather easy, since a random search can have a CSF of about 29% when using the same maximum number of evaluations as the GE experiment;
- when wrapping is on, the small change between these two grammars imply a noticeable change in the difficulty.

We compute the so-called effort defined in Koza [2]: let p be the probability of finding a solution, the number of evaluations needed to ensure we find a solution,

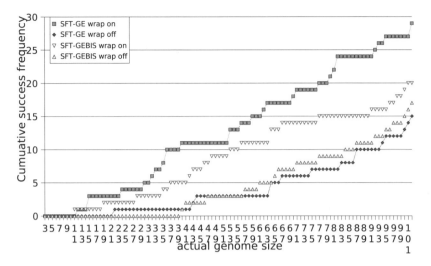

Fig. 2. Randomly solving the benchmark defined by grammars SFT-GE and SFT-GEBIS

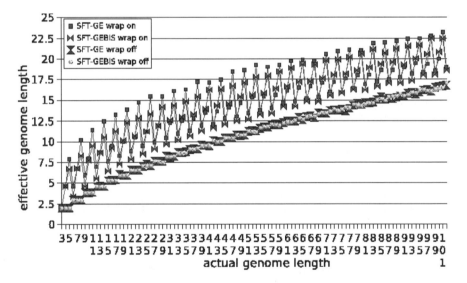

Fig. 3. Effective length versus actual length for SFT-GE and SFT-GEBIS random solutions

with probability at least $1 - \epsilon$, is: Effort $= -\log(\epsilon)/p$. We have drawn $2,500,000$ independent solutions with sizes from 3 to 102, and obtained 37 perfect solutions with wrapping. This gives an effort of around $310,000$ for our random search to solve SFT-GE with a 99% confidence.

A possible hint for the gap in performance between SFT-GE and SFT-GEBIS with wrapping may be given by the difference in average effective length of solutions (remember all codons are not always used). As it is shown in Figure 3, random SFT-GE solutions with wrapping have a slightly greater average length, and this certainly allows to explore a more promising part of the search space. The mapping process in combination with wrapping is also at the origin of the quite astonishing "stair" effect that can be observed on Figure 3: the bias introduced in the sampling of programs is rather complex and hard to foresee using the grammar alone.

4.4   Semantic Equivalence Is Not Enough

Even when the competing heuristics work on semantically equivalent search spaces, as is the case between SFT-BAP and SFT, a matter of concern raises from the different probabilities of randomly sampling a given tree. This impinges the run at least during the random initialization phase, between grammar-based systems using the SFT-BAP grammar and standard GP "grow" or "ramped half & half" mechanisms, but also during the evolution which is a stochastic process:

- through random mutation;
- through crossover in GP and GE, since crossover points are chosen at random; moreover in the GE scheme exchanged codons are often interpreted differently in their new context, and it can also be seen as a kind of random sampling;
- through the BAP generation phase, where the new population is drawn randomly with the biases obtained from the Bayesian network;

Sampling biases are of course a well-known cause for different search performances, and such discrepancies are certainly very hard to avoid in many cases, but above all it can fairly be argued that these are indeed desirable, because new paradigms are designed precisely to introduce new biases in the exploration process. Then a refined definition of the SFT could state that:

"The search space of programs must be semantically equivalent to the set of programs possible within the original SFT definition".

This would be enough to discard the SFT-GE grammar and keep the SFT-BAP one.

But then, if a new paradigm such as SFT-BAP is allowed to change the exploration bias, presumably to its best, one must also seek to give GP a good bias if both are to be compared. As can be seen on Figures 4, 5 and 6, simply dropping the progn3 statement from the function set gives a much more favourable bias to random search and to GP, while the search space remains semantically equivalent to the original SFT. From Figure 6, and from the plots and the reference to Koza appearing in [13], we can suppose that both progn2 and progn3 were used in their BAP versus GP experiment, in which case GP performance was not measured at its best.

One can also notice that the ramped half and half random search effort on Figure 4 incurs a sudden and dramatic drop when the time steps limit is

Effort to solve SFT with ramped–half&half random search

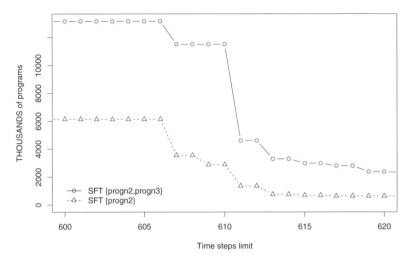

Fig. 4. Effort needed to solve with 99% confidence the SFT using Ramped Half and Half random search, program depth in the range 2 to 6, time steps limits ranging from 600 to 620 and **progn2** versus **progn2 + progn3** operators

Effort to solve SFT with Uniform random search

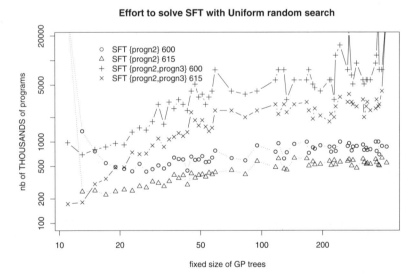

Fig. 5. Effort needed to solve with 99% confidence the SFT using uniform random search, time steps limits 600 and 615, and **progn2** versus **progn2 + progn3** operators

pushed from 600 towards 615 steps. The 15,000,000 programs already reported by Langdon *et al.* in [5] drops to around 2,400,000 with **progn3**, while it moves from 6,000,000 to 620,000 with **progn2** alone. This seems to be a sensitive

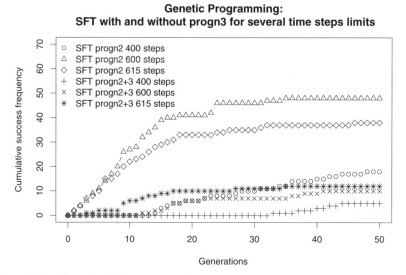

Fig. 6. GP solves the SFT easier without `progn3` for several time steps limits (other parameters are taken from [5])

parameter in this range of values, even if there is no significant difference in the way GP solves the problem as illustrated on Figure 6.

5   Conclusion

This paper shows how references to an *a priori* standard benchmark, the Santa Fe Trail, can indeed cover different settings associated to various difficulty levels, up to the point where comparing results is impossible. Let us say that this problem is more likely to be encountered when the new paradigm to be tested does not allow a straight re-use of fundamental concepts such as genotype and phenotype representations, as it was the case in grammar-based systems due to the translation phase via the derivation rules.

In the meantime, new paradigms also question whether some previously defined parts of a benchmark should be considered as open to legitimate changes, somehow contributing to define the search space associated to the problem. We think it is the case when tackling the SFT: from a problem-oriented point of view, it is enough to preserve the semantic of programs, whatever the biases introduced by the representation. In turn, when comparing search methods with different biases, it is of course necessary to push all competing heuristics to their best and this may imply at least to drop the `progn3` statement from the GP function set.

Re-visiting some recent works, we have shown that the SFT-GE grammar define a search space that is different from the SFT one, contrary to the SFT-BAP grammar which is a fair equivalent as far as the semantic of programs is concerned. We have pointed at the sensitivity of the grammar choice when wrapping is allowed for GE. The work from Langdon et al. has also been extended by computing the effort with the sequence operator `progn2` alone, and showing a sudden drop in the effort needed to solve SFT with ramped half and half random search when the time limit moves from 600 to 615 steps.

References

[1] David Jefferson, Robert Collins, Claus Cooper, Michael Dyer, Margot Flowers, Richerd Korf, Charles Taylor, and Alan Wang. Evolution as a theme in artificial life: The genesys/tracker system. In Langton, Taylor, Farmer, and Rasmussen, editors, *Proceedings of Artificial Life 2*, volume 10 of *Proceedings volume in the Santa Fe Institute studies in the sciences of complexity*, pages 549–578. Addison Wesley, 1992.

[2] John Koza. *Genetic Programming: On the Programming of Computers by Means of Natural Selection*. The MIT Press, 1992.

[3] John Koza. Genetic evolution and co-evolution of computer programs. In Langton, Taylor, Farmer, and Rasmussen, editors, *Proceedings of Artificial Life 2*, volume 10 of *Proceedings volume in the Santa Fe Institute studies in the sciences of complexity*, pages 603–629. Addison Wesley, 1992.

[4] W. B. Langdon. Better trained ants. In Riccardo Poli, W. B. Langdon, Marc Schoenauer, Terry Fogarty, and Wolfgang Banzhaf, editors, *Late Breaking Papers at EuroGP'98: the First European Workshop on Genetic Programming*, pages 11–13, Paris, France, 14-15 April 1998. CSRP-98-10, The University of Birmingham, UK.

[5] W. B. Langdon and Riccardo Poli. *Foundations of Genetic Programming*. Springer-Verlag, 2002.

[6] Michael O'Neill and Conor Ryan. Evolving multi-line compilable C programs. In Riccardo Poli, Peter Nordin, William B. Langdon, and Terence C. Fogarty, editors, *Genetic Programming, Proceedings of EuroGP'99*, volume 1598 of *LNCS*, pages 83–92, Goteborg, Sweden, 26-27 May 1999. Springer-Verlag.

[7] Michael O'Neill and Conor Ryan. Grammatical evolution. *IEEE Transactions on Evolutionary Computation*, 5(4):349–358, 2001.

[8] Maarten Keijzer, Michael O'Neill, Conor Ryan, and Mike Cattolico. Grammatical evolution rules: The mod and the bucket rule. In James A. Foster, Evelyne Lutton, Julian Miller, Conor Ryan, and Andrea G. B. Tettamanzi, editors, *Genetic Programming, Proceedings of the 5th European Conference, EuroGP 2002*, volume 2278 of *LNCS*, pages 123–130, Kinsale, Ireland, 3-5 April 2002. Springer-Verlag.

[9] Conor Ryan, Miguel Nicolau, and Michael O'Neill. Genetic algorithms using grammatical evolution. In James A. Foster, Evelyne Lutton, Julian Miller, Conor Ryan, and Andrea G. B. Tettamanzi, editors, *Genetic Programming, Proceedings of the 5th European Conference, EuroGP 2002*, volume 2278 of *LNCS*, pages 278–287, Kinsale, Ireland, 3-5 April 2002. Springer-Verlag.

[10] Michael O'Neill and Conor Ryan. Grammatical evolution by grammatical evolution: The evolution of grammar and genetic code. In Maarten Keijzer, Una-May O'Reilly, Simon M. Lucas, Ernesto Costa, and Terence Soule, editors, *Genetic Programming 7th European Conference, EuroGP 2004, Proceedings*, volume 3003 of *LNCS*, pages 138–149, Coimbra, Portugal, 5-7 April 2004. Springer-Verlag.

[11] Michael O'Neill and Conor Ryan. Automatic generation of high level functions using evolutionary algorithms. In Conor Ryan and Jim Buckley, editors, *Proceedings of the 1st International Workshop on Soft Computing Applied to Software Enineering*, pages 21–29, University of Limerick, Ireland, 12-14 April 1999. Limerick University Press.

[12] Michael O'Neill and Conor Ryan. *Grammatical Evolution: Evolutionary Automatic Programming in a Arbitrary Language*, volume 4 of *Genetic programming*. Kluwer Academic Publishers, 2003.

[13] Evandro Nunes Regolin and Aurora Trinidad Ramirez Pozo. Bayesian automatic programming. In Collet and Tomassini, editors, *Proceedings of EuroGP'05*, volume 3000 of *LNCS*, pages 39–48, Lausanne, 2005. Springer.

[14] R. P. Salustowicz and J. Schmidhuber. Probabilistic incremental program evolution. *Evolutionary Computation*, 5(2):123–141, 1997.

[15] Sean Luke and Liviu Panait. Is the perfect the enemy of the good? In W. B. Langdon, E. Cantú-Paz, K. Mathias, R. Roy, D. Davis, R. Poli, K. Balakrishnan, V. Honavar, G. Rudolph, J. Wegener, L. Bull, M. A. Potter, A. C. Schultz, J. F. Miller, E. Burke, and N. Jonoska, editors, *GECCO 2002: Proceedings of the Genetic and Evolutionary Computation Conference*, pages 820–828, New York, 9-13 July 2002. Morgan Kaufmann Publishers.

Size Control with Maximum Homologous Crossover

Michael Defoin Platel, Manuel Clergue, and Philippe Collard

Laboratoire I3S, CNRS-Université de Nice Sophia Antipolis

Abstract. Most of the Evolutionary Algorithms handling variable-sized structures, like Genetic Programming, tend to produce too long solutions and the recombination operator used is often considered to be partly responsible of this phenomenon, called *bloat*. The Maximum Homologous Crossover (MHC) preserves similar structures from parents by aligning them according to their homology. This operator has already demonstrated interesting abilities in bloat reduction but also some weaknesses in the exploration of the size of programs during evolution. In this paper, we show that MHC do not induce any specific biases in the distribution of sizes, allowing size control during evolution. Two different methods for size control based on MHC are presented and tested on a symbolic regression problem. Results show that an accurate control of the size is possible while improving performances of MHC.

1 Introduction

One of the major research areas in Genetic Programming (GP) is the management of the size of programs. Indeed, the "natural" trend of GP systems is to quickly increase the size of individuals until they reach the maximum allowed size, a phenomenon commonly known as *bloat*.

1.1 Bloat

This uncontrolled growth of programs is one of the weaknesses of GP as a problem-solver: resources needed by the system to address a problem are not adapted to the difficulty, the system consumes all the resources provided, leading generally to a waste of computing time and memory. Moreover, this behavior may dramatically influence the efficiency of the system in terms of solution quality and it works against the assumption [14] that between two equally fit programs, we should retain the smaller, which is often more robust and more evolvable.

Many authors have proposed explanations for bloat. To name a few, Altenberg [1] notes that bloat arises during evolution as the population attempts to protect useful subtrees from the crossover effects. This is the *protection hypothesis*. On the other hand, in [8], Langdon and Poli argue that fitness causes bloat. The idea is that the search starts from short genotypes with a given fitness. Then after a while, since the chance of finding better solutions is low, the

E. Talbi et al. (Eds.): EA 2005, LNCS 3871, pp. 13–24, 2006.

process becomes neutral and only equally fit solutions can be retained. But the search space contains many more long genotypes than short ones with the same fitness. This is the *drift hypothesis*. We note that recent work on Exact Schemata Theorems [14] tends to confirm this hypothesis, while giving a theoretical explanation for bloat. In [16], authors give another explanation for bloat by pointing out the asymmetric effects on the fitness of subtrees deletions and insertions. Indeed, they show that when a subtree is removed, the effects on the fitness depend on its size (strong effects for large subtrees) but not in case of a subtree insertion. This is the *removal bias hypothesis*. Another important aspect of the bloat problem is the presence in programs of inviable code, called the introns. Most of bloat theories suggest that the phenomenon is due to the propagation of introns. However, some interesting work [9][10] tends to contradict the *introns hypothesis*.

Various methods have been investigated to solve the size problem. Maybe the widespread idea to control the size, is to modify the fitness of programs and so the selection process. For examples, we can quote : the variable fitness [17], the parsimony pressure [16], the multi-objective evaluation [5] and the Tarpeian method [12]. Another way to tackle the size control problem is based on specific genetic operators, in particular more homologous crossover operators, [13] and [4].

1.2   Maximum Homologous Crossover

The Maximum Homologous Crossover (MHC) [7] is a recombination mechanism mimicking natural crossover that maximally preserves homology between parents. The MHC ensures that the genetic material exchanged during crossover is chosen, according to an *edit distance*[1] , in the most dissimilar regions of parents, and so leaves unchanged their nearly identical parts, *ie* the homologous regions. Thus, offspring can not be very different from their parents.

MHC was originally designed for Linear GP (LGP), where programs are sequences of instructions of an imperative language (C, machine code, ...). Our study is based on a stack-based GP implementation [11] and [3], where a sequence of instruction is evaluated using an operand stack. Figure 1 gives an example of MHC recombination between two programs P_x and P_y in stack-based representation. We see that during Step 1, an alignment $(\overline{P}_x, \overline{P}_y)$ of the two parents is computed, see [7] for details, to identify homologous regions. We note that aligned programs may contain some gaps (ε) and that they always have the same size. Thus, a crossover site can be chosen in $(\overline{P}_x, \overline{P}_y)$, here at position 5, and the classical 1-point crossover used in GA can be used, see Step 2. Finally, in Step 3, the inserted gaps are removed, producing offspring P'_x and P'_y. In a previous study [6], authors have shown, on the Even-N Parity Problem, that MHC is a less destructive operator than the Standard Crossover (SC) used in LGP[2]. Moreover the performances of the two crossover operators were very

---

[1] The *edit distance* corresponds to the minimal number of elementary operations (deletion, insertion or substitution) required to change one program into the other.

[2] In LGP, the SC operator randomly exchanges prefixes (or suffixes) between linear sequences.

P_x	P_y		$(\overline{P_x}, \overline{P_y})$			$Xo(\overline{P_x}, \overline{P_y})$			P'_x	P'_y
DIV	X		ε	X		ε	X		DIV	X
SUB	COS		ε	COS		ε	COS		-1	COS
ADD	DIV		DIV	DIV		DIV	DIV		SUB	DIV
X	ADD		ε	ADD		ε	ADD		X	ADD
0.56	-1	1	ε	-1	2	-1	ε	3	-0.10	SUB
MUL	SUB	⟹	SUB	SUB	⟹	SUB	SUB	⟹	MUL	ADD
	X		ADD	ε		ε	ADD		SIN	X
	-0.10		X	X		X	X			0.56
	MUL		0.56	-0.10		-0.10	0.56			MUL
	SIN		MUL	MUL		MUL	MUL			
			ε	SIN		SIN	ε			

Fig. 1. MHC of programs P_x and P_y in stack-based representation : Step 1, alignment and Xover site selection (here 5); Step 2, swapping sequences ; Step 3, deletion of gaps

similar but MHC has demonstrated a significant tendency in bloat reduction. The fact that using less desctructive operators allows a kind of reduction in bloating behaviours tends to confirm the *protection hypothesis*. However, an unexpected consequence of this size growth limitation was the need to accurately tune initial sizes in the population. The hypothesis was that MHC is unable to properly manage the size of individuals. In this context, the size may be viewed as a new dimension of the search space that needs some specific operators to be explored. Some experiments, on a flat landscape and on the Even-N Parity Problem, have demonstrated the possibility of controlling the size of programs using MHC.

In this paper, we investigate further, with two methods based on MHC, how to control the size of program during evolution but also how to improve the performances of MHC, as a fully functional recombination mechanism.

2   Size Control with MHC

In [14][15], authors have shown the biases introduced by SC in the exploration of the size of programs. They concluded that, without selective pressure, the distribution of the size converges toward a gamma distribution, *ie* SC does not modify the average length of individuals but leads to an oversampling of shorter programs of the search space and also to the creation of very long programs compared to the average size. To compare the effects of SC and MHC on size distribution, we have performed, for both crossover operators and without mutatin, 200 experiments on a flat landscape with a population of 1000 individuals during 1000 generations. The initial size of programs was randomly chosen between 1 and 50 instructions and the instruction set was defined with 10 different symbols.

We can see, in Figures 2 and 3, the expected gamma distribution obtained with SC, while with MHC, the distribution seems to converge much more slowly toward a gamma distribution. More precisely, at the last generation, when SC

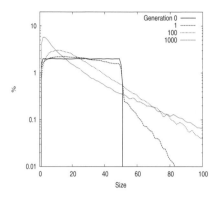 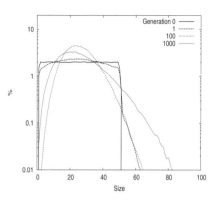

Fig. 2. Distribution of programs size using SC and without selective pressure

Fig. 3. Distribution of programs size using MHC and without selective pressure

is used, the most numerous programs, around 7% of the population, have only 2 instructions, while for MHC, they represent 4% of the population and have 23 instructions. So MHC is less biased than SC what explains its ability to reduce bloat and at the same time its difficulty to explore efficiently the size dimension.

We have already mentioned that two main strategies have been investigated to fight the bloat phenomenon. With the first one, the idea is to work on fitness to modify the search space in order to eliminate too long programs. For example, with the Tarpeian Method (TM) (see [12] for pseudo-code), some "holes" are dynamically introduced in the fitness landscapes by assigning, with a given probability, a very low fitness to programs whose size is higher than average in the population. With the second strategy, the approach consist in designing unbiased operators that prevent the creation of too long programs. For example, the size-fair operators (cf. [4]) ensure that the amount of genetic material exchanged during recombination is comparable between parents and so they modify little the size of programs. In this case, the goal is to control the distribution of size of programs that undergo the selection process. We focus on two different ways to control the distribution of size with MHC.

Firstly, we propose to use the mutation operator to modify the average size of programs during evolution. In our stack-based system, mutation consists either of an insertion, a deletion or a substitution of one instruction, each operator having its own application rate. We define an operator $MHC+INS_r$ to be MHC combined with an insertion rate of r higher than deletion and substitution rates. This unbalanced setting of the mutation rates must enable the system to increase the average size of programs and so to increase the chances of visiting areas of high performances. We note that, in [2], a similar setting was used in the context of LGP with homologous recombination to improve performances. We have plotted, in Figure 4, the size distribution obtained with $MHC+INS_{1.0}$, *ie* insertion rate equal to 1.0 and deletion and substitution rates fixed to 0.0. We see that $MHC+INS_r$ allows a translation of the size distribution reported when using MHC alone.

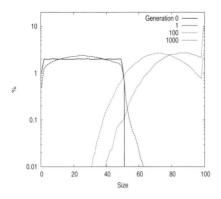 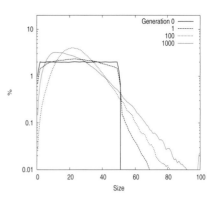

Fig. 4. Distribution of programs size using MHC+INS$_{1.0}$ and without selective pressure

Fig. 5. Distribution of programs size using MHC+SC$_{0.2}$ and without selective pressure

Secondly, we propose to use SC to modify the average size of programs during evolution. An operator MHC+SC$_p$ is defined where MHC is used to perform recombination but with p being the probability that SC will be used instead. Thus, MHC+SC$_{0.5}$ corresponds to an equally use of both MHC and SC. We note that in [4], authors have already speculated that judicious mixing of size-fair and standard operators could be the best way to encourage robust problem solving performances. We have plotted, in Figure 5, the size distribution obtained with MHC+SC$_{0.2}$. We see that MHC+SC$_p$ allows a transformation of the size distribution reported when using MHC alone.

3   Experimental Results

3.1   Problem and Parameters Settings

In this section, we aim to verify the ability to control the size of programs on a Symbolic Regression Problem. We choose the Poly-10 problem [12], where the target function is the 10-variate cubic polynomial $x_1x_2 + x_3x_4 + x_5x_6 + x_1x_7x_9 + x_3x_6x_{10}$, because it was introduced as a benchmark for the study of the TM. In this study, the fitness is the classical Root Mean-Square Error. The dataset contains 50 test points and is generated by randomly assigning values to the variables x_i in the range $[-1, 1]$.

We want to compare the performance between the operators SC with TM, MHC and the two alternatives MHC+INS and MHC+SC. For the TM, we call n the parameter giving the probability that programs whose size is higher than average will receive a very bad fitness. We test different value for n varying from 0.05 to 0.9. The GP system has very distinct behavior according to the operator used, this is why to perform a fair comparison, the evolutionary parameters tuning must be extensively investigated. For each operator and for each tuning of the size control parameters (n, p and r), we perform 50 independent runs with various mutation and crossover rates. Let us notice that a mutation rate of 1.0 means that each program involved in reproduction will undergo, on average, one insertion, one deletion and one substitution.

Populations of 500 individuals are randomly created according to a maximum creation size of 50. The instruction set contains: the four arithmetic instructions ADD, SUB, MUL, DIV, the ten variables $X_1 \ldots X_{10}$ and one stack-based GP specific instruction DUP which duplicates the top of the operand stack. The evolution, with elitism, maximum program size of 500, 16-tournament selection, and steady-state replacement, takes place over 100 generations[3]. We use a statistical unpaired, two-tailed t-test with 95% confidence to determine if results are significantly different.

3.2   Best Results

In what follows, SC stands for SC without TM ($n=0$), MHC+INS stands for MHC+INS$_{2.0}$ and MHC+SC stands for MHC+SC$_{0.1}$. In Table 1, the best results, in terms of average fitness of the best program found, among all the parameters settings tested, are reported (crossover rate varying from 0 to 1.0 and mutation rate from 0 to 2.0). As expected, using MIIC, the system has found less

Table 1. Best Results

Xover Type	Fitness	Size	Effective Size
SC	$0.13_{(\sigma=0.03)}$	$457.42_{(\sigma=79.07)}$	$457.14_{(\sigma=79.25)}$
MHC	$0.25_{(\sigma=0.05)}$	$92.28_{(\sigma=31.39)}$	$91.74_{(\sigma=31.45)}$
MHC+INS	$0.14_{(\sigma=0.03)}$	$247.18_{(\sigma=90.36)}$	$245.00_{(\sigma=90.75)}$
MHC+SC	$0.11_{(\sigma=0.02)}$	$419.12_{(\sigma=96.90)}$	$418.80_{(\sigma=96.82)}$

fit but smaller programs than using other operators. This is unsurprising since the optimization of the "maximum initial size" parameter, needed by MHC (see Section 1.2), has not been performed. Statistical analysis of the results of SC, MHC+INS and MHC+SC shows that their average fitness does not differ significantly. Conversely, the average size of the best solution found varies greatly. The operator MHC+INS seems to give a good trade-off between fitness and size since, in this case, the average size is almost 2 times smaller than with SC. We note that an increase of the n parameter has always led to fitness degradation for the SC operator.

3.3   Application Rates

In what follows, SC stands for SC without TM ($n=0$), MHC+INS stands for MHC+INS$_{2.0}$ and MHC+SC stands for MHC+SC$_{0.1}$. We have gone to great

---

[3] In a steady state system, the generation concept is somewhat artificial and is used only for comparison with generational systems. Here, a generation corresponds to a number of replacement equal to the number of individual in the population, *ie* 500.

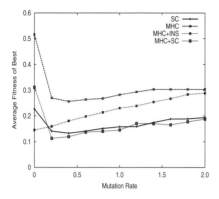

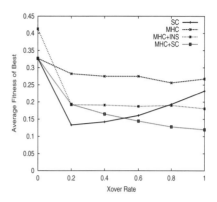

Fig. 6. Average fitness of best as a function of the mutation rate on Poly-10

Fig. 7. Average fitness of best as a function of the crossover rate on Poly-10

effort to determine the appropriate setting for each operator studied. Figure 6 depicts the average fitness of the best program found as a function of the mutation rate for the best crossover rate found. In other words, each point of the plot corresponds, for a given mutation rate, to the best result found among all crossover rates. All operators demonstrate a similar behavior according to the mutation rate, except for MHC+INS, which has obtained better performances without mutation. However, we know that it performs, by construction, at least 2.0 insertions on average per individual. The use of the mutation operator is critical but with low rates (the optimal is less than 0.4). Let us recall that a rate of 0.4 corresponds to, on average, 0.4 mutations of each type (insertion, deletion and substitution) per individual, so to a little more than one change per individual.

In Figure 7, we have plotted the best results found according to the crossover rate for a mutation rate set to 0.2. We see that SC obtains its best result with a small crossover rate but that its performances tend to worsen when too many recombinations are performed. On the other hand, the performances of MHC,

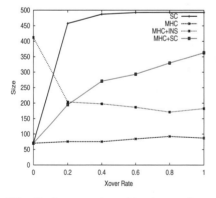

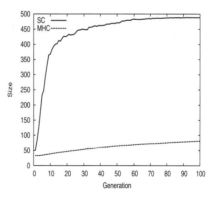

Fig. 8. Average size of best as a function of the crossover rate on Poly-10

Fig. 9. Evolution of the average size of best on Poly-10

MHC+INS and MHC+SC operators do not vary so much according to the
crossover rate, but with a small tendency to increase for high rates. Figure 8 represents the average size of the best program found as a function of the crossover
rate for a mutation rate of 0.2. We see that the size of the programs found using
SC, MHC and MHC+INS does not depend on the crossover rate. More precisely,
for SC, the size is limited by the "maximum allowed size" parameter, here 500 instructions. Whereas for MHC, the "maximum creation size", here 50 instructions,
is the major parameter influencing size (see also Figure 9). Finally for MHC+INS,
we see that the insertion of instructions, here 2.0 on average, in each individual
of the population leads to an increase of more than 100 instructions compared to
MHC. It is obvious that in the case of MHC+INS, when no recombination is performed, size control does not work (around 400 instructions) since there is nothing
to compensate the unbalanced mutation setting. Conversely, the MHC+SC operator finds programs of different sizes according to the crossover rate. This means
that the size of programs does not depend only on the proportion of MHC and SC
(the parameter p) but rather on the number of SC recombinations performed per
generation, which increases with the crossover rate.

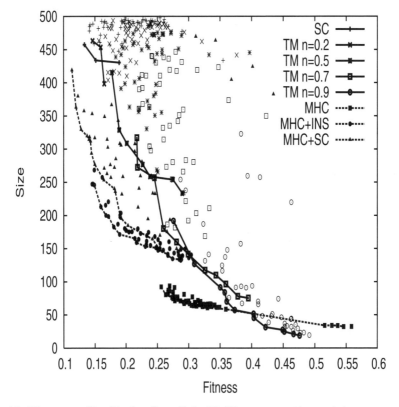

Fig. 10. Fitness vs Size Trade-off on Poly-10. Lines connecting points correspond to
Pareto frontiers.

3.4   Fitness *vs* Size Trade-Off

In order to visualize the fitness *vs* size trade-off, we have plotted, in Figure 10 a scatter plot of the average fitness and the average size of the best solutions found. Each point corresponds to one of the setting of the parameters (of both mutation and crossover rates) tested in this study. Lines connecting points depict the Pareto frontiers. We can see that the trade-off between size and fitness differs for the four operators. We see that when SC with $n=0$ or MHC are used, variations in the size dimension are very small. On the other hand, frontiers for SC with TM and for both alternatives of MHC cover larger ranges in the fitness vs size space. However, excepted for SC with $n=0.9$ that gives the shorter programs, the size control methods using MHC report better trade-off than SC with TM. Let us recall that results presented here do not correspond to a multi-objective approach since our goal was not to minimize, in words of Pareto optimality, both size and fitness criteria. We next investigate further the influence of the parameters r and p on fitness and size for both MHC+INS$_r$ and MHC+SC$_p$ operators. We have performed some specific experiments with a mutation rate of 0.2 and a crossover of 0.80. Figures 11 and 12 show the variations of, respectively, the average fitness and size of the best program found as a function of r for MHC+INS$_r$. We see that the insertion of instructions, controlled by r, always leads to an improvement in fitness but that for r greater than 2.0, no gains can be obtained. The average size is strongly correlated to the parameter r and all the allowed sizes in the search space can be reached. Compared to the performances of MHC, using MHC+INS$_{2.0}$, we obtain programs around two times more fitter but also two times bigger (see Table 1 above).

Figures 13 and 14 show the variations of, respectively, the average fitness and size of the best program found as a function of p for MHC+SC$_p$. We see an improvement of the fitness, compared to MHC and SC, for all the values of p. This means that the combination of both MHC and SC performed better

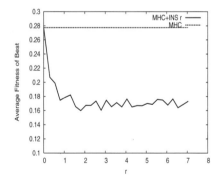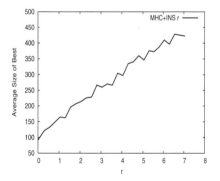

Fig. 11. Average fitness of best as a function of r on Poly-10 with MHC+INS$_r$

Fig. 12. Average size of best as a function of r on Poly-10 with MHC+INS$_r$

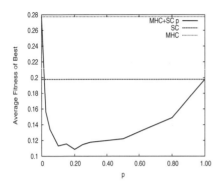
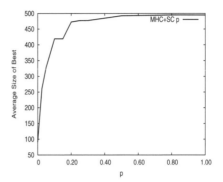

Fig. 13. Average fitness of best as a function of p on Poly-10 with MHC+SC$_p$

Fig. 14. Average size of best as a function of p on Poly-10 with MHC+SC$_p$

than when MHC and SC are used separately. However the size of the programs increases quickly with parameter p and seems to reach the maximum size when p is greater than 0.5. Moreover, we note a minimum of the fitness curve, when p is equal to 0.2. This implies that the p parameter must be carefully fixed. In the size control context, we can define, for the Poly-10 problem, a "region of interest" of the MHC+SC$_p$ operator for p in the range [0,0.2].

4   Conclusion and Perspectives

Contrary to SC, MHC do not induce any specific biases in the distribution of sizes and so an accurate control of the distribution during evolution is possible and have to be investigated.

In this paper, two methods for controlling the distribution with MHC are introduced and tested. The first one, called MHC+INS$_r$, where MHC works in conjunction with the mutation operator, directly modifies the number of genes in the population, *ie* the total amount of available instructions. In the second one, called MHC+SC$_p$, the MHC works in conjunction with SC to allow the creation of much bigger programs than the average size in the population. As expected, we note a significant increase in the average size and in the average fitness of the solution found. This reinforces our first assumption: to be efficient with MHC, the size of programs has to be explored as a new dimension of the search space. Nevertheless, the two methods presented here are static and so require a specific tuning that may depend on the problem addressed. Hopefully, the first steps in the study of size control methods, and more generally of MHC behavior, allow us to believe that dynamic control of the size is possible, according to some exogenous or endogenous properties.

MHC understanding, thanks to experimental results, is improved. For various benchmarks, the performance of this operator is equivalent but with an accurate management of the size. Future work should consist in a study of much more complex problems and then to real-world applications where an uncontrolled growth of the size of programs is a strong limitation for GP. For this purpose, the use and design of new dynamic methods for size control with MHC, taking into account some exogenous or endogenous additional informations, will undoubtedly be required.

References

[1] L. Altenberg. The evolution of evolvability in genetic programming. In *Advances in Genetic Programming*. MIT Press, 1994.

[2] M. Brameier and W. Bhanzhaf. Explicit control of diversity and effective variation distance in linear genetic programming. In *Genetic Programming, Proceedings of the 5th European Conference, EuroGP 2002*, volume 2278 of *LNCS*, pages 37–49, Kinsale, Ireland, 3-5 April 2002. Springer-Verlag.

[3] W. S. Bruce. The lawnmower problem revisited: Stack-based genetic programming and automatically defined functions. In *Genetic Programming 1997: Proceedings of the Second Annual Conference*, pages 52–57, Stanford University, CA, USA, 13-16 1997. Morgan Kaufmann.

[4] R. Crawford-Marks and L. Spector. Size control via size fair genetic operators in the PushGP genetic programming system. In *GECCO 2002: Proceedings of the Genetic and Evolutionary Computation Conference*, pages 733–739, New York, 9-13 July 2002. Morgan Kaufmann Publishers.

[5] E. D. de Jong, R. A. Watson, and J. B. Pollack. Reducing bloat and promoting diversity using multi-objective methods. In *Proceedings of the Genetic and Evolutionary Computation Conference (GECCO-2001)*, pages 11–18, San Francisco, 2001. Morgan Kaufmann.

[6] M. Defoin Platel, M. Clergue, and P. Collard. Homolgy gives size control in genetic porgramming. In *Proceedings of the 2003 Congress on Evolutionary Computation CEC2003*, pages 281–288. IEEE Press, 2003.

[7] M. Defoin Platel, M. Clergue, and P. Collard. Maximum homologous crossover for linear genetic programming. In *Genetic Programming, Proceedings of EuroGP'2003*, volume 2610 of *LNCS*, pages 194–203, Essex, 14-16 April 2003. Springer-Verlag.

[8] W. B. Langdon and R. Poli. Fitness causes bloat. In *Second On-line World Conference on Soft Computing in Engineering Design and Manufacturing*, pages 13–22. Springer-Verlag London, 23-27 1997.

[9] S. Luke. Code growth is not caused by introns. In *Late Breaking Papers at the 2000 Genetic and Evolutionary Computation Conference*, pages 228–235, Las Vegas, Nevada, USA, 8 2000.

[10] S. Luke. Modification point depth and genome growth in genetic programming. *Evol. Comput.*, 11(1):67–106, 2003.

[11] T. Perkis. Stack-based genetic programming. In *Proceedings of the 1994 IEEE World Congress on Computational Intelligence*, volume 1, pages 148–153, Orlando, Florida, USA, 27-29 1994. IEEE Press.

[12] R. Poli. A simple but theoretically-motivated method to control bloat in genetic programming. In *Genetic Programming, Proceedings of EuroGP'2003*, volume 2610 of *LNCS*, pages 204–214, Essex, 14-16 April 2003. Springer-Verlag.

[13] R. Poli and W. B. Langdon. Genetic programming with one-point crossover. In *Soft Computing in Engineering Design and Manufacturing*, pages 180–189. Springer-Verlag London, 23-27 June 1997.

[14] R. Poli and N. F. McPhee. Exact schema theorems for GP with one-point and standard crossover operating on linear structures and their application to the study of the evolution of size. In *Genetic Programming, Proceedings of EuroGP'2001*, volume 2038, pages 126–142. Springer-Verlag, 18-20 2001.

[15] J. E. Rowe and N. F. McPhee. The effects of crossover and mutation operators on variable length linear structures. In *Proceedings of the Genetic and Evolutionary Computation Conference (GECCO-2001)*, pages 535–542, San Francisco, California, USA, 7-11 July 2001. Morgan Kaufmann.

[16] T. Soule and J.A. Foster. An analysis of the causes of code growth in genetic programming. *Genetic Programming and Evolvable Machines*, 3(1):283–309, 2002.

[17] R.E. Keller W. Banzhaf, P. Nordin and F.D. Francone. *Genetic Programming - An Introduction*. Morgan Kaufmann, 1998.

A New Classification-Rule Pruning Procedure for an Ant Colony Algorithm

Allen Chan and Alex Freitas

Computing Laboratory, University of Kent,
Canterbury, CT2 7NF, UK
{ac207, A.A.Freitas}@kent.ac.uk

Abstract. This work proposes a new rule pruning procedure for Ant-Miner, an Ant Colony algorithm that discovers classification rules in the context of data mining. The performance of Ant-Miner with the new pruning procedure is evaluated and compared with the performance of the original Ant-Miner across several datasets. The results show that the new pruning procedure has a mixed effect on the performance of Ant-Miner. On one hand, overall it tends to decrease the classification accuracy more often than it improves it. On the other hand, the new pruning procedure in general leads to the discovery of classification rules that are considerably shorter, and so simpler (more easily interpretable by the users) than the rules discovered by the original Ant-Miner.

1 Introduction

Ant-Miner [3] is an Ant Colony algorithm that discovers classification rules in the context of data mining. The basic goal of data mining is to extract, from data, knowledge that is not only accurate but also comprehensible to the user [9], [5]. Knowledge comprehensibility is important because in many applications of data mining the user should validate and interpret discovered knowledge, rather than blindly trust the result provided by an algorithm. A typical example of an application where rule comprehensibility is crucial is medical diagnosis, where rules suggesting a diagnosis for a patient must be interpreted and validated by a medical doctor.

Ant-Miner has been shown to be competitive with a well-known classification algorithm in [3], in experiments across several datasets. However, those experiments did not involve datasets with a large number of attributes, where the rule pruning procedure of Ant-Miner tends to be very time consuming. In order to improve Ant-Miner's scalability to data sets with a larger number of attributes, this paper proposes a faster rule pruning procedure for Ant-Miner. The proposed procedure is essentially a hybrid pruning procedure. It combines Ant-Miner's original pruner with a faster pruning based on the information gain of individual attributes. (See [5] for a review of information gain in general.) The basic idea is that, if the candidate rule to be pruned is a long one, instead of applying Ant-Miner's original pruner the algorithm first applies the faster information gain-based pruner, as a first step to reduce the rule length. In terms of computational cost, this first step "comes for free", since the required value of the information gain is already computed by another procedure of Ant-Miner. Once the rule has been so reduced, Ant-Miner's original pruner – slower but more effective – can be applied to the rule, further reducing its length.

E. Talbi et al. (Eds.): EA 2005, LNCS 3871, pp. 25–36, 2006.

The proposed hybrid rule pruner is evaluated across several datasets, most of them with more than 100 attributes. The results are evaluated with respect to the classification accuracy and the comprehensibility of the discovered rules.

The remainder of this paper is organized as follows. Section 2 reviews the Ant-Miner algorithm. Section 3 describes the proposed hybrid rule pruner. Section 4 reports the results of computational experiments, and section 5 concludes the paper.

2 The Original Ant-Miner Classification Algorithm

A single ant within a colony is normally seen as a highly unintelligent individual, but collectively, as a colony, ants exhibit what is known as swarm intelligence. While ants forage for a food source they deposit on their paths a certain amount of pheromone, a chemical substance to which other ants are attracted. It turns out that over time shorter routes between two points (such as the colony's nest and some food source) will have more pheromone than longer routes, because in a fixed period of time there will be more ants completing a shorter path than a longer path. When selecting between multiple paths, ants will in general be attracted to those paths with the highest concentration of pheromone. As a result, the ants will in general prefer to follow the shortest route within a network of paths, which will further increase the concentration of pheromones in the shortest path, attracting more ants to that path. Therefore, over time ants will converge and follow the shortest route within a network of paths. This has been shown by experiments performed by Deneubourg et al. [2].

Dorigo et al, inspired by this interesting behaviour of ant colonies, first developed Ant Colony Optimization (ACO) to solve difficult combinatorial optimization problems like the classic travelling salesman problem [1], [8]. This idea was then taken from solving optimization problems and applied in the field of data mining for discovering classification rules. The Ant-Miner algorithm, developed by Parpinelli et al. [3], is an adaptation of the ACO paradigm especially for the classification task of data mining. The algorithm implements the basic idea of awarding the best attributes (used by the ants to construct the best rules) with pheromone, which increases the probability of those attributes being selected by the next ants to construct other rules. A simple high-level pseudocode of Ant-Miner is shown in Pseudocode 1, adapted from [7]. A more detailed description of Ant-Miner can be found in [3].

The Ant-Miner algorithm uses a sequential covering approach to discover a list of classification rules which will cover all or most of all the examples in the training set. Rules discovered are in the form of: IF <term1> AND ... AND <term-n> THEN <class>. Each term takes the form <attribute=value>, where value belongs to the domain of the attribute. The training set holds examples that are used for discovering a list of classification rules. The discovered rule list is empty to start with. Every iteration of the outer REPEAT-UNTIL loop of Pseudocode 1 discovers one classification rule and adds it to the list of discovered rules. Each iteration of the inner REPEAT-UNTIL loop corresponds to the trail of an ant that constructs one candidate rule. At the end of the inner REPEAT-UNTIL loop, the best rule from the set of rules constructed by all ants (i.e., in all iterations of the inner REPEAT-UNTIL loop) is added to the discovered rule list. Examples correctly covered by this rule are removed from the training set before the next iteration of the outer REPEAT-UNTIL loop begins to discover the next rule. (An example is correctly covered by a rule if the example satisfies all conditions of the rule and it has the class predicted by the rule.)

The first procedure of the inner REPEAT-UNTIL loop consists of incrementally constructing a candidate rule. This procedure starts with an empty rule and then adds one term at a time to the current rule. This incremental rule construction will terminate when one of the following two stopping criteria is met: any term added to the current rule would make the rule cover a number of examples less than a user specified threshold, or when all attributes have already been used in the current rule being generated, so that no other attribute is available. (A rule cannot have two occurrences of the same attribute, because this would lead to invalid rules such as "IF <gender = male> AND <gender = female>…".)

The inner REPEAT-UNTIL loop will terminate when one of two stopping criteria is met: the number of constructed rules is equal or greater than the maximum number of ants specified by the user, or the rule constructed by an ant is exactly the same as the rule constructed by the a certain number of previous ants. The latter criterion is checked via a convergence test. These stopping criteria are controlled by parameters, which are discussed in detail in [3]. Finally, the outer REPEAT-UNTIL loop terminates when the number of examples in the training set becomes lower than a predefined threshold.

```
TrainingSet = {all training examples};
DiscoveredRuleList = {} /* initialized with empty list */
REPEAT
    Initialize all trails with the same amount of
    pheromone;
    REPEAT
        An ant incrementally constructs a
            candidate classification rule;
        Prune the just-constructed rule;
        Update the pheromone of all trails;
    UNTIL (stopping criteria)
    Choose the best rule out of all candidate
        rules constructed by all ants;
    Add the best rule to DiscoveredRuleList;
    TrainingSet = TrainingSet - {examples correctly
        covered by best rule};
UNTIL (stopping criteria)
```

Pseudocode 1. A high-level description of the original Ant-Miner

For the purpose of this paper, the most important part of Ant-Miner is its rule pruning procedure. This procedure is computationally expensive and it can be considered the bottleneck of the algorithm with respect to processing time and scalability to large data sets, as discussed in the next section.

3 Extending Ant-Miner with a Faster Rule Pruning Procedure

3.1 The Motivation for Rule Pruning

Rule pruning is a commonplace data mining technique, used in the vast majority of rule induction algorithms [5]. Pruning can improve the quality of a rule by removing irrelevant terms from the rule antecedent. As a result, pruning can improve both the predictive accuracy and the comprehensibility of the rule.

It should be noted that Ant-Miner, like the majority of rule induction algorithms, can potentially discover rules with a long rule antecedent (with many terms), hindering the comprehensibility of the rule. Indeed, rules take the form of IF <antecedent> THEN <consequent> where the rule antecedent is a conjunction of n terms, where the value of n can potentially be close to the total number of attributes in the dataset. This means that a rule can become too long for a user to be able to interpret it. Hence, there is a preference for shorter, more comprehensible rules.

3.2 Ant-Miner's Original Rule Pruner

Ant-Miner's original rule pruner takes a freshly generated rule by the current ant and tries to improve its quality (measured by the rule's predictive accuracy), by removing irrelevant terms from the rule antecedent. This is done by iteratively removing one term at a time while it improves on the rule's quality. This iterative process stops when no term removal will further increase the quality of the current rule undergoing pruning. The entire rule pruning process is described in Pseudocode 2.

```
Execute_pruning - true;
WHILE (Execute_pruning = true) AND
      (Number of terms in current rule antecedent > 1)
   FOR EACH (term t_i in the current rule to be pruned)
      Temporarily remove t_i and assign to
      the rule consequent the most frequent class among
      the examples covered by the rule antecedent;
      Evaluate rule quality;
      Reinstate term t_i in rule antecedent;
   END FOR
   IF (rule quality was improved w.r.t. original rule's
      Quality in some iteration of the FOR loop) THEN
      Remove permanently the term whose removal improves
      current rule most;
   ELSE
      Execute_pruning = false;
   END IF-THEN-ELSE
END WHILE
```

Pseudocode 2. A high-level description of Ant-Miner's original rule pruner

Initial experiments conducted by Parpinelli et al. [3] showed that Ant-Miner produces rules that have a good predictive accuracy and are relatively short on average. However, that work also presented an analysis of computational time complexity showing that rule pruning is the most time consuming part of the algorithm, and that the time taken by Ant-Miner's rule pruner is quite sensitive to the number of attributes in the data being mined. This is due to the fact that the larger the number of attributes in the data being mined, in general the larger the number of terms in a constructed rule before pruning, and so the larger the number of iterations in the loops of Pseudocode 2. In each iteration of the FOR EACH loop a term is temporarily removed and the quality of the reduced candidate rule has to be computed by formula (1).

Rule Quality = Sensitivity $\times$ Specificity = (TP / (TP+FN)) $\times$ (TN / (TN+FP))   (1)

where:

- TP (true positives) is the number of examples that are covered by the rule and have the same class as predicted by the rule;
- FP (false positives) is the number of examples that are covered by the rule and have a class different from the class predicted by the rule;
- TN (true negatives) is the number of examples that are not covered by rule and have a class different from the class predicted by the rule;
- FN (false negatives) is the number of examples that are not covered by the rule but have the same class as predicted by the rule.

The computation of formula (1) is computationally expensive because it requires scanning the entire current training set in order to compute the values of TP, FP, TN and FN. For rules generated with a small number of terms in its antecedent, the pruning method shown in Pseudocode 2 is relatively quick, as there are not a large number of terms to temporarily remove and evaluate rule quality. But for rule antecedents containing a large number of terms, this type of pruning is very computationally expensive. This is because the WHILE loop of the Pseudocode 2 is potentially performed a large number of times (in the worse case the number of terms in the original rule), each iteration of the while loop involves a FOR EACH loop over all current terms in the rule, and each FOR EACH iteration involves a scan of the training set.

It should be noted that the computational time taken by Ant-Miner was not a significant problem in the experiments reported by Parpinelli et al. for the following reason: those experiments involved datasets where the number of attributes was *not* very large. However, in addition to the previously-mentioned theoretical analysis of the computational time complexity of Ant-Miner identifying the rule pruner as the bottleneck of the algorithm [3], there is empirical evidence that the computational time taken by Ant-Miner becomes very long when the data being mined contains a large number of attributes. This empirical evidence consists of recent experiments trying to apply Ant-Miner to a large bioinformatics data set containing 33,079 examples and 854 attributes [10]. In that project Ant-Miner turned out to be so slow that it was not viable to use it to discover classification rules, and a much faster hybrid ACO/PSO (Particle Swarm Optimization) algorithm was developed and used instead. To quote [10]: "...the unusually large amount of attributes and classes associated with this problem mean an *extremely* large amount of computation time is required [by Ant-Miner]." Therefore, there is a clear motivation for developing a considerably faster rule pruning procedure for Ant-Miner and investigate its performance, which is the focus of the remainder of the paper.

3.3 Proposed Hybrid Rule Pruner for Ant-Miner

After an analysis of Ant Miner's original rule pruner, the following is a proposal to a new hybrid rule pruner, combining the original Ant-Miner's rule pruner with a rule pruner based on information gain – the latter somewhat inspired by the rule pruner proposed in [4]. (For a review of information gain in general, see [5].)

```
INPUT:

a)   information   gain   of   all   terms   individually,
calculated using the entire current training set;
/* previously done by another procedure of Ant Miner */
b) value of r /* user-defined parameter: number of terms
in the current rule which will be given to Ant-Miner's
original rule pruner */

Reduced_rule = {};
Num_terms_selected = 0;
IF (number of terms in current rule's antecedent > r)
THEN
   WHILE (Num_terms_selected < r)
      FOR EACH (term t_i in current rule's antecedent)
         Calculate probability of selecting a term t_i as:
            prob(t_i) =      InfoGain(t_i)
                        ─────────────────────
                        T∑_{i=1}(InfoGain(t_i))
         /*T = number of terms in the rule antecedent */
      END FOR
      Create roulette wheel for selection and select one
      Term, called selected_term, by spinning the wheel;
      Reduced_rule = Reduced_rule ∪ selected_term;
      Remove selected_term from current rule's antecedent
      to avoid reselection;
      Num_terms_selected = Num_terms_selected + 1;
   END WHILE
   Assign to the consequent of the Reduced_rule the most
   frequent class among all examples covered by the rule;
   Run Ant-Miner's original rule pruner on Reduced_rule;
ELSE
   Run Ant-Miner's original rule pruner on current rule;
END IF-THEN-ELSE
```

Pseudocode 3. A high-level description of the proposed hybrid rule pruner

First of all, the motivation for this new hybrid rule pruner is to significantly reduce the computational time taken by Ant-Miner, and hopefully do it without unduly reducing the accuracy of the discovered rules, by comparison with the original Ant-Miner. In other words, the basic idea is to combine the effectiveness of the original Ant-Miner pruner (in terms of maximizing predictive accuracy) with the speed of a rule pruner based on information gain. This latter is very fast, because it does not require any scan of the training set, as explained below.

The way this hybrid rule pruner functions is described in Pseudocode 3. The information gain of each term has already been computed by Ant-Miner – in order to compute the values of the heuristic function [3], and is re-used in this hybrid procedure. The parameter r represents the number of terms in the rule antecedent that will be subject to the original Ant-Miner's rule pruner.

As shown in Pseudocode 3, the hybrid pruner selects r terms, out of all the terms in the current rule, before applying Ant-Miner's original rule pruner. If the number of terms in the rule antecedent of a freshly generated rule exceeds the value of r, the rule first undergoes reduction of the number of terms to the value of parameter r. This reduction is obtained as follows. For each term within the rule antecedent, the rule pruner computes a measure of the probability of selecting that term. This probability measure is based on the pre-computed value of that term's information gain with respect to the class attribute. Then the rule pruner selects r number of terms using the roulette wheel selection technique (commonplace in genetic algorithms), with the probability of selecting each term proportional to the information gain of that term. Once r terms have been selected by spinning the roulette wheel r times, the resulting reduced rule is placed back into Ant-Miner's original rule pruner.

If the original rule does not contain a number of terms in its rule antecedent exceeding the value of r parameter, then it gets placed straight into Ant-Miner's original rule pruner, with no need to apply the information gain-based pruning.

Intuitively it is difficult to specify an ideal value for parameter r, since the best value of this parameter tends to be dataset-dependent. Therefore we have conducted experiments to investigate the influence of different values of this parameter in the performance of the proposed hybrid rule pruner, as discussed in the next section.

4   Computational Results

4.1   Experimental Setup and Datasets Used in the Experiments

As discussed earlier, the proposed hybrid rule pruner has a parameter, r, which can have a significant influence in the performance of Ant-Miner. To investigate this issue, we conducted experiments with different values of r, varying from 3 to 10. These experiments have two main goals. First, evaluating how sensitive the performance of Ant-Miner with the hybrid pruner is with respect to different values of the parameter r. Second, comparing the performance of the hybrid rule pruner with the performance of the original Ant-Miner's rule pruner.

The experiments used mainly 5 datasets – as summarised in Table 1, detailing key statistics (the number of examples, attributes and classes) for each dataset. The *Chess* and the *House-votes* datasets have been taken from the well-known UCI Machine Learning dataset repository – see [6] for details about these data sets. They have a small number of attributes, and were included in the experiments as control datasets. By contrast, the three *Web-mining* datasets are more challenging, because they have a considerably larger number of attributes, varying from 159 to 339. In addition, note that these data sets are very "sparse", in the sense that the number of examples is even smaller than the number of attributes. (Such challenging datasets are commonplace in text/web mining and bioinformatics applications, and therefore it is important to investigate the performance of Ant-Miner in this kind of very sparse dataset.)

In the *Web-mining* datasets, each example is a web page, and the goal is to classify each example into one of three classes: 'Technology', 'Sport' and 'Education'. These classes represent the general subject of the web page. These datasets were harvested from a small selection of BBC and Yahoo web pages relating to the above named subjects. All attributes within these datasets are binary, where each attribute denotes whether or not a given word occurs in a given web page (example). These datasets have been collected by and previously been experimented with Ant-Miner by Holden & Freitas [7].

In addition to the above datasets, we also did experiments with 2 bioinformatics datasets using a single value of r. Both datasets have 1872 examples (proteins) and the same values of 102 binary predictor attributes. Each attribute indicates whether or not a protein has a given Prosite pattern. The datasets differ in the class to be predicted: whether or not a protein is involved in DNA repair (first dataset) or in DNA damage (second dataset). The creation of these datasets is explained in [11].

Table 1. Summarized details of the 5 main datasets used in the experiments

Dataset	No. of examples	No. of attributes	No. of classes
Chess	3196	36	2
House-votes-84	434	16	2
Web-mining 1	124	159	3
Web-mining 2	124	293	3
Web-mining 3	124	339	3

Ant-Miner takes on several parameters besides the one we have discussed for the hybrid pruning procedure. The values of all those other parameters were maintained at their default values, specified in [3].

All the experiments were conducted using a stratified 5-fold cross validation procedure [5]. In essence, the dataset was partitioned into five folds with each fold retaining as closely as possible the class distribution of the whole dataset being mined. Each version of Ant-Miner (with original pruner and with the new hybrid pruner) is then run 5 times. In these runs, each fold was used four times as the training set and once as a test set. All results reported in this paper were averaged over the five iterations of the cross validation procedure.

4.2 Results on Classification Accuracy

Reported in Table 2 are the average classification accuracies on the test set and the corresponding standard deviations for each value of r in the range of 3 to 10. The classification accuracy is the number of correctly classified test examples divided by the total number of test examples. The first line with results in the table shows the classification accuracy of Ant-Miner using only its original rule pruner. The results for this original version of Ant-Miner are provided as a comparison to see how well it performs against the new hybrid rule pruner. For each dataset, the best result (out of the trials using the hybrid pruner) is highlighted in bold, in order to indicate which value of r yielded the highest accuracy.

From Table 2, in general there was a considerable variation in the classification accuracy across different values of the parameter r. The only exception was the *House-votes* dataset, where accuracies varied only in the range 93.1–95.4%. In the *Chess* dataset, most values of r led to an accuracy higher than the accuracy obtained with the original Ant-Miner's rule pruner, although in general the differences are not statistically significant (considering the standard deviations). On the other hand, in the three *Web-mining* datasets the accuracies obtained with the hybrid rule pruner were lower than the accuracies obtained with Ant-Miner's original rule pruner. This drop in accuracy associated with the use of the hybrid rule pruner in the *Web-mining* datasets can be explained as follows.

Table 2. Classification accuracy rate (%) on the test set (5-fold cross validation)

Value of *r*	Dataset				
	Chess	House-votes	Web-mining 1	Web-mining 2	Web-mining 3
Original	72.18±9.61	94.23±1.75	68.53±4.29	57.26±7.25	55.90±4.79
3	78.79±7.84	**95.38±1.33**	50.76±3.57 (-)	44.89±5.95	48.83±1.00 (-)
4	83.77±7.77	94.23±1.75	40.26±4.07 (-)	43.32±6.56 (-)	48.95±5.02
5	74.00±9.80	93.07±1.95	50.86±3.24 (-)	38.55±5.48 (-)	50.4±1.91
6	67.40±8.82	94.23±1.75	49.16±2.81 (-)	43.20±6.58 (-)	**52.24±8.17**
7	78.75±9.42	94.00±1.61	51.86±4.62 (-)	49.56±6.37	51.87±5.23
8	79.85±7.99	94.23±1.75	51.10±2.37 (-)	44.92±1.78 (-)	50.27±5.67
9	76.91±8.33	94.00±1.69	**62.83±3.86**	40.06±4.11 (-)	45.84±5.90
10	**84.13±8.88**	93.53±2.19	53.96±2.39 (-)	**55.26±5.92**	49.50±6.00

As mentioned earlier, the *Web-mining* datasets are particularly challenging because they are very "sparse". Each of those datasets contains a number of attributes greater than the number of examples. Recall that the hybrid rule pruner selects *r* number of terms, and terms are selected with probability based on their information gain. In very sparse datasets such as the *Web-mining* datasets, the values of the information gain of the attributes are not very "reliable", since they are prone to overfitting issues. As a result, the hybrid rule pruner has difficulty in selecting *r* relevant terms based on the computed information gain values. Of course, the issue of overfitting also occurs with the other component of the hybrid rule pruner, i.e., the original Ant-Miner's rule pruner. However, the latter is a more direct and more reliable measure of the relevance (predictive power) of the terms, since it is based on evaluating a candidate pruned rule as a whole, taking into account term interactions. By contrast, the heuristic of selecting terms based on the information gain of individual attributes seems more sensitive to overfitting issues, since the quality of each term is estimated by ignoring term interactions, i.e., ignoring the actual effect of the term in the current candidate rule. As a result, in the *Web-mining* datasets the accuracy obtained with the hybrid rule pruner is consistently lower than the accuracy obtained with the Ant-Miner's original rule pruner; a phenomenon that is not observed in the much less sparse *Chess* and *House-votes* datasets.

In any case, the difference of accuracy between Ant-Miner's original rule pruner and the new hybrid rule pruner is not significant in the majority of the cases in Table 2, taking into account the standard deviations. More precisely, in Table 2 the cells where the accuracy of the hybrid pruner is significantly lower than the accuracy of Ant-Miner's original rule pruner – in the sense that the corresponding standard deviation intervals do not overlap – are marked with the symbol "(-)". The drop in accuracy associated with the hybrid pruner was significant in 13 out of the 40 cells with hybrid pruner results in Table 2. In the other 27 cells the difference in accuracy is not significant, and as mentioned earlier the hybrid rule pruner even obtains a somewhat higher accuracy in the majority of the cases for the Chess data set.

We have also applied the hybrid rule pruner with a single value of r, viz. $r = 5$, to a couple of bioinformatics datasets with 102 attributes and 1872 examples, as mentioned in section 4.1, to evaluate the performance of the method in less sparse datasets. In the DNA repair dataset the hybrid rule pruner obtained a predictive accuracy of 97.69% ± 0.81%, against the original Ant-Miner's accuracy of 98.50% ± 0.58%. In the DNA damage dataset the hybrid rule pruner obtained an accuracy of 95.30% ± 4.03%, against the original Ant-Miner's accuracy of 93.25% ± 3.49%. Hence, in these datasets the hybrid pruner did not significantly reduce the accuracy.

4.3 Results on Rule Comprehensibility

We now turn to another criterion of performance often used in data mining, namely the comprehensibility of the discovered rules. We emphasize that rule comprehensibility is an important performance criterion in the context of data mining [5], [9] where the goal usually is to discover knowledge that can be interpreted and validated by human beings, to support intelligent decision making. As usual in the literature, we measure rule comprehensibility by the average number of terms in the discovered rules. The basic idea is that in general the shorter a rule is (i.e., the fewer terms it has in its antecedent), the simpler and more easily interpretable the rule is to the user. In this spirit, Table 3 reports the average number of terms per discovered rule when using the original Ant-Miner's rule pruner and when using the hybrid rule pruner – again, with values of r varying from 3 to 10. Similarly to Table 2, the numbers after the symbol "±" are standard deviations. For each dataset, the best result (i.e., the smallest number of terms per rule) is shown in bold.

Table 3. Average number of terms per discovered rule

| Value of r | Dataset | | | | |
	Chess	House-votes	Web-mining 1	Web-mining 2	Web-mining 3
Original	3.35±0.49	0.95±0.05	10.07±0.65	10.02±1.76	7.33±1.52
3	**1.50±0.14**	0.96±0.07	**1.77±0.23**	**1.32±0.15**	**1.30±0.24**
4	1.85±0.15	**0.86±0.06**	2.42±0.28	2.00±0.15	2.00±0.15
5	1.97±0.24	1.12±0.12	3.13±0.08	2.30±0.13	2.60±0.24
6	2.07±0.14	0.95±0.05	3.62±0.19	3.20±0.36	3.13±0.27
7	2.62±0.24	1.27±0.17	3.97±0.42	3.38±0.20	3.65±0.49
8	2.22±0.65	0.95±0.05	4.67±0.47	3.77±0.34	4.67±0.27
9	2.48±0.15	1.00±0.08	5.67±0.38	3.80±0.34	3.87±0.54
10	2.51±0.30	0.95±0.05	5.53±0.49	4.33±0.46	4.37±0.38

The only dataset in which the hybrid rule pruner did not significantly lower rule length, by comparison with Ant-Miner's original rule pruner, was the *House-votes* dataset. In this dataset the original Ant-Miner already obtained a very short average rule length, close to 1, and so it is perfectly acceptable that this result cannot be significantly improved. In the other four datasets, the hybrid rule pruner has significantly lowered the rule length, and so significantly improved rule comprehensibility, taking into account the standard deviations, for all tested values of r. The results were particularly good in the *Web-mining* datasets, as can be seen in Table 3, where rule length is reduced to less than half the rule length associated with the original Ant-Miner in most cases.

As expected, the shortest rule lengths were in general obtained with the smallest tried value of r, i.e., $r = 3$. With this value of r, in the *Web-mining* datasets rule length is reduced from about 10 or 7 to less than 1.5, a very significant improvement in rule comprehensibility. In addition, in all datasets but the *House-votes* one, there is a clear correlation between the value of r and the average length of the discovered rules. That is, in four out of the five datasets, in general the larger the value of r the larger the average rule length, and so the less comprehensible the discovered rules are. This result can be explained by the fact that the hybrid rule pruner's component based on information gain is more "aggressive" than the other component – Ant-Miner's original pruner. The latter is more "conservative" in the sense that it will only remove a term from a candidate rule if that removal improves the rule quality. By contrast, the pruner based on information gain always reduces the rule to r terms as the first step of the hybrid pruner, and so the hybrid pruner as a whole tends to produce shorter rules as the value of r is reduced.

5   Conclusions and Future Work

This work has proposed a new hybrid rule pruner for the Ant-Miner algorithm. The hybrid pruner combines Ant-Miner's original pruner with a faster pruning based on information gain. The basic idea is that, if the candidate rule to be pruned is a long one, instead of applying Ant-Miner's original pruner the algorithm first applies the faster information gain-based pruner, as a first step to reduce the rule length. In terms of computational cost, this first step "comes for free", since the required value of the information gain is already computed by another procedure of Ant-Miner. Once the rule has been so reduced, Ant-Miner's original pruner – slower but more effective – can be applied to the rule, further reducing its length.

Experiments were performed with several data sets, comparing the performance of the proposed hybrid rule pruner with the performance of Ant-Miner's original rule pruner. In general the hybrid pruner significantly reduced the computational time of Ant-Miner, by comparison with the computational time taken with the original rule pruner. In the datasets with the largest numbers of attributes (the *Web-mining* datasets), in most cases the computational time was significantly reduced, by comparison with the original Ant-Miner's computational time. In particular, in the *Web-mining-2* data set, the use of the hybrid rule pruner reduced Ant-Miner's computational time to a fraction of the original Ant-Miner's time in all cases, and this fraction varied from 11.6% in the best case to 65.0% in the worst case. A larger computational time reduction is expected in a dataset with a much larger number of attributes and examples. Concerning the quality of the classification rules discovered by Ant-Miner with the new hybrid rule pruner, there are three main conclusions.

First, the predictive accuracy of Ant-Miner is quite sensitive to values of a parameter of the hybrid rule pruner that determines how aggressive the information gain-based rule pruner is. Hence, when using the hybrid rule pruner in important real-world problems, it is recommended to carry out experiments optimizing the value of this parameter for the target dataset. Such parameter optimization is, of course, normally recommended in the context of data mining in general, where the performance of the algorithm is typically considerably dependent on the dataset being mined. Second, with respect to the comprehensibility of the discovered rules, the hybrid rule pruner in general led to the discovery of rules considerably shorter (and so more easily interpretable by users) than the rules discovered with the original Ant-Miner's rule pruner. Hence, the hybrid rule pruner is particularly recommended in applications where rule comprehensibility is very important, such as medical applications – where discovered rules should be carefully interpreted by experts before they are actually used to diagnose a patient or suggest a medical treatment. Third, the results suggest that, as long as the main parameter of the hybrid rule pruner is suitably adjusted for the target data set, it is possible to obtain a good trade-off between accuracy and comprehensibility. In each of the three web mining data sets – where accuracy was overall most reduced by using the hybrid rule pruner – the hybrid rule pruner with its best parameter value obtained a rule set with no significant drop in accuracy and with a significant gain in comprehensibility. However, to be on the safe side it is recommended to use Ant-Miner's original rule pruner whenever possible, in order to avoid the potential loss of accuracy associated with the hybrid rule pruner.

A future research direction is to develop a more adaptive version of the proposed hybrid rule pruner, where the value of r is automatically adapted by the algorithm on-the-fly, rather than being statically determined by the user.

References

[1] M. Dorigo, and L. M. Gambardella, "Ant colonies for the travelling salesman problem", *BioSystems, 43:* 73 – 81, 1997.
[2] J. L. Deneubourg, S. Aron, S. Goss, and J. M. Pasteels, "The self-organizing exploratory pattern of the argentine ant.", *Journal of Insect Behaviour, 3*:159 – 168, 1990.
[3] R.S. Parpinelli, H.S. Lopes and A.A. Freitas, "Data Mining with an Ant Colony Optimization Algorithm", *IEEE Trans. on Evolutionary Comput., 6(4)*, Aug 2002, 321-332
[4] Deborah R. Carvalho and A.A. Freitas, "A hybrid decision tree/genetic algorithm method for data mining" *Information Sciences 163(1-3)*, pp. 13-35. June 2004.
[5] I. H. Witten and E. Frank, *Data Mining – Pratical Machine Learning Tools and Techniques with Java Implementations*, Morgan Kaufmann 2000.
[6] UCI Machine Learning Repository (University of California at Irvine) – http://www.ics.uci.edu/~mlearn/MLSummary.html (visited 14/10/2004)
[7] N.Holden and A.A.Freitas, "Web Page Classification with an Ant Colony Algorithm" *Proc. 2004 Parallel Problem Solving from Nature, LNCS 3242,* 1092-1102. Springer, 2004.
[8] M. Dorigo and T. Stuetzle. *Ant Colony Optimization.* MIT Press, 2004.
[9] U.M. Fayyad, G. Piatetsky-Shapiro and P. Smyth. From data mining to knowledge discovery: an overview. In: U.M. Fayyad et al (Eds.) *Advances in Knowledge Discovery and Data Mining*, 1-34. AAAI/MIT, 1996.
[10] N. Holden and A.A. Freitas. "A Hybrid Particle Swarm/Ant Colony Algorithm for the Classification of Hierarchical Biological Data". *Proc. 2005 IEEE Swarm Intelligence Symposium*, 100-107. IEEE, 2005.
[11] A. Chen. *Ant Colony Optimisation for High-Dimensional and Multi-Label Classification in Data Mining.* Master Thesis (in preparation). University of Kent, UK. Sep. 2005.

Swarm-Based Distributed Clustering in Peer-to-Peer Systems

Gianluigi Folino, Agostino Forestiero, and Giandomenico Spezzano

Institute for High Performance Computing and Networking (ICAR),
Via P. Bucci 41c, I-87036 - Rende (CS), Italy
{folino, forestiero, spezzano}@icar.cnr.it

Abstract. Clustering can be defined as the process of partitioning a set of patterns into disjoint and homogeneous meaningful groups, called clusters. Traditional clustering methods require that all data have to be located at the site where they are analyzed and cannot be applied in the case of multiple distributed datasets. This paper describes a multi-agent algorithm for clustering distributed data in a peer-to-peer environment. The algorithm proposed is based on the biology-inspired paradigm of a flock of birds. Agents, in this context, are used to discovery clusters using a density-based approach. Swarm-based algorithms have attractive features that include adaptation, robustness and a distributed, decentralized nature, making them well-suited for clustering in p2p networks, in which it is difficult to implement centralized network control. We have applied this algorithm on synthetic and real world datasets and we have measured the impact of the flocking search strategy on performance in terms of accuracy and scalability.

1 Introduction

Clustering algorithms have been applied to a wide range of problems, including exploratory data analysis, data mining, image segmentation and information retrieval. In all of these disciplines the common problem is that of grouping similar objects according to their distance, connectivity, or their relative density in space [5] [9].

Traditional clustering methods require that all data have to be located at the site where they are analyzed and cannot be applied where there are multiple distributed datasets unless all data are transferred in a single node and then clustered. Today's large-scale data sets are usually logically and physically distributed, requiring a distributed approach to mining. Huge amounts of data are stored in autonomous, geographically distributed sources over networks with limited bandwidth and large number of computational resources. Such computing systems include Grid computing platforms, sensor networks and peer-to-peer (P2P) computing environments. The scale of these systems poses several difficulties, such as the impracticability of global communications and global synchronization, dynamic topology changes of the network, on-the-fly data updates and the frequent failure and recovery of resources. In last years, many effective

E. Talbi et al. (Eds.): EA 2005, LNCS 3871, pp. 37–48, 2006.
© Springer-Verlag Berlin Heidelberg 2006

and scalable clustering methods have been developed [8] [13] [14] [7] but they cannot manage with efficiency large-scale distributed clustering problem.

Recently, other data mining paradigms based on biological models [10] [11] have been introduced to solve the clustering problem. These paradigms are characterized by the interaction of a large number of simple agents that sense and change their environment locally. Ants' colonies, flocks of birds, termites, swarms of bees etc. are agent-based insect models that exhibit a collective intelligent behavior (*swarm intelligence*) [1] that may be used to define new distributed clustering algorithms. In these models, the emergent collective behavior is the outcome of a process of self-organization, in which insects are engaged through their repeated actions and interaction with their evolving environment. Intelligent behavior frequently arises through indirect communication between the agents using the principle of **stigmergy** [3]. This mechanism is a powerful principle of cooperation in insect societies. According to this principle an agent deposits something in the environment that makes no direct contribution to the task being undertaken but it is used to influence the subsequent behavior that is task related. Swarm intelligence (SI) models have many features in common with Evolutionary Algorithms (EA). Like EA, SI models are population-based. The system is initialized with a population of individuals (i.e., potential solutions). These individuals are then manipulated over many iteration steps by mimicking the social behavior of insects or animals, in an effort to find the optima in the problem space. Unlike EAs, SI models do not explicitly use evolutionary operators such as crossover and mutation. A potential solution simply 'flies' through the search space by modifying itself according to its past experience and its relationship with other individuals in the population and the environment.

We believe that this biology-inspired paradigm could serve as a basis for supporting the design of completely distributed clustering algorithms in large scale systems with a dynamic environment as P2P systems. The advantages of SI are twofold. Firstly, it offers intrinsically decentralized algorithms that can use P2P systems quite easily. Secondly, these algorithms show a high level of robustness to change by allowing the solution to dynamically adapt itself to global changes by letting the agents self-adapt to the associated local changes.

In this paper, we present P-SPARROW a novel algorithm which uses the concepts of a flock of birds that move together in a complex manner with simple local rules, to cluster spatial data in P2P systems. P-SPARROW assumes that the objects to be clustered are created and located at local sites. Each local site situates its own objects in a local 2D cellular space. P-SPARROW clusterizes data independently on the different local sites by a smart exploratory strategy based on a colored flock of birds combined with a density-based clustering method. On each cellular space, a set of agents, equipped with a set of attributes, will be working with the goal to discover local clusters. The flocking algorithm determines a local model that consists of a set of representative agents (RAs). Each RA represents a *skeletal* object in which the cardinality of the neighborhood exceeds some threshold. At intervals, as RAs discover skeletal points, each node transfers the agents to neighboring nodes. In the receiving nodes, all the objects that are in the neighborhood of these agents are considered belonging to the same cluster. Furthermore, as the clusters are discovered they are merged using an iterative distributed labeling strategy to generate global labels with which identify the clusters of all nodes. P-SPARROW has a number of nice properties. It has the advantages of being easily implementable on distributed systems as P2P networks and it is robust compared to the failure of individual agents. It can also be applied to perform efficiently *approximate clustering* since the points that are, to each iteration, visited and analyzed by the flock of agents represent a significant (in *ergodic* sense) subset of the entire dataset. The subset reduces the execution time since reduces the space of solutions that a clustering algorithm has to search keeping the accuracy loss as small as possible. P-SPARROW has no centralized coordinator. Each node acts independently from each other and intermediate results may be overturned as new data arrives. P-SPARROW behaves as an *anytime* algorithm in which the quality of results improves as computational time increases. Each node maintains an assumption of the correct result and updates it whenever new skeletal objects are discovered. During the execution of the algorithm, if the system remains static, then the solution will quickly converge toward an exact solution. However, in a dynamic system, where nodes dynamically join or depart and the data changes over time, the changes are quickly and locally adjusted to, and the solution continues to converge. This property is particularly interesting if continuous data are analyzed. Furthermore, each node communicates only with its immediate neighbors. Locality implies that the algorithm is scalable to very large networks. Another outcome of the algorithm's locality is that the communication load it produces is small and decreases with the time. We have implemented P-SPARROW in a P2P network to investigate the interaction of the parameters that characterize the algorithm. The remainder of this paper is organized as follows. Section 2 introduces P-SPARROW, presents the principles of the density-based algorithms and describes the heuristics of the proposed method. Section 3 discusses the obtained results and Section 4 draws some conclusions.

2   A Flocking Algorithm for Distributed Clustering

In this section we present P-SPARROW a multi-agent distributed clustering algorithm implemented in a P2P network which combines the stochastic search of an adaptive flocking with a density-based clustering method and an iterative self-labeling strategy to generate global labels with which identify the clusters of all nodes. Since P-SPARROW utilizes the principles of the conventional density-based algorithms, some of them are described first. After, the algorithm proposed is described.

2.1   Density-Based Clustering

Density-based clustering methods try to find clusters based on the density of points in regions. Dense regions that are reachable from each other are merged to formed clusters. The key idea is that for each point of a cluster the neighborhood of a given radius has to contain at least a minimum number of points, i.e. the density in the neighborhood has to exceed some threshold. The shape of a neighborhood is determined by the choice of a distance function for two points p and q, denoted by *dist(p,q)*. Density-based clustering methods excel at finding clusters of arbitrary shape. Examples of density-based clustering methods include DBSCAN [2] and DBRS [15]. DBSCAN is one of the most popular density-based spatial clustering algorithms. A complete description of the algorithm and its theoretical basis is presented in the paper by Ester et al. [2]. In the following we briefly present the main principles of DBSCAN. The algorithm is based on the idea that all points of a data set can be regrouped into two classes: *clusters* and *noise*. Clusters are defined as a set of dense connected regions with a given radius (*Eps*) and containing at least a minimum number (*MinPts*) of points. Data are regarded as noise if the number of points contained in a region falls below a specified threshold. The two parameters, Eps and MinPts, must be specified by the user and allow to control the density of the cluster that must be retrieved. The algorithm defines two different kinds of points in a clustering: *core points* and *non-core points*. A core point is a point with at least MinPts number of points in an Eps-neighborhood of the point. The non-core points in turn are either *border points* if they are not core points but they are *density-reachable* from another core point or *noise points* if they are not core points and are not density-reachable from other points. To find the clusters in a data set, DBSCAN starts from an arbitrary point and retrieves all points with the same density-reachable from that point using Eps and MinPts as controlling parameters. A point *p* is density reachable from a point *q* if the two points are connected by a chain of points such that each point has a minimal number of data points, including the next point in the chain, within a fixed radius. If the point is a core point, then the procedure yields a cluster. If the point is on the border, then DBSCAN goes on to the next point in the database and the point is assigned to the noise. DBSCAN builds clusters in sequence (that is, one at a time), in the order in which they are encountered during space traversal. The retrieval of the density of a cluster is performed by successive spatial queries. Such queries are supported efficiently by spatial access methods such as R*-trees. DBSCAN is not suitable for finding *approximate* clusters in very large datasets. DBSCAN starts to create and expand a cluster from a randomly picked point. It works very thoroughly and completely accurately on this cluster until all points in the cluster have been found. Then another point outside the cluster is randomly selected and the procedure is repeated. This method is not suited to stopping early with an approximate identification of clusters.

DBRS modifies DBSCAN introducing an approximate clustering method which can produce approximate purity density-based clusters with far fewer region queries. The intuition behind DBRS is that a cluster can be viewed as a minimal number of core points (called *skeletal* points) and their neighborhoods. In a dense cluster, a neighborhood may have far more than *MinPts* points, but examining the neighborhoods of these points in detail is not worthwhile, because we already know that these points are part of a cluster. If an unclassified point in a neighbor's neighborhood should be part of this cluster, we are very likely to discovery this later when we select it or one of its other unclassified neighbors.

To find cluster, it is sufficient to perform region queries on the skeletal points. However, identifying skeletal points is NP-complete. Instead, it is possible randomly select sample points, find their neighborhoods, and merge their neighborhoods if they intersect. If enough samples are taken, a close approximation to the cluster without checking every point can be found. The sample points may not be the skeletal points, but the number of region queries can be significant fewer than for DBSCAN for datasets with widely varying densities. Likewise of DBSCAN also DBRS is a centralized clustering algorithm.

Recently, DBDC a distributed version of DBSCAN algorithm has been presented in [6]. DBDC uses DBSCAN to make local clustering and determinate a local model after the local clustering is finished. All information which is comprised within the local model, i.e. the representatives and their corresponding e-ranges, is sent to a global server site, where a global clustering representation is produced from local representations. Based on this small number of representatives, the global clustering can be done very efficiently. After having created a global clustering, the complete global model is sent back to all client sites. The client sites relabel all objects located on their site independently from each other. DBDC does not scale-up well since, in a P2P scenario, no centralized coordinator and limited communications are required.

2.2   The P-SPARROW Clustering Algorithm

As in DBRS, P-SPARROW finds cluster performing region-queries on skeletal points but it replaces the random search of the skeletal points with a stochastic multi-agent search that discovers in parallel skeletal points. P-SPARROW is constituted of two phases: a local phase for the discovery of the skeletal points on each node and a iterative phase that concerns a global relaxation process in which nodes exchange cluster labels with nearest neighbors until a fixed point (i.e. all nodes detect no change in the labels) is reached.

On each peer, P-SPARROW uses a 2D cellular space to situate objects that must be clustered. In the cellular space, objects have a global position. All objects are partitioned among the cellular spaces without replication. Each node uses a flocking algorithm constituted by a fixed number of agents that occupy a randomly generated position to explore its own cellular space. Each agent moves around the cellular space testing the neighborhood of each object (point) it visits in order to verify if the point can be identified as a *skeletal point*. Then, P-SPARROW uses a flocking algorithm with an exploring behavior in which individual members (agents) to first explore the environment searching for goals whose positions were not know *a priori*, and then, after the goals are located, all the flock members should move to the goals. Agents search the goals in parallel and signal the presence or the lack of significant patterns into the data

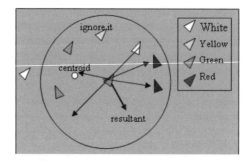

Fig. 1. Computing the direction of a green agent

to other flock members, by changing color. The entire flock then moves towards the agents (*attractors*) that have discovered interesting regions to help them, avoiding the uninteresting areas that are instead marked as obstacles. The color is assigned to the agents by a function associated with the data analyzed during the exploration according to the DBSCAN density-based rules with the same parameters: *MinPts*, the minimum number of points to form a cluster and *Eps*, the maximum distance that the agents can look. In practice, the agent computes the local density (*density*) in a circular neighborhood (with a radius determined by its limited sight, i.e. *Eps*) and then it chooses the color (and the speed) in accordance to the following simple rules:

$$
\begin{aligned}
density > MinPts &\Rightarrow mycolor = red \quad (speed = 0) \\
\tfrac{MinPts}{4} < density \leq MinPts &\Rightarrow mycolor = green \quad (speed = 1) \\
0 < density \leq \tfrac{MinPts}{4} &\Rightarrow mycolor = yellow \quad (speed = 2) \\
density = 0 &\Rightarrow mycolor = white \quad (speed = 0)
\end{aligned}
$$

So *red*, reveals a high density of interesting patterns in the data, *green*, a medium one, *yellow*, a low one, and white, indicates a total absence of patterns. The color is used as a communication mechanism among flock members to indicate them the roadmap to follow. The roadmap is adaptively adjusted as the agents change their color moving to explore data until they reach the goal. The main idea behind our approach is to take advantage of the colored agent in order to explore more accurately the most interesting regions (signaled by the red agents) and avoid the ones without clusters (signaled by the white agents). Red and white agents stop moving in order to signal this type of regions to the others, while green and yellow ones fly to find more dense clusters. Indeed, each flying agent computes its heading by taking the weighted average of alignment, separation and cohesion (as illustrated in figure 1).

Green and yellow agents compute their movement observing the positions of all other agents that are at most at some fixed distance (*dist_max*) from them and applying the rules of Reynolds' [12] with the following modifications:

- *Alignment* and *cohesion* do not consider yellow agents, since they move in a not very attractive zone.

- *Cohesion* is the resultant of the heading towards the average position of the green flockmates (centroid), of the attraction towards red agents, and of the repulsion by white agents.
- A *separation* distance is maintained from all the agents, whatever their color is.

Agents will move towards the computed destination with a speed depending from their color: green agents will move more slowly than yellow agents since they will explore denser zones of clusters. An agent will speed up to leave an empty or uninteresting region whereas it will slow down to investigate an interesting region more carefully. The variable speed introduces an adaptive behavior in the algorithm. In fact, agents adapt their movement and change their behavior (speed) on the basis of their previous experience represented from the red and white agents.

P-SPARROW assumes that the objects to be clustered are created and located at local sites. On each node (as described in the pseudocode of figure 2), the set of red agents determinates the local model of clustering. The local models must be incrementally combined within the cellular spaces of the neighbors and transferred to the entire network. To this end, red agents create clone agents. At intervals, when a fixed number of clone agents is achieved (i.e. a bag of agents has reached the desired dimension) or a certain number of iterations are performed, each node sends new clone agents only to a small group of neighbor peers. The number of messages exchanged is reduced by grouping several agents within one message. In the receiving nodes, the clone agents will occupy the same correspondent position in the local cellular spaces and will continue their execution.

```
for i=1..Max Generations
    foreach agent (yellow, green)
        age=age+1;
        if (age>Max_Life)
            generate_new_agent(); die();
        endif
        if (not visited (current_point))
            density = compute_local_density();
            mycolor= color_agent(density);
    endif
    end foreach
    foreach agent (yellow, green)
                dir= compute_dir();
    end foreach
    foreach agent (all)
        switch (mycolor){
                case yellow, green:    move(dir, speed(mycolor)); break;
                case white:            stop (); generate_new_agent();break;
                case red:              stop (); merge (); if (new_red()) clone_agent(); break;
                }
        end foreach
        if (bag_out dimension()> threshold1)
            send_bag();
        if (bag_in_full())
            notify_changes ();
    end for
```

Fig. 2. The pseudo-code of P-SPARROW executed on every node

Temporary labels will be given to red and clone agents and to all the objects contained in their neighborhood. If any objects have already a label then the label of the agents is set to the minimum of the two. Subsequently, red and clone agents will run the merge procedure. Agents, during the merge phase, continuously update the labels as multiple clusters take shape concurrently. On each node, changes in the label of the agents will be communicated to the neighboring nodes. So, the labels are continuously set making a repetitive comparison between the label of the agents and those of the objects belonging to the neighborhood. This continues until nothing changes, by which time all the clusters will have been labeled with the minimum initial label of all the sites containing the data. At the end all sites are labeled with their "local representative" labels which are then globalized (i.e. made unique over the whole system).

During simulations a cage effect, was observed; in fact, some agents could remain trapped inside regions surrounded by red or white agents and would have no way to go out, wasting useful resources for the exploration. So, a limit on their life was imposed to avoid this effect; hence, when their age exceeded a determined value (maxLife) they were killed and were regenerated in a new randomly chosen position of the space.

3 Experimental Results

In this section, we want to analyze the goodness of our algorithm in the task of performing approximate clustering and we want to verify some interesting properties of our distributed system (i.e. accuracy, scalability, etc..). In our experiments, we used three two-dimensional spatial datasets (showed in 3): two synthetic (called GEORGE and DS1-CURE) and one real (SEQUOIA). GEORGE consists of 8000 points and it is characterized from a large number of noise points; DS1-CURE (used in [4]) contains 100000 points in three circles and two ellipsoids connected by a chain of outliers and random noise scattered in the entire space; SEQUOIA was composed by 62556 names of landmarks (and their coordinates), and was extracted from the US Geological Survey's Geographic Name Information System.

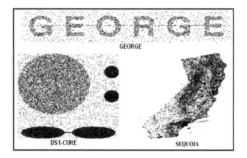

Fig. 3. The three datasets used in our experiments

Where not differently specified, we run our algorithm on five peers using 50 agents working until they explore the 2%, 10%, 15% and 30% of the entire data set. All the experiments were averaged over 20 tries.

3.1 Accuracy

First of all, we want to evaluate the capacity of our algorithm in finding approximate clusters, exploring only a portion of the entire dataset. Our algorithm uses the same parameters of DBSCAN. Therefore, if we visited all the points of the dataset, we would obtain the same results as DBSCAN. Then, in our experiments we consider as 100% the cluster points found by DBSCAN (note DBSCAN visits all the points). We want to verify how we come close to this percentage visiting only a portion of the entire dataset. In figure 4 we can observe the experimental results. Exploring the 10% of points we are able to individuate more than 50% (and in some cases 70%)of the points of the clusters (with the exception of the Big-Circle cluster of the DS1-Cure dataset, as it has a really low density compared to the others) and with the 15% we succeed in discovering more than 70% and even 80% of the points. Next, we want to determine the effect of using P-SPARROW search strategy as opposed to a random-walk search strategy in order to identify clusters; so we implemented a version of the algorithm changing only the phase of search, i.e. replacing the flock strategy with the random-walk one. Results are illustrated in figure 5 (a) for the SEQUOIA dataset, but the same considerations are valid for the other two datasets. At the beginning, the random strategy overcomes P-SPARROW, but, after about 300 visited points, the flock presents a superior behavior because of the adaptive behavior of the algorithm that allows agents to learn on their previous experience. An interesting property of our strategy consists in finding more points belonging to clusters

Clustering (GEORGE)	Perc. of data points for cluster found by P-SPARROW			
	2%	10%	15%	30%
G	14,50%	50,18%	70,58%	93,38%
E	12,34%	64,25%	73,47%	94,25%
O	16,44%	54,21%	74,95%	93,16%
R	16,20%	56,22%	71,12%	93,43%
G	15,62%	51,82%	73,17%	94,55%
E	16,03%	58,16%	79,14%	94,92%

Clustering (DS1-Cure)	Perc. of data points for cluster found by P-SPARROW			
	2%	10%	15%	30%
Big Circle	14,09%	32,45%	65,28%	91,10%
Circle 1	57,98%	71,23%	80,66%	94,55%
Circe 2	58,14%	77,29%	81,04%	93,87%
Ellipse 1	58,31%	75,78%	86,30%	94,11%
Ellipse 2	54,63%	78,93%	86,58%	93,99%

Clustering (Sequoia)	Perc. of data points for cluster found by P-			
	2%	10%	15%	30%
Philadelphia	39,53%	58,52%	78,81%	95,17%
New York	37,72%	63,46%	77,35%	94,37%
Boston	42,53%	67,77%	83,68%	95,23%

Fig. 4. Number of clusters and number of points for cluster for George, DS1-Cure, Sequoia (percentage in comparison to the total number of points for cluster) when P-SPARROW analyse 2%, 10%,15% and 30% of total points, using 5 peers

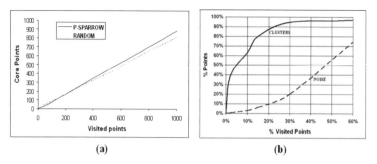

Fig. 5. a) Number of core points found for P-SPARROW and Random strategy vs. total number of visited points for the SEQUOIA dataset. b) Number of noise and cluster points found for P-SPARROW vs. number of visited points for the SEQUOIA dataset.

than noise points. In fact, if we observe the figure 5 (b), we can note that until the 30% of visited points, the algorithm is able to find a large percentage of cluster points. After this threshold a few new cluster points are discovered and more noise points are found. So it is not convenient go on searching beyond this value.

3.2 Scalability

In order to evaluate the performance of our algorithm, we image to have a certain number of P2P networks, with different configurations. Each configuration consists of a variable number of nodes, where the data are located. For large networks, the density of points for cluster for peer necessarily decreases; so we

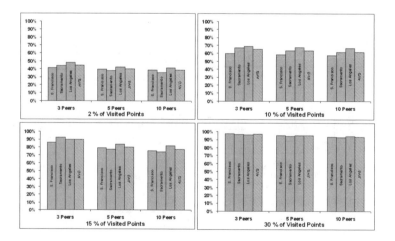

Fig. 6. Number of points for cluster for Sequoia (percentage in comparison to the total number of points for cluster) when P-SPARROW analyses 2%, 10%, 15% and 30% of total points, using 3, 5 and 10 peers

have to choose a different value of the parameter MinPts to keep into account this aspect. In practice, we choose a value of MinPts inversely proportional to the number of peers (i.e. if we fix MinPts as 4 on 10 peers, we have to fix as 8 on 5 peers). In figures 6, we show the experimental results concerning the scalability of the algorithm by varying the number of peers for the SEQUOIA dataset. We obtained a reduction from 89% to 77%. Note the scalability is quite good for all datasets, since if we consider the DS1-Cure dataset, exploring only the 15% of the data points, we are able to find on average more than 70% of the clusters, also using 10 peers, with a reduction from 87% (3 peers) to 71% (10 peers). This larger reduction is due to the cluster representing the big circle; in fact, for it, we have a reduction in accuracy from the 88% with 3 peers to the 42% with 10 peers, because the low density of this cluster, after the decomposition in ten parts, and consequentially the number of points on every peer is not sufficiently large to make the algorithm work with a discrete approximation. Also using the George dataset (for the 15% case), we had a reduction from 77% to 71% on average, as the clusters are composed of few points, so the reduction is considerable.

4   Conclusions

This paper describes the P-SPARROW algorithm for distributed clustering of data in peer-to-peer environments. The algorithm combines a smart exploratory strategy based on a flock of birds with a density-based strategy to discover clusters of arbitrary shape, size in spatial data. The algorithm has been implemented in a peer-to-peer system and evaluated using two synthetic datasets and one real word dataset. Measures of accuracy of the results show that P-SPARROW can be efficiently applied as a data reduction strategy to perform approximate clustering. Moreover, the adaptive search strategy of P-SPARROW is more efficient than that of the random-walk search strategy. Finally the algorithm shows a good scalable behavior.

Acknowledgments

This work has been supported by MIUR programme L.449/97 "Enabling ITC distributed complex platforms" and by FIRB strategic project on "Enabling Technologies for Information Society, Grid.it(RBNE01KNFP)".

References

[1] Eric Bonabeau, Marco Dorigo, and Guy Theraulaz. Swarm intelligence: From natural to artificial systems. *J. Artificial Societies and Social Simulation*, 4(1), 2001.

[2] Martin Ester, Hans-Peter Kriegel, Jorg Sander, and Xiaowei Xu. A density-based algorithm for discovering clusters in large spatial databases with noise. In *Proc. 2nd Int. Conf. on Knowledge Discovery and Data Mining*, pages 226–231, 1996.

[3] P.P. Grass. *La Reconstruction du nid et les Coordinations Inter-Individuelles chez Beellicositermes Natalensis et Cubitermes sp. La Thorie de la Stigmergie : Essai d'interprtation du Comportement des Termites Constructeurs in Insect. Soc. 6.* Morgan Kaufmann, 1959.

[4] Sudipto Guha, Rajeev Rastogi, and Kyuseok Shim. CURE: an efficient clustering algorithm for large databases. In *ACM SIGMOD International Conference on Management of Data*, pages 73–84, June 1998.

[5] Jiawei Han and Micheline Kamber. *Data Mining: Concepts and Techniques (The Morgan Kaufmann Series in Data Management Systems).* Morgan Kaufmann, September 2000.

[6] Eshref Januzaj Hans-Peter. Towards effective and efficient distributed clustering, 2003.

[7] M. Kamber J. Han and a. K. H. Tung. *Spatial Clustering Methods in Data Mining: A Survey, Geographic Data Mining and Knowledge Discovery.* Morgan Kaufmann, 2001.

[8] George Karypis, Eui-Hong Han, and Vipin Kumar. Chameleon: Hierarchical clustering using dynamic modeling. *IEEE Computer*, 32(8):68–75, 1999.

[9] L. Kaufman and P. J. Rousseeuw. *Finding Groups in Data: An Introduction to Cluster Analysis.* John Wiley, 1990.

[10] Deneubourg J. L., Goss S., Franks N., Sendova-Franks, A. C. Detrain, and L. Chretien. The dynamics of collective sorting robot-like ants and ant-like robots. In *From Animals to Animats: Proc. of the 1st Int. Conf. on Simulation of Adaptive Behaviour.* MIT Press/Bradford Books, 1990.

[11] Nicolas Monmarché, M. Slimane, and Gilles Venturini. On improving clustering in numerical databases with artificial ants. In *ECAL '99: Proceedings of the 5th European Conference on Advances in Artificial Life*, pages 626–635, London, UK, 1999. Springer-Verlag.

[12] Craig W. Reynolds. Flocks, herds and schools: A distributed behavioral model. In *SIGGRAPH '87: Proceedings of the 14th annual conference on Computer graphics and interactive techniques*, pages 25–34, New York, NY, USA, 1987. ACM Press.

[13] Jörg Sander, Martin Ester, Hans-Peter Kriegel, and Xiaowei Xu. Density-based clustering in spatial databases: The algorithm gdbscan and its applications. *Data Min. Knowl. Discov.*, 2(2):169–194, 1998.

[14] Wei Wang, Jiong Yang, and Richard R. Muntz. Sting: A statistical information grid approach to spatial data mining. In *VLDB'97, Proceedings of 23rd International Conference on Very Large Data Bases, August 25-29, 1997, Athens, Greece*, pages 186–195. Morgan Kaufmann, 1997.

[15] Xin Wang and Howard J. Hamilton. Dbrs: A density-based spatial clustering method with random sampling. In *PAKDD*, pages 563–575, 2003.

Simultaneous Optimization of Weights and Structure of an RBF Neural Network

Virginie Lefort, Carole Knibbe, Guillaume Beslon, and Joël Favrel

INSA-IF/PRISMa, 69621 Villeurbanne CEDEX, France
virginie.lefort@insa-lyon.fr

Abstract. We propose here a new evolutionary algorithm, the RBF-Gene algorithm, to optimize Radial Basis Function Neural Networks. Unlike other works on this subject, our algorithm can evolve both the structure and the numerical parameters of the network: it is able to evolve the number of neurons *and* their weights.

The RBF-Gene algorithm's behavior is shown on a simple toy problem, the 2D sine wave. Results on a classical benchmark are then presented. They show that our algorithm is able to fit the data very well while keeping the structure simple – the solution can be applied generally.

1   Introduction

Radial Basis Function Neural Networks (RBF NN) [1] are widely used in regression and classification tasks. They are often coupled with evolutionary algorithms, especially Genetic Algorithms (GAs) [2, 3, 4]. The GA is used to find the optimal parameters of the network.

However, in a neural network, all parameters cannot be considered to be equivalent. In particular, one can distinguish structural parameters from scalar parameters. The former define the general structure of the network: number of layers, number of neurons by layer and the topology of the network; they directly influence the capabilities of the network. The later, namely the neural weights and bias parameters, define the precise input-output mapping of the network. The scalar parameters clearly strongly depend on the structural parameters.

A classical GA can easily encode the scalar parameters, as their number is known once the structure is chosen, and so we would expect it to perform well in this regard. However, a more sophisticated evolutionary algorithm will be required to optimize the structural parameters in conjunction with the scalar parameters.

In this context, we present here a new Genetic Algorithm that can optimize simultaneously both the structural and scalar parameters of a feed-forward RBF NN. We have named it the RBF-Gene algorithm.

In the next section, we will briefly present some work on how to optimize an RBF-NN using Genetic Algorithms. We will then present our algorithm. In the last section, we will present the results obtained with it and compare with other published results on some benchmark tests.

E. Talbi et al. (Eds.): EA 2005, LNCS 3871, pp. 49–60, 2006.

2   Evolving Neural Networks Using Genetic Algorithms

Radial Basis Function Neural Networks are classical neural models with one layer of hidden neurons. They are used in classification or regression tasks as they are universal approximators.

In an RBF NN, the output of the network o is a weighted sum of the output of H hidden neurons. The transfer function of each neuron is a Gaussian function $g(\boldsymbol{X})$:

$$o(\boldsymbol{X}) = \sum_{j=0}^{H} w_j.g_j(\boldsymbol{X}) = \sum_{j=0}^{H} w_j.e^{\frac{-\|\boldsymbol{X}-\boldsymbol{\mu}_j\|^2}{\sigma_j^2}}$$

with $\boldsymbol{X}$ the input vector of the network, $\boldsymbol{\mu}_n$ the vector representing the center of the Gaussian for the nth hidden neuron and σ_n its standard deviation.

Implementing an RBF NN is a two-stages process. First, we have to choose the structure of the network; second, we must find its free parameters. Once the structure is fixed, the free parameters can be chosen manually, often the case for the Gaussian parameters; analytically, with high sensitivity to the quality of the dataset; or by a learning algorithm, generally applied only on the output weights w_{ij}. However the free parameters, and therefore the behavior of the network, strongly depend on the network's structure.

In "difficult" cases, with few examples or significant noise, an evolutionary algorithm (like GAs) is a good solution. GAs can be used to optimize the parameters of the neurons directly: The structure is fixed at the beginning of the run and the algorithm finds strong values for the fixed number of parameters.

Goldberg [5] used GAs in this manner as early as 1989. It is easy to use and is therefore rather popular (e.g. [2].)

In another approach, GAs are used to optimize the structure of the network [3, 6, 7]. Then another algorithm such as a learning algorithm optimizes the scalar parameters for each structure. This method is time consuming since for each step in the structure algorithm we have to perform a complete optimization on the scalar parameters. Moreover the fitness associated to a specific structure strongly depends on the optimization algorithm used to compute the weights.

To overcome the limits of these approaches, "integrative" GAs have been proposed. The idea is to optimize simultaneously the structure and the weights of the network. The main problem is then the encoding of the chromosome. Two different solutions exist: the "direct encoding" scheme, where the structure and the weights are directly encoded in the same string; and the "indirect encoding" scheme, where the chromosome contains generative instructions used to build the network.

The direct encoding method was used successfully in several works (e.g. [4, 8]) with variable-length chromosomes. But the encoding introduces difficulties in creating offspring from parents[1]. One solution is to use species, where each

---

[1] Chromosomes are typically made of a structural part and a parameter part. Since the size of the parameter sequence depends on the values encoded in the structural sequence, the crossover operator often doesn't make sense.

species represent networks with the same structure. Since genetic exchange may not be possible between species, such an algorithm is closely related to and has many of the problems of two-phases optimization.

The direct encoding scheme also suffers from the "permutation problem"[2] that leads to the failure of the crossover. Some work has been done on this problem [9] but extra computations on the individuals, such as reordering the genes on the chromosome or ignoring duplicated genes, are required.

On the other hand, the indirect encoding scheme (see [3] for a review) is based on generative strategies: It doesn't directly encode the network but rather how to construct it from a seed. Generative strategies can be based on a grammar with a seed and rules or cellular growth, with instructions to add neurons and edges from one progenitor cell.

At first glance, this scheme seems to be better than direct encoding. However, much of the work has been done on simplified networks, such as binary weighted networks. Moreover mutations have counter-intuitive effects in these systems and the fitness landscape can be much more complicated than in a simple GA.

3   The RBF-Gene Model

In this context, we have developed an evolutionary algorithm based on GAs whose main goal is to optimize both the structure and the weights of a feed-forward RBF neural network. Our central idea is to build a chromosome in which each gene is an atomic entity that can be combined to the others to produce the phenotype (i.e. the RBF NN.) In order to avoid the difficulties encountered by generative approach, the product of one particular gene must not depend on the values of the other ones, nor on their relative positions along the chromosome.

Moreover, in order to avoid the "direct encoding" problem, we need to forbid any hierarchy between the genes. In other words, the genes encoded onto the chromosome must be homogeneous, i.e. they all code for the same kind of "basic block" of the answer.

3.1   Principles

This idea of homogeneous units is partly inspired by molecular biology and particularly by the encoding of the proteins on the chromosome. In biology, the translation process is only based on local rules: promoters, terminators, Shine-Dalgarno sequences, start/stop codons and so on. Consequently, adding, deleting or changing a gene (i.e. a protein) will have no consequences on the other proteins: The effect will only be visible at the global, phenotypic level. We wanted to have the same property for our algorithm as we wanted to make structural changes easy by adding, deleting or removing neurons or connections without changing the entire chromosome.

---

[2] The same genes are encoded in a different order in two individuals and so the one-point crossover leads to individuals without a special gene or with two copies of it.

In the RBF-Gene model, our basic block will be a "kernel": that is a complete hidden neuron together with all its numeric parameters, namely the mean and the standard deviation of the Gaussian, and the output weight.

Each kernel will be encoded on the chromosome as a "gene"; so the number of genes will indicate the number of kernels. In order to allow all the possible structures, we have to allow a variable number of kernels, i.e. a variable number of genes encoded onto the chromosome. The simplest way to do it is to have a variable-length chromosome.

As in biology, our genes will be located on the chromosome using purely local rules. In particular, we include two special sub-sequences indicating the beginning and the end of each gene. The chromosome is then a succession of coding and non-coding sequences of different size and purely local rules are required to decode it: if a coding sequence appears or disappears elsewhere in the genome, there will be no influence at all on the present genes. Some work has been done showing the interest of variable length chromosome and the use of coding and non-coding sequences, for instance [10, 11, 12].

Since all the kernels are equivalent, the order or the position of the corresponding gene on the chromosome doesn't matter and the permutation problem vanishes. Moreover, this property enables us to introduce rearrangements (i.e large scale mutation operators) that will help avoiding local optima by significantly changing the genetic code. Specifically, we will show that the sequence of copy-and-edit used by natural evolution of genes can be used by the RBF-Gene algorithm as well.

To create the next generation, we need to compute the fitness of each individual. This is done in three steps : First, using special sequences, we find the genes on the chromosome; second, using a genetic code, we extract all the parameters of the corresponding kernel; third, as we now have all the hidden neurons and their links, we construct the NN and test it on the dataset.

Once all the individuals are evaluated, we use a standard evolutionary process to compute the next generation. For the recombination/mutation step, we introduce rearrangement operators that change the structure of the chromosome. These operators modify the chromosome on a large scale independently of the nature of the region (coding or non-coding sequence).

3.2   Encoding

In the RBF-Gene algorithm, each gene encodes for the parameters of a hidden neuron: its mean vector μ and its standard deviation σ. Moreover, it also encodes for the output links w_i. So, if we have n input values and m output weights, we have $n + m + 1$ real values to define for each neuron (n for the mean vector, 1 for the standard deviation and m for the output weights).

The simplest way to encode a value onto a chromosome is to use a binary encoding. So we need a "0" base and a "1" base for each parameter. This gives us an artificial genetic code: Our chromosome will be a string of characters (A, B, C...) and each parametric value has two characters associated with: one for the "0" and one for the "1". As we want a homogeneous chromosome, using purely local signals to detect genes, we add two special characters for the start and the stop[3]. So our chromosome is a variable length string built with a $2(n+m+1)+2$ character alphabet.

Since we have a homogeneous chromosome, there are no special regions on the chromosome or on the gene. The different characters are mixed together at random (i.e. the parameters are not encoded sequentially). In order to compute a parameter, we only have to extract the corresponding characters: thanks to the genetic code, the mixed character string (the gene) is converted into three ordered binary strings (one per parameter). Each of them is then transformed into a numerical value using a Gray code [13].

One of the major advantages of our model is that gene's length is no longer fixed: it only depends on the relative position of a start character and the next stop character. Thus the number of bits per parameter is not fixed at all. This overcomes a classical drawback of binary encoding in GAs since the precision of each parameter is able to evolve by simply modifying the length of the gene. Consequently, the algorithm is able to generate rough solutions at first, with a small number of kernels or with numeric values of low precision. It can then progressively refine the input/output mapping by either adding new kernels or enhancing the precision of existing ones.

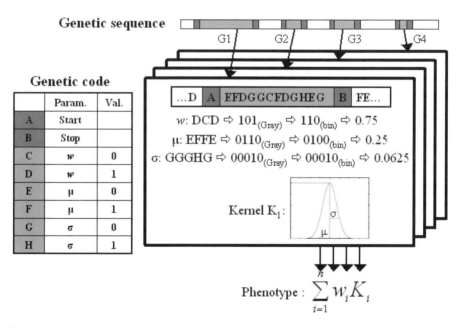

Fig. 1. A simple example of the mapping from the chromosome to the neurons. Here we only have one input value ($n = 1$) and one output value ($m = 1$). Thus the genetic code is composed of $2 + 2(1 + 1 + 1) = 8$ letters.

---

[3] Of course, we can find start signals inside a gene or stop signals between genes. In such a case they are ignored.

The overall translation algorithm can be summarized as:

1. Find the genes using the alphabetic start and stop signals
2. For each gene
   (a) Extract each parameter sequence using the "genetic" code translation
   (b) Decode each parameter using a variable-length Gray code
   (c) Build the associated Gaussian kernel K
3. Construct the neural network
4. Evaluate the individual on the data

A simple example of the first two steps is shown in figure 1.

3.3   Evolution

Our chromosomes are homogeneous: they are a simple string of characters. More-over the structure of the chromosome is free to evolve without any influence on the phenotype: a gene can move from one locus to an other or two genes can be swapped. So the chromosome length can vary and so does the number of genes and we have more biologically inspired operators available to us than the two classic ones: point mutation and crossover.

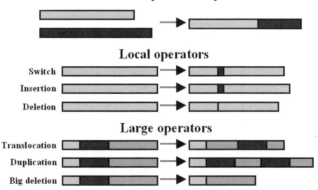

Fig. 2. A schematic view of the different operators

To create the next generation, we use two types of operators: first, the re-combination operator if needed, and then the mutation operators. The selection of the fittest is done by a roulette-wheel based on the rank[4].

The recombination operator used here is a classic one-point crossover and the crossing point is chosen randomly. The probability of using it can vary between 0% for clonal reproduction to 100% for sexual recombination.

---

[4] However, the RBF-Gene algorithm can be used with any selection operator without loss of its properties.

We have two families of mutation operators (see figure 2):

- Local operators: They modify only one base. As the structure can change, we can use the traditional switch operator but also the local insertion or deletion.
- Large operators (rearrangements): They modify the global structure either by translocation, which move one part elsewhere on the chromosome; duplication, which copy and paste a sequence elsewhere; or large deletion, which erase a part of the chromosome.

Notice that the local mutation rates are given per base while the rearrangements rates are given per chromosome. Thus, the mutation effect remains constant whatever the genome size: since the "cut and paste" points are chosen randomly, the average number of bases affected by one rearrangement directly depends on the genome size [14].

The structure of the chromosome can then change and genes can be added, deleted or moved along the genome. So, the algorithm can adapt:

- **the complexity of the solution** by changing the number of genes
- **the local precision of each parameter** by changing the number of bases in each individual gene
- **the structure of the genetic sequence** by changing the order of the genes or the length of the non-coding sequences. It can be seen as the evolution of the robustness and the evolvability of the solution: the algorithm can change the structure in order to resist best to deleterious mutations or to help evolution to find optima more quickly.

4   Simulations and Results

In order to illustrate the behavior of the algorithm, we will first present results on a simple toy problem (the 2D sine wave, a $\mathbb{R}^2 \to \mathbb{R}$ problem). Then, we will present a real regression problem, the Boston dataset (a $\mathbb{R}^{13} \to \mathbb{R}$ problem), and compare our results with other results on the same benchmark[5].

4.1   2D Sine Wave: Graphical Results

The 2D sine wave is a straightforward problem which is interesting because the results can easily be visualized. The goal is to approximate the curve:

$$y(\boldsymbol{x}) = 0.8sin(\frac{x_1}{4})sin(\frac{x_2}{2}), x_1 \in [0; 10], x_2 \in [-5; 5]$$

This test has been proposed by Orr [17] and we use the same protocol for our experiment. We generate a training set of 200 patterns sampled at random. We add a normally-distributed noise with $\sigma = 0.1$ and zero mean. The test set contains 400 noiseless samples arranged as a 20*20 grid covering the input ranges. The fitness used by the algorithm is the total squared error (SE) on the

---

[5] The algorithm has been tested on other datasets as well [15, 16]: the $sin(12x)$ problem, a $\mathbb{R} \to \mathbb{R}$ problem or the abalone dataset, a $\mathbb{R}^9 \to \mathbb{R}$ problem.

learning set (the fitness does not depend on the genetic parameters like the size or the number of genes). The test set, independent from the training set, is used to evaluate the generality of the solution as we want to fit the 2D sine curve at all points, not just at the training points.

We have done 5 runs of 5000 generations with the following parameters:

- Population size: 100 individuals of initially 200 random bases
- Local mutation rates: $5e - 4$ per base
- Rearrangements rates: 0.05 per genome
- Crossover rate: 0.6 to create an offspring from two parents (and so 0.4 from one parent)

Figure 3 summarizes some indicators of evolution: the fitness (total squared error) on the training set and the test set, the size and the number of neurons at the 5000th generation for the best individual of each run. The training fitness ranges from 1.7762 to 2.1736, the test fitness from 1.0499 to 1.9613, the size from 614 characters to 1342 and the number of neurons from 8 to 15. However, beyond the given results, this problem is of low enough dimension to see the structure of the proposed solution.

	Mean	Std. dev.
Learning fitness	1.902	0.156
Test fitness	1.432	0.339
Size	1066.0	276.6
Nb. of neurons	11.2	2.6

Fig. 3. Statistical results and indicators on the 2D Sine Wave problem for the best individual of each simulation after 5000 generations

The figure 4 shows the original points, the final surface and the different Gaussian functions. Each function is equivalent to a neuron. The individual shown is the best individual after 5000 generations.

4.2   The Boston Dataset

In order to be able to compare the performance of the RBF-Gene algorithm with other models, we have performed experiments on the Boston dataset. This dataset is a well-known benchmark that can be downloaded on the UCI Machine Learning Repository [18]. It is a real dataset made by the U.S Census Service concerning housing in the area of Boston.

There are 13 inputs and 1 output. Since the inputs have different dimensionality, we have normalized them before applying the algorithm (mean of 0 and standard deviation of 1). The output is between 0 and 50 and we have kept it unchanged in order to compare our results. There are 506 points in the dataset, and no data is missing.

We have done 25 simulations by set of parameters using different seeds and/or different partitions of the data (keeping a ratio of 5 learning points for 1 validation point). The fitness function used by the algorithm is the mean square error

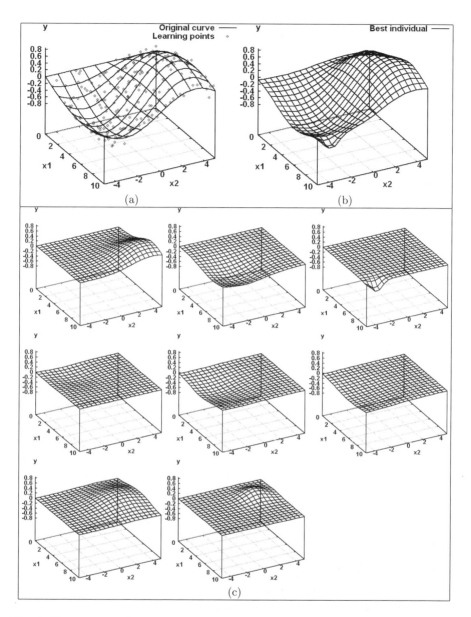

Fig. 4. The target surface and one generated individual after 5000 generations: (a) the desired curve and the training points; (b) one individual from the last generation; (c) the different Gaussian functions of the individual. Note that the noise in the data is partly extracted by the algorithm (kernel nb. 3).

(MSE) on the training set and we compare our results using the MSE on the validation set. The parameters are the same than in the 2D sine wave, save that the local mutation rates is $1e - 4$ per base here.

	Mean	Std. dev.	Median
Learning fitness	13.41	3.92	13.06
Validation fitness	16.70	4.59	16.60
Nb. of neurons	29.2	39.2	18.0

Fig. 5. Statistical results and indicators on the Boston dataset for the best individual of each simulation after 5000 generations

	Learning fitness	Validation fitness	Nb. of neurons
Minimum	7.35	10.19	4
Maximum	22.21	27.48	178

Fig. 6. Minimum and maximum values obtained by the best individual of each simulation after 5000 generations

Figure 5 summarizes our results on the dataset. We can compare them with [19] in which different methods are tested. We see that we have similar results to the best of the tested methods. However, since Madigan et al. don't provide enough statistical information to run a statistical test, we cannot make comparative claims with any certainty. Indeed, we have an average MSE of 16.70 ± 4.59 (best individual: 10.19) while the results in [19] range from 14.1 for the GBM two-way method to 25.8 for the Stagewise method.

Figure 6 shows the minimum and maximum value of each of learning fitness, validation fitness and number of neurons. When the standard deviation on the number of kernels is more than the mean, it is because we have a highly skewed distribution. 2 simulations have more than 100 kernels (126 and 178 respectively) while the other ones all have less than 60 kernels.

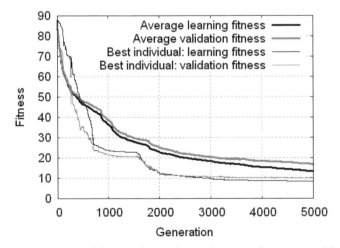

Fig. 7. Average fitness and best individual: the learning fitness is in black and the validation fitness in gray

Figure 7 shows the evolution of the fitness over the course of the runs (average fitness and the evolution of the fitnesses for the individual with the best validation fitness). As we can see, there is no over-learning as the validation fitness is still decreasing and the gap between the learning fitness and the validation fitness remains slight.

Moreover, during evolution, the structure of the genome changes and the number of genes increases progressively from 1 at the first generation to about 30 in the last generation. Of course, while the number of genes increases, so does the genome size (from 200 bases initially to a median size of about 6000 bases) but we also observe that the coding proportion (proportion of coding sequences on the genome) grows from about 20% to 60% without any hard-coded limit to the genome size.

5 Discussion and Future Works

The preliminary results we have obtained with the RBF-Gene algorithm are very encouraging. Quantitatively the algorithm performance is quite good. Qualitatively, the chromosome structure obviously evolves during the evolutionary process thus showing a genuine *simultaneous* evolution of the neural network structure and the scalar parameters.

However, more work has to be done. We would like to study the influence of different parameters such as the mutations rates. A better understanding of the evolution of the genome structure would be quite instructive, especially on the chromosome size, the individual gene size, and the genes' order. This work is in progress; early results suggest that the algorithm is very robust – that is it gives similar results over a wide range of parameters. We are also studying the possibility of using a real code instead of the binary Gray code for the parameter encoding.

References

1. Haykin, S.: Neural Networks - A Comprehensive Foundation. 2nd edn. Prentice Hall (1999)
2. Blanco, A., Delgado, M., Pegalajar, M.: A real-coded genetic algorithm for training recurrent neural networks. Neural Networks **14** (2001) 93–105
3. Kuşçu, I., Thornton, C.: Design of artificial neural networks using genetic algorithms: review and prospect. Technical Report 319, Cognitive and Computing Sciences, University of Sussex, Falmer, Brighton, Sussex, UK (1994)
4. Arotaritei, D., Negoita, M.G.: Optimization of Recurrent NN by GA with Variable Length Genotype. LNAI (Springer-Verlag) **AI 2002** (2002) 681–692
5. Goldberg, D.E.: Genetic Algorithms in Search Optimization and Machine Learning. Addison-Wesley (1989)
6. MacLeod, C., Maxwell, G.M.: Incremental Evolution in ANNs: Neural Nets which Grow. Artificial Intelligence Review **16** (2001) 201–224
7. Barrios, D., Manrique, D., Plaza, R.M., Juan, R.: An Algebraic Model for Generating and Adapting Neural Networks by Means of Optimization Methods. Annals of Mathematics and Artificial Intelligence **33** (2001) 93–111

8. Cliff, D., Harvey, I., Husbands, P.: Incremental evolution of neural network architectures for adaptive behaviour. Technical Report Cognitive Science Research Paper CSRP256, University of Sussex - School of Cognitive and Computing Science, Brighton BN1 9QH, England, UK (1992)
9. Thierens, D.: Non-Redundant Genetic Coding of Neural Networks. In: International Conference on Evolutionary Computation. (1996) 571–575
10. Levenick, J.R.: Inserting introns improves genetic algorithm success rate: Taking a cue from biology. In Belew, R., Booker, L., eds.: Proceedings of the Fourth International Conference on Genetic Algorithms, San Mateo, CA, Morgan Kaufman (1991) 123–127
11. Wu, A.S., Lindsay, R.K.: A survey of intron research in genetics. In Voigt, H.M., Ebeling, W., Ingo, R., Hans-Paul, S., eds.: Parallel Problem Solving From Nature IV. Proceedings of the International Conference on Evolutionary Computation. Volume 1141., Berlin, Germany, Springer-Verlag (1996) 101–110
12. Burke, D.S., De Jong, K.A., Grefenstette, J.J., Ramsey, C.L., Wu, A.S.: Putting more genetics into genetic algorithms. Evolutionary Computation 6 (1998) 387–410
13. Whitley, D.L., Barbulescu, L., Watson, J.P.: Local Search and High Precision Gray codes: convergence Results and Neighborhoods. (2000)
14. Knibbe, C., Beslon, G., Lefort, V., Chaudier, F., Fayard, J.: Self-adaptation of Genome Size in Artificial Organisms. In Capcarrere, M.S., al., eds.: Advances in Artificial Life, Proceedings of the 8th European Conference ECAL 2005. Volume 3630 of Lecture Note in Artificial Life (LNAI)., Springer-Verlag (2005) 423–432
15. Lefort, V., Knibbe, C., Beslon, G., Favrel, J.: The RBF-Gene Model. In: Proceedings of GECCO 04, Late Breaking Papers. (2004)
16. Lefort, V., Knibbe, C., Beslon, G., Favrel, J.: A bio-inspired genetic algorithm with a self-organizing genome: The RBF-Gene Model. In: Proceedings of GECCO 04. Lecture Notes in Computer Science, Springer-Verlag (2004)
17. Orr, M.J.L., Hallam, J., Takezawa, K., Murray, A.F., Ninomiya, S., Oide, M., Leonard, T.: Combining Regression Trees and Radial Basis Function Networks. International Journal of Neural Systems 10 (2000) 453–465
18. Blake, C., Merz, C.: UCI repository of machine learning databases (1998)
19. Madigan, D., Ridgeway, G.: Discussion of "Least Angle Regression" by Efron et al. The Annals of Statistics 32 (2004) 465–469

An Exponential Representation in the API Algorithm for Hidden Markov Models Training

Sébastien Aupetit[1], Nicolas Monmarché[1], Mohamed Slimane[1], and Pierre Liardet[2]

[1] Université François-Rabelais de Tours, Laboratoire d'Informatique, Polytech'Tours, 64, Av Jean Portalis, Tours 37200, France
{aupetit, monmarche, slimane}@univ-tours.fr
[2] Université de Provence, CMI, Laboratoire ATP, UMR-CNRS 6632, 39 rue F. Joliot-Curie, 13453 Marseille cedex 13, France
liardet@cmi.univ-mrs.fr

Abstract. In this paper, we show how an efficient ant based algorithm, called API and initially designed to perform real parameter optimization, can be adapted to the difficult problem of Hidden Markov Models training. To this aim, a transformation of the search space that preserves API's vectorial moves is introduced. Experiments are conducted with various temporal series extracted from images.

1 Introduction

The pattern recognition domain involves a wide class of problems which often require to learn and recognize temporal data. To perform this task, many various tools are available such as hidden Markov models (HMM). These statistical tools allow to model almost all temporal data so complex they are. They have been applied successfully in various research domains like speech or image recognition, but they suffer of a great disadvantage: the training problem is not completely solved. Indeed, there are only automatic training procedures that converge to local optima according to the training criterion. In some cases, local optima can be used with success while global optimum could achieve better recognition. That is precisely the question we are concerned in: finding the best or at least a good optimum for the training task. In the literature, many researches have been conducted to tackle this problem. Simulated annealing [1], Tabu search [2] and genetic algorithm [3, 4] have been proposed. These methods are often considered to be useful to learn HMMs but they have not been extensively benchmarked. However, those metaheuristics have been constructed to optimize discrete problems, while HMMs training mainly concerns problems with continuous parameters. So that, we may expect that continuous metaheuristics could provide better results.

A subfamily of these methods concerns artificial ant-based algorithms. Most of these algorithms require discrete search spaces and are applied to combinatorial problems. Recently, a new ant algorithm has been introduced in [5]: the API

E. Talbi et al. (Eds.): EA 2005, LNCS 3871, pp. 61–72, 2006.

algorithm, which is inspired from the foraging behavior of *Pachycondyla apicalis* ants. This algorithm is independent, in its general description, of the search space and consequently can be easily adapted to a continuous search space. Our goal is to modify the API algorithm in order to apply it to HMMs training task.

This paper is organized as follows: the first part recalls basic definitions concerning usual hidden Markov models, the second part presents the API algorithm, and the third part is devoted to its adaptation to the training of HMMs. In the last part, experimental results obtained from image data are presented and discussed.

2   Hidden Markov Models

HMMs are essentially stochastic finite state automata that extend Markov models by associating to each state a random output variable which produces the observed outcomes. They are statistical models designed for learning, recognition and analysis of temporal data. They are involved in numerous domains. For instance, they are used intensively for speech recognition and synthesis, and even in biology. A good review including many references can be found in [6]. Several kinds of HMMs have been proposed such as multidimensional HMMs with independent processes or hierarchical HMMs [7] to tackle specificities of the problem to solve. In this paper, we only consider first order HMMs in discrete time and discrete symbols.

Recall that a Markov model is a stochastic process consisting of a set $S = \{S_1, S_2, \ldots, S_N\}$ of N states and a S-valued Markov chain $(q_t(\cdot))_{t \geq 0}$ of order 1, that is to say, for all $t \geq 1$ the Markov property

$$\mathbf{P}(q_t = s_t | q_1 = s_1, \ldots, q_{t-1} = s_{t-1}) = \mathbf{P}(q_t = s_t | q_{t-1} = s_{t-1}) \quad s_t \in S \quad (1)$$

holds. A Markov chain is thus defined by a state transition matrix $A = (a_{ij})$ with $1 \leq i \leq N$ and $1 \leq j \leq N$ where a_{ij} is the transition probability $\mathbf{P}(q_t = S_j | q_{t-1} = S_i)$. For a stationary model, this probability depends neither on t nor on states prior to q_t. In addition to the stochastic matrix A, initial state probabilities (π_i) are given, where π_i is the probability that the chain starts from the state S_i i.e., $\pi_i = \mathbf{P}(q_0 = S_i)$. This model is useful when the current state is known explicitly. However, in many cases, states S_j are not directly observed but they emit symbols randomly from a given alphabet $V = \{V_1, V_2, \ldots, V_M\}$ according to emission probabilities $b_j(k)$. Let $(o_t)_{t \geq 1}$ denote the sequence of observed random variables. By definition $b_j(k) = \mathbf{P}(o_t = V_k | q_t = S_j)$, with $1 \leq j \leq N$ and $1 \leq k \leq M$. Moreover, the random variables o_t are mutually independent and independent of the $q_{t'}$ for $t \neq t'$. Finally, an HMM is thus defined by the triplet $\lambda = (A, B, \Pi)$ with $B = (b_j(k))$ and will be denoted by HMM(λ). For our purpose, the parameter λ will be viewed as a row vector of row vectors. More explicitly

$$\lambda = (A_1, \ldots, A_N, B_1, \ldots, B_N, (\pi_1, \ldots, \pi_N)) \quad (2)$$

where $A_i = (a_{i1}, \ldots, a_{iN})$ and $B_k = (b_k(1), \ldots, b_k(M))$ are row vectors respectively in $\mathbb{R}^N$ and $\mathbb{R}^M$ with non negative entries. Notice that here $\mathbb{R}^k$ denotes the space of k-dimensional real row vectors, equipped with the Euclidean norm $||\cdot||_2$. Hence, λ belongs to the vector space $(\mathbb{R}^N)^N \times (\mathbb{R}^M)^N \times \mathbb{R}^N$ identified to $\mathbb{R}^{N(N+M+1)}$. The set of all possible values of λ is the subset Λ of $[0, 1]^{N(N+M+1)}$ defined by

$$\sum_{j=1}^{N} \pi_j = 1, \quad \sum_{j=1}^{N} a_{ij} = 1, \quad \sum_{j=1}^{M} b_i(j) = 1, \qquad (1 \leq i \leq N). \tag{3}$$

For a given $\lambda \in \Lambda$, let $\mathbf{P}_\lambda$ be the probability of the underlying probability space related to $HMM(\lambda)$, and suppose that the sequence $\mathcal{O} = (O_1, \ldots, O_T) \in V^T$ is observed through HMM(λ). The likelihood $L_{\mathcal{O}}(\lambda)$ that the model HMM(λ) generates $\mathcal{O}$ is by definition

$$L_{\mathcal{O}}(\lambda) = \sum_{s \in S^T} \mathbf{P}_\lambda(\mathcal{O}, s) \tag{4}$$

where $\mathbf{P}_\lambda(\mathcal{O}, s)$ is set for $\mathbf{P}_\lambda(o_1 = O_1, \ldots, o_T = O_T, q_1 = s_1, \ldots, q_T = s_T)$. From the independent assumption

$$\mathbf{P}_\lambda(\mathcal{O}, s) = \pi_{s_1} \prod_{t=1}^{T-1} a_{s_t, s_{t+1}} \prod_{t=1}^{T} b_{s_t}(O_t) = \mathbf{P}_\lambda(\mathcal{O}|s)\mathbf{P}_\lambda((q_1, \ldots, q_T) = s) \tag{5}$$

is the probability, under HMM(λ), to observe $\mathcal{O}$ when the successive hidden states are $s = (s_1, \ldots, s_T)$. Now, four basic problems arise if we want to apply HMMs efficiently:

1. **Evaluation**: given a series of T observations $\mathcal{O} = (O_1, \ldots, O_T)$ and a model HMM(λ) with N hidden states, how to evaluate the likelihood $L_{\mathcal{O}}(\lambda)$? The solution is given by the Forward/Backward algorithm [8], which computes this value in polynomial time, precisely in complexity $\theta(N^2 T)$.
2. **Most probable path**: the Viterbi algorithm can find the sequence $s = (s_1, \ldots, s_T)$ of hidden states which have the highest probability of generating the series $\mathcal{O}$ of observations [9]. This algorithm has a complexity $\theta(N^2 T)$.
3. **Optimal adaptation**: this is the problem of learning a HMM from a series of observations $\mathcal{O}$. For a given number of hidden states N, how to find a model λ that maximizes $L_{\mathcal{O}}(\lambda)$? The Baum-Welch algorithm [10], denoted BW for short, provides a method that iteratively improves an initial model. Notice that BW suffers from drawbacks, inherent to gradient-like algorithm: it may converge to critical points of $L_{\mathcal{O}}(\lambda)$ like local minima or inflection points. It is also sensitive to the chosen initial model. The complexity of this algorithm is $\theta(TN(N + M))$.
4. **Optimal number of hidden states**: how many hidden states N we need for a given HMM(λ) in order to maximize $L_{\mathcal{O}}(\lambda)$? This problem is of course linked to the previous one and remains difficult. Most of learning algorithms,

like BW, need to have this number fixed. In this work, we consider an arbitrary fixed number of hidden states since our problem is only to find optimum of $L_O(\lambda)$, knowing N.

To explain empirically the difficulty of HMMs training, it is enough to notice that the likelihood $L_O(\lambda)$ is a polynomial in $N(N + M + 1)$ variables and of degree $\leq 2T$. However, stochastic constraints reduce the search space to the set Λ, that reduces the number of free parameters to $N^2 + N(M - 1) - 1$. As we can notice, the likelihood is a high degree polynomial and thus can have several local optima. This is the reason why it is very difficult to find a global optimum. The only simple way to find a solution near the global optimum is then to use metaheuristics.

3   The API Algorithm

3.1   Ant Algorithms

Ants have recently inspired new researches in computer science and many successful works deal with combinatorial optimization (see the review in [11]). In most of cases, a global memory is used to guide the search agents toward promising solutions. This is achieved in the same way that real ants spread volatile substances, known as pheromones, on their path leading to food sources. In our case, artificial pheromones are real values which are used with a positive feedback mechanism that reinforce promising solutions. Such a mechanism is often described as stigmergy [12]. A very successful metaheuristic, called ACO (for Ant Colony Optimization), have been applied to a large variety of optimization problems [13] such as, but far from limited to, the Traveling Salesman Problem [14, 15], the Quadratic Assignment Problem [16],or scheduling problems [17, 18].

Whereas many of ant algorithms have discrete search spaces, a little number of ant inspired methods can be found where continuous search spaces are used [19, 20, 21]. But, as we shall see in the following, not all of ant species use pheromones and so, in opposite to all optimization heuristics cited above, we do not need to use artificial pheromones to build ant algorithms.

3.2   The Foraging Strategy of *Pachycondyla apicalis* Ants

Pachycondyla apicalis [22] ants have two main characteristics regarding their foraging behavior: they memorize explorations of hunting sites around their nest and they systematically go back to the last site where a prey was found. Moreover the nest is regularly moved and represents a central point from which ants perform their explorations. When the location of the nest is changed, the ants are able to perform recruitments: an ant brings one of its colleagues to the new nest. This behavior is called tandem-running recruitment. Notice that these ants do not use pheromones to find their path in their natural environment, but they use visual landmarks instead. Since these ants demonstrate good performances in their prey research, they furnish a relevant model (called API) to solve optimization problems [23].

3.3   The API Algorithm for Continuous Search Spaces

The API algorithm is directly inspired by the foraging behavior of *Pachycondyla apicalis* ants. We have adapted the natural behavior of these ants to the general problem of optimization: a hunting site corresponds to a point in the solution space and an ant performs a prey capture when it succeeds to improve the value of the evaluation function defined on hunting sites. Hence, more ants improve their memorized hunting sites, better will be the results. This general principle can be applied with various search spaces like searching for neural networks weights or TSP [24] and for example, in [23], the results obtained by API on standard multimodal functions is compared against random hill climbing and a genetic algorithm. A previous model of *Pachycondyla apicalis* foraging behavior has been proposed in [25] and was focusing on the probabilities that one ant would leave its nest to search preys and about the learning process involved. However, our model does not take into account this learning process.

To apply the API algorithm on a search space $\mathcal{S}$, we only need to define two operators which are random generators:

- $\mathcal{O}_{\mathrm{rand}}(\mathcal{S})$, which generates a random point in $\mathcal{S}$ according to a uniform distribution. This operator is used to create a new nest location.
- $\mathcal{O}_{\mathrm{explo}}(s, A^i)$, which generates a point s' in the neighborhood of s according to the amplitude A^i of the ant a_i. This operator is used to create new hunting sites around the nest $\mathcal{N}$ (with $A^i = A^i_{\mathrm{site}}$ and $s = \mathcal{N}$) and to perform an exploration (with $A^i = A^i_{\mathrm{local}}$ and $s = s^i_j$, where s^i_j is a hunting site of ant a_i). If the new point s' is a better solution than s then the new hunting site position becomes s' ($s^i_j \leftarrow s'$). This improvement corresponds to a prey discovery.

Algorithm 1 describes the main principles of API algorithm seeking the global minimum of a function f. This algorithm is sequential but it can be easily transformed into a parallel algorithm where each ant is assigned to a different processor. The search space is Λ (see *supra*), which is a product of convex subsets, and we introduce the following parameters:

- m, the number of ants. The *Pachycondyla apicalis* colonies being small (from 40 to 100 workers) we often use a small number of ants.
- T_{max}, the number of iterations of the algorithm. A large value of T_{max} gives more chances to find better solutions.
- T_{nest}, the number of times one ant leaves its nest to search for a prey in between two nest moves. T_{nest} can also be fixed automatically to the value $T_{\mathrm{nest}} = m * 2 \times (P^i_{\mathrm{local}} + 1) \times p^i$ in order to give each ant enough iterations to explore all the sites in its memory.
- for each ant a_i:
  - p^i, the number of hunting sites memorized by ant a_i.
  - P^i_{local}, the patience of the ant, *i.e.* the number of unsuccessful explorations of a hunting site before its deletion of ant's memory.

- A_{site}^i the hunting site creation amplitude.
- A_{local}^i, the hunting site exploration amplitude.

These two last parameters are relative to the search space size.

—— **Algorithm 1.** Main API algorithm ——

1: **Choose** the initial nest location: $\mathcal{N} \leftarrow \mathcal{O}_{\text{rand}}(\mathcal{S})$
2: **Initialize** each ant's memory M_i: $\forall i \in \{1, \ldots, m\}$, $M_i = \emptyset$
3: **for** $t = 1$ to $T_{\max}$ **do**
4:    **for all** ant a_i **do**
5:       {Ant a_i has less than p^i sites in memory ?}
6:       **if** card$(M_i) < p^i$ **then**
7:          **Create** a new site in the neighborhood of $\mathcal{N}$: $s \leftarrow \mathcal{O}_{\text{explo}}(\mathcal{N}, A_{\text{site}}^i)$
8:          **Explore** this new site: $s' \leftarrow \mathcal{O}_{\text{explo}}(s, A_{\text{local}}^i)$
9:          **if** $f(s') < f(s)$ **then**
10:             $s \leftarrow s'$
11:          **end if**
12:          $M_i \leftarrow M_i \cup \{s\}$
13:       **else**
14:          **if** the previous exploration performed by a_i was successful **then**
15:             **Explore** the same site s;
16:          **else**
17:             **Explore** a randomly selected site s (among the p^i sites in M_i)
18:          **end if**
19:          $s' \leftarrow \mathcal{O}_{\text{explo}}(s, A_{\text{local}}^i)$
20:          {successful exploration}
21:          **if** $f(s') < f(s)$ **then**
22:             $M_i \leftarrow M_i \setminus (\{s\} \cup \{s'\})$
23:          **else**
24:             **if** a_i has explored its site s unsuccessfully more than P_{local}^i consecutive times **then**
25:                $M_i \leftarrow M_i \setminus \{s\}$ {Remove site from a_i's memory}
26:             **end if**
27:          **end if**
28:       **end if**
29:    **end for**
30:    **if** the nest is moved **then**
31:       **Change** the nest location to the best solution found and **Reset** ants' memories
32:    **end if**
33: **end for**

4   Learning HMMs with API Algorithm

In a previous study, we have experimented the hybridization of API with HMMs [26] in order to solve continuous optimization problems. HMMs were used to generate binary strings and ants were used to improve HMMs. With the present problem, ants are introduced to perform an exploration of HMMs' search space Λ corresponding to the set of all possible HMMs for a fixed number of hidden states. Nest and locations of hunting sites are support of HMMs and ants perform moves in this space. The objective is now to maximize the map $\lambda \mapsto L_{\mathcal{O}}(\lambda)$. To adapt API to HMMs training, we need to introduce a suitable search space representation.

In the present work, two natural operators $\mathcal{O}_{\mathrm{rand}}(\mathcal{S}_{\mathrm{HMM}})$ and $\mathcal{O}_{\mathrm{explo}}(\lambda, A^i)$ are introduced by the way of a new linear representation of HMMs. In order to define these operators, we first need some definitions and notations.

For a given dimension k, let $\mathbf{1}_k$, denote the column vector in $\mathbb{R}^k$ with all entries equal to 1 and let E_k be the quotient vector space $\mathbb{R}^k/\mathbb{R}\cdot\mathbf{1}_k$. We denote by $c_k : \mathbb{R}^k \to E_k$ the canonical map and define the so-called regularization map $r_k : \mathbb{R}^k \mapsto \mathbb{R}^k$ by

$$r_k(\mathbf{x})_i = x_i - \max_{j=1..k} x_j \qquad (6)$$

where $r_k(\mathbf{x})_i$ represents the i-th component of the vector $r_k(\mathbf{x})$. Obviously $r_k(\mathbf{x}) \equiv \mathbf{x} \pmod{\mathbb{R}\cdot\mathbf{1}_k}$ for all $\mathbf{x} \in \mathbb{R}^k$. We usually drop the index k if the reference to the dimension k is obvious. The subset $\Omega_k = r(\mathbb{R}^k)$ is also the set of points in $\mathbb{R}^k$ such that $r(\mathbf{x}) = \mathbf{x}$. In fact, Ω_k is a cone with vertex $\mathbf{0}$ (the null vector), which is the union of the faces of the k-dimensional polyhedron $]-\infty, 0]^k$. The map $r(\cdot)$ is the projection of $\mathbb{R}^k$ onto Ω_k, parallel to the vector space $\mathbb{R}\cdot\mathbf{1}$. In other words, $r(x)$ is the unique element in the intersection $(x + \mathbb{R}\cdot\mathbf{1}) \cap \Omega_k$. Moreover r is one-to-one on Ω_k. Consequently, there exists a bijection $\gamma_k : E_k \to \Omega_k$ (also denoted simply by γ) determined by the relation $r = \gamma \circ c$. We use $\gamma(\cdot)$ to carry the vector structure of E_k to the cone Ω_k. We obtain a vector space $(\Omega_k, \oplus, \odot)$ where the addition law $\oplus$ and the scalar law $\odot$ are respectively defined by $\mathbf{x} \oplus \mathbf{y} = \gamma(c(\mathbf{x}) + c(\mathbf{y}))\ (= r(\mathbf{x} + \mathbf{y}))$ and $t \odot \mathbf{x} = \gamma(t \cdot c(\mathbf{x}))\ (= r(t \cdot \mathbf{x}))$. One main interest in introducing the linear space $(\Omega_k, \oplus, \odot)$ is to be able to define transformations on Ω_k using linear operators.

Straightforward computations lead to the following properties of r:

(i) $r \circ r = r$ (in particular $r(\mathbf{x}) = \mathbf{x}$ for any $\mathbf{x} \in \Omega_k$),
(ii) $r(r(\mathbf{u}) + r(\mathbf{v})) = r(\mathbf{u} + \mathbf{v})$,
(iii) $r(t\cdot\mathbf{u}) = r(t\cdot r(\mathbf{u}))$ and $r(t\cdot\mathbf{u}) = t\cdot r(\mathbf{u})$ if $t \geq 0$,
(iv) $r(\mathbf{u} + t\cdot\mathbf{1}_k) = r(\mathbf{u})$,

where $\mathbf{u}$ and $\mathbf{v}$ are any vectors in $\mathbb{R}^k$, and t any real number. Let Q_k be the set of probability vectors $\{\mathbf{x} \in]0,1]^k; \sum_{i=1}^{k} x_i = 1\}$ and define the maps $\phi_k : \mathbb{R}^k \to Q_k$, $\psi_k : Q_k \mapsto \Omega_k$ (or ψ, ϕ for short) by $\phi(\mathbf{x})_i = \exp x_i / \sum_{1 \leq j \leq k} \exp x_j$ and $\psi(\mathbf{x})_i = \log x_i - \max_{j \in \{1,\dots,k\}} \log x_j$.

Notice that $\Lambda = (Q_N)^N \times (Q_M)^N \times Q_N$. Consequently, we introduce the vector spaces $E = (E_N)^N \times (E_M)^N \times E_N$, $\Omega = (\Omega_N)^N \times (\Omega_M)^N \times \Omega_N$ and the maps $C : (\mathbb{R}^N)^N \times (\mathbb{R}^M)^N \times \mathbb{R}^N \to E$, $\Gamma : E \to \Omega$, $\Phi : \mathbb{R}^{N^2} \times \mathbb{R}^{NM} \times \mathbb{R}^N \to \Lambda$ and $\Psi : \Lambda \to \Omega$. By definition, $C = (c_N, \dots c_N, c_M, \dots, c_M, c_N)$ and similarly for Γ, Φ and Ψ (the letter c being replaced by γ, ϕ, ψ respectively). In addition, let $R : (\mathbb{R}^N)^N \times (\mathbb{R}^M)^N \times \mathbb{R}^N \to \Omega$ be given by $(r_N, \ldots r_N, r_M, \ldots, r_M, r_N)$ and remark that R is one-to-one on Λ with $R((\mathbb{R}^N)^N \times (\mathbb{R}^M)^N \times \mathbb{R}^N) = \Omega$. From these definitions, we also derive that Γ is an isomorphism between E and Ω verifying $R = \Gamma \circ C$. Moreover, $\Phi(\mathbf{x} + \mathbf{y}) = \phi(\mathbf{x})$ for any $\mathbf{x} \in \mathbb{R}^{N(N+M+1)}$ and $\mathbf{y} \in \mathbb{R} \cdot \mathbf{1}_{N^2} \times \mathbb{R} \cdot \mathbf{1}_{NM} \times \mathbb{R} \cdot \mathbf{1}_N$, $\Phi \circ R = \Phi$ and, both $\Phi \circ \Psi = Id_{|_\Lambda}$, $\Psi \circ \Phi_{|_\Omega} = Id_{|_\Omega}$. Now, the map Ψ allows to identify the parameter $\lambda = (A, B, \Pi)$ (in Λ) defining $HMM(\lambda)$ with $\Psi(\lambda)$ in the vector space Ω $(= \Psi(\Lambda))$ and so, the representation space for HMMs with N hidden states and M symbols is now $\mathcal{E}_{\mathrm{HMM}} = \Omega$.

We are ready to defined the API's operators considered in our experiments. The first one is

$$\mathcal{O}_{\mathrm{rand}}(\mathcal{E}_{\mathrm{HMM}}) = R\big(\mathcal{U}([-30, 0[^{N^2} \times [-30, 0[^{NM} \times [-30, 0[^{N})\big) \qquad (7)$$

where $\mathcal{U}(H)$ returns a uniformly random value in the box H. The next one is given by

$$\mathcal{O}_{\mathrm{explo}}(\lambda, A^i) = \Phi\left(\tilde{\Psi}(\lambda) \oplus \left(\frac{\mathcal{U}([0, 1[)A^i}{||W||_2} \odot W\right)\right) \qquad (8)$$

where $W = \mathcal{O}_{\mathrm{rand}}(\mathcal{E}_{\mathrm{HMM}})$ (used independently of the previous one) is now the direction of the ant's exploration (we ensure that $||W||_2 > 0$) and $\tilde{\Psi}$ is a modification of Ψ in order to deal with null probabilities. In our case, we choose $\tilde{\Psi} = (\tilde{\psi}_{N^2}, \tilde{\psi}_{NM}, \tilde{\psi}_N)$ such that, for any $\mathbf{x} \in Q_k$,

$$\tilde{\psi}_k(\mathbf{x})_i = \begin{cases} \log x_i - \max\{\ \max\limits_{j \in \{1, \ldots, k\}} \log x_j, -100\} \text{ if } x_i > e^{-100} \\ -100 - \max\{\ \max\limits_{j \in \{1, \ldots, k\}} \log x_j, -100\} \text{ otherwise.} \end{cases} \qquad (9)$$

Notice that, even if null probabilities do not occur in theory, they may appear in practice during computations, due to the numerical precision of the implementation.

An easier parameter setting strategy should be to choose ants' parameters identical for all ants (homogeneous case), but it is showed in [23] that, for numerical optimization, best results are obtained with heterogeneous ants' parameters (heterogeneous API). In order to simplify ants' parameter settings in this later case, we introduce the following automatic settings:

- $A^i_{\mathrm{nest}} = \frac{15i}{m}$;
- the amplitude of the local search is set to $A^i_{\mathrm{local}} = A^i_{\mathrm{nest}}/10$;
- The number of sites memorized by real ants being unknown, we arbitrarily set the default number of sites memorized by each ant's memory to $p^i = 2$, $(1 \leq i \leq m)$.

This choice leads to an API algorithm that is more robust for a wider range of problems. Finally, the classical Baum-Welch algorithm (BW) should be used to locally improve models built by ants' explorations. For this goal, BW is applied at the final step, replacing $\mathcal{O}_{\mathrm{explo}}$ by $\mathcal{O}_{\mathrm{exploBW}}(\lambda, A^i) = \mathrm{BW}(\mathcal{O}_{\mathrm{explo}}(\lambda, A^i))$.

5   Experimental Study

For our experiments, we have chosen to do two iterations of the Baum-Welch algorithm. We consider 10 ants (*i.e.*, $m = 10$), 2 hunting sites (*i.e.*, $p^i = 2$) and a local patience of 3 (*i.e.*, $P^i_{\text{local}} = 3$). We allow 1000 moves in the search space so that $T_{\max} = 1000/10 = 100$ (each ant can build 100 HMMs) and the nest patience is 20 (*i.e.*, $T_{\text{nest}} = 20$).

As testbed, we consider series of observations built from a set of images using JPEG encoding but details are omitted here: the only important thing to keep in mind is that those series correspond to a real application of image recognition. Images are extracted from [27]. The observations have different length (T) and are encoded by various number of symbols (M). For the training, 11 hidden states is a quite reasonable value.

To emphasize the interest of the API algorithm for HMMs training, we compare it with a random search algorithm denoted by `Random` which consists in generating randomly 1000 HMMs and returning the best one. The interest of the hybridization with the Baum-Welch algorithm is exhibited by performing a `Random+BW` algorithm which consists in randomly generating 1000 HMMs, followed by 2 iterations of Baum-Welch algorithm and then returning the best model. The API algorithm is stated in four versions: two with homogeneous parameters, denoted as `APIhomo` and `APIhomo+BW`, and two with heterogeneous parameters, denoted as `APIhete` and `APIhete+BW`. Site amplitudes for

Table 1. Mean performance results in log-likelihood

Image	T	M	Random	Random+BW	APIhomo	APIhomo+BW	APIhete	APIhete+BW
1	400	16	-1043.75	-776.92	-1053.61	-699.67	-1041.74	-525.78
2	400	16	-1045.88	-776.78	-1063.75	-727.62	-1039.16	-575.60
3	400	16	-1067.61	-916.18	-1075.82	-872.18	-1071.30	-699.42
4	400	16	-1072.06	-927.36	-1075.74	-859.78	-1078.03	-649.35
1	400	32	-1340.36	-1040.92	-1347.66	-967.39	-1347.85	-728.64
2	400	32	-1343.27	-1061.93	-1362.36	-1008.18	-1349.06	-813.94
3	400	32	-1358.32	-1183.37	-1371.11	-1137.57	-1366.94	-939.58
4	400	32	-1357.70	-1191.59	-1372.87	-1132.08	-1364.61	-881.84
1	400	64	-1630.28	-1319.15	-1648.86	-1252.77	-1645.66	-975.68
2	400	64	-1633.48	-1319.21	-1650.49	-1263.06	-1647.22	-1038.66
3	400	64	-1642.02	-1451.16	-1656.34	-1403.15	-1652.02	-1177.02
4	400	64	-1644.36	-1458.69	-1657.52	-1398.58	-1655.37	-1117.29
1	2000	16	-5269.79	-3887.00	-5292.09	-3551.14	-5192.61	-2497.94
2	2000	16	-5259.94	-3926.22	-5314.93	-3700.36	-5283.02	-2911.64
3	2000	16	-5344.28	-4475.89	-5411.76	-4140.47	-5359.73	-3164.02
4	2000	16	-5359.13	-4541.35	-5396.17	-4224.90	-5362.49	-3241.26
1	2000	32	-6708.70	-5324.81	-6748.03	-4999.43	-6783.95	-3642.75
2	2000	32	-6719.60	-5377.93	-6794.59	-5141.07	-6790.99	-4122.03
3	2000	32	-6786.96	-5915.23	-6865.31	-5639.53	-6825.65	-4420.70
4	2000	32	-6802.65	-5988.95	-6844.56	-5715.88	-6837.17	-4472.57

Table 2. Means cpu time utilization in second

T	M	Random	Random+BW	API	API+BW
400	16	0.7	3	7	45
400	32	0.7	3	8	49
400	64	0.7	4	10	54
2000	16	1.6	30	32	390
2000	32	1.6	32	33	394

homogeneous versions are $A_{\text{nest}}^i = 8.25$ and $A_{\text{local}}^i = 0.825$. These constants correspond to the averaged values of the corresponding site amplitudes in the heterogeneous case. For each algorithm and each series of observations, we average the log-likelihood of best output models obtained with 15 runs.

Results are given in table 1 and 2. We notice that, algorithms which involve Baum-Welch (BW) perform significantly better than the ones without BW, strengthening the usefulness of the hybridization with BW. Moreover, table 2 shows that BW also increases significantly the computing time. When BW is not used, our experiments point out that the random search performs better than API algorithms. This behavior seems to be surprising but is in accordance with the fact that a uniform random search furnishes, in averaged, better results than a directed search. This is due to the small number of iterations compared with the search space size. However, notice that the heterogeneous API gives results which are better or equal than those obtained with the homogeneous API. Notice that BW algorithm in both versions of API performs significantly better than Random+BW and the heterogeneous API version performs much better than the corresponding homogeneous version. Moreover, the parameter settings could probably be improved. Also, we confirm the tendency observed in [5]: heterogeneous parameters improve results especially when the search space dimension increases. To conclude, we can say that depending of the available computing time, we can significantly improve the training. For fast training, we can use Random. For better performance and reasonable computing time, we can use Random+BW. For optimal training, we can use more computing time and use the APIhete+BW algorithm.

6   Conclusion

In this paper, we have introduced a new application of the API algorithm for the training of HMMs involving suitable search space representations. Two kinds of parameter settings are considered: one is homogeneous, independent of the ants, and the other one is heterogeneous.

To complete this approach, these algorithms are hybridized with the Baum-Welch algorithm. Our experiments show that hybridized API algorithms outperforms a straightforward random search associated to BW and the best performances are obtained if we choose heterogeneous parameters in the API algorithm.

In future works, we plan to study more precisely the effect of parameter settings and propose to compare such algorithms with other implementations involving the operator $\mathcal{O}_{\text{explo}}(\lambda, A^i)$ already used in other metaheuristics.

References

1. Douglas, B.P.: Training of hmm recognizers by simulated annealing. In: Proceedings of ICASSP'85. (1985) 13–16
2. Chen, T.Y., Mei, X.D., Pan, J.S., Sun, S.H.: Optimization of HMM by the tabu search algorithm. Journal of Information science and engineering **20** (2004) 949–957
3. Slimane, M., Venturini, G., Asselin de Beauville, J.P., Brouard, T., Brandeau, A.: Optimizing hidden Markov models with a genetic algorithm. In: proceedings of Artificial Evolution conference. Volume 1063 of Lecture Notes in Computer Science., Springer Verlag (1996) 384–396
4. Slimane, M., Venturini, G., Asselin de Beauville, J.P., Brouard, T.: Hybrid genetic learning of hidden markov models for time series prediction. Biomimetic approaches in management science, Kluwer Academics (1998)
5. Monmarché, N., Venturini, G., Slimane, M.: On how *Pachycondyla apicalis* ants suggest a new search algorithm. Future Generation Computer Systems **16** (2000) 937–946
6. Cappé, O.: Ten years of HMM. http://ww.tsi.enst.fr/~cappe/docs/hmmbib.html (2001)
7. Fine, S., Singer, Y., Tishby, N.: The Hierarchical Hidden Markov Model: Analysis and applications. Machine Learning **32** (1998) 41–62
8. Rabiner, L.: A tutorial on hidden Markov models and selected applications in speech recognition. Proceedings of IEEE **77** (1989) 257–286
9. Forney, G.: The Viterbi algorithm. Proceedings of IEEE **61** (1973) 268–278
10. Baum, L., Petrie, T., Soules, G., Weiss, N.: A maximization technique occuring in the statistical analysis of probabilistic functions of markov chains. Ann. Math. Stat. **41** (1970) 164–171
11. Bonabeau, E., Dorigo, M., Theraulaz, G.: Swarm Intelligence: From Natural to Artificial Systems. Oxford University Press, New York (1999)
12. Dorigo, M., Bonabeau, E., Theraulaz, G.: Ant algorithms and stygmergy. Future Generation Computer Systems **16** (2000) 851–871
13. Dorigo, M., Di Caro, G.: The Ant Colony Optimization Meta-Heuristic. In Corne, D., Dorigo, M., Glover, F., eds.: New Ideas in Optimisation. McGraw-Hill, London, UK (1999) 11–32 Also available as Tech.Rep.IRIDIA/99-1, Université Libre de Bruxelles, Belgium.
14. Dorigo, M., Gambardella, L.: Ant Colony Sytem: A cooperative learning approach to the Travelling Salesman Problem. IEEE Transactions on Evolutionary Computation **1** (1997) 53–66
15. Stützle, T., Hoos, H.: $\mathcal{MAX} - \mathcal{MIN}$ Ant System and local search for the Traveling Salesman Problem. In IEEE, ed.: Proceedings of the fourth International Conference on Evolutionary Computation (ICEC), IEEE Press (1997) 308–313
16. Stützle, T., Dorigo, M.: ACO algorithms for the Quadratic Assignment Problem. In Corne, D., Dorigo, M., Glover, F., eds.: New Ideas in Optimisation. McGraw-Hill, London, UK (1999) 33–50

17. T'Kindt, V., Monmarché, N., Tercinet, F., Laügt, D.: An Ant Colony Optimization algorithm to solve a 2-machine bicriteria flowshop scheduling problem. European Journal of Operational Research **142** (2002) 250–257
18. Ying, K.C., Liao, C.J.: An ant colony system for permutation flow-shop sequencing. Computers & Operations Research **31** (2004) 791–801
19. Bilchev, G., Parmee, I.: The ant colony metaphor for searching continuons design spaces. In Fogarty, T., ed.: XX. Volume 993 of Lecture Notes in Computer Science. Springer Verlag (1995) 24–39
20. Li, S., Liu, Z.: General CAC approach using novel ant algorithm training based neural network. In: Proceedings of the International Joint Conference on Neural Networks. Volume 3. (1999) 1885–1888
21. Dréo, J., Siarry, P.: Continuous interacting ant colony algorithm based on dense heterarchy. Future Generation Computer Systems (2004) In Press.
22. Fresneau, D.: Individual foraging and path fidelity in a ponerine ant. Insectes Sociaux, Paris **32** (1985) 109–116
23. Monmarché, N., Venturini, G., Slimane, M.: On how *Pachycondyla apicalis* ants suggest a new search algorithm. Future Generation Computer Systems **16** (2000) 937–946
24. Monmarché, N.: Algorithmes de fourmis artificielles : applications à la classification et à l'optimisation. Thèsc de doctorat, Laboratoire d'Informatique, Université de Tours (2000)
25. Deneubourg, J., Goss, S., Pasteels, J., Fresneau, D., Lachaud, J.: Self-organization mechanisms in ant societies (ii): learning in foraging and division of labor. In Pasteels, J., Deneubourg, J., eds.: From individual to collective behavior in social insects, Experientia supplementum. Volume 54. Bikhauser Verlag (1987) 177–196
26. Soukhal, A., Monmarché, N., Laügt, D., Slimane, M.: How hidden markov models can help artificial ants to optimize. In: Proceedings of the Optimization by Building and Using Probabilistic Models workshop, Genetic and Evolutionary Computation Conference. (2001) 226–229
27. Samaria, F., Harter, A.: Parameterisation of a stochastic model for human face identification. In: IEEE workshop on Applications of Computer Vision, Florida (1994)

Memetic Algorithms for the MinLA Problem*

Eduardo Rodriguez-Tello[1], Jin-Kao Hao[1], and Jose Torres-Jimenez[2]

[1] LERIA, Université d'Angers,
2 Boulevard Lavoisier, 49045 Angers, France
{ertello, hao}@info.univ-angers.fr
[2] Mathematics Department, University of Guerrero,
54 Carlos E. Adame, 39650 Acapulco Guerrero, Mexico
jose.torres.jimenez@acm.org

Abstract. This paper presents a new Memetic Algorithm designed to compute near optimal solutions for the MinLA problem. It incorporates a highly specialized crossover operator, a fast MinLA heuristic used to create the initial population and a local search operator based on a fine tuned Simulated Annealing algorithm. Its performance is investigated through extensive experimentation over well known benchmarks and compared with other state-of-the-art algorithms.

Keywords: Memetic Algorithms, Linear Arrangement, Heuristics.

1 Introduction

Evolutionary algorithms (EAs), as general purpose optimization procedures, have been successfully applied in a broad spectrum of areas in physics, chemistry, engineering, management science, biology and computer science [22].

It is well recognized that it is essential to incorporate some form of domain knowledge into EAs to arrive at highly effective search [1,4,10]. There are many ways to achieve this, for example by the combination of EAs with other efficient problem-dependent heuristics, or by using encodings and genetic operators that are tailored to the problem to be solved. Memetic algorithms (MAs) follow such an approach and have demonstrated recently to be very efficient [3,7,8,12,16,23]. Under different contexts and situations, MAs are also known as hybrid EAs or genetic local searchers.

In this paper, we are interested in tackling with the use of MAs a well-known combinatorial optimization problem: the Minimum Linear Arrangement problem (MinLA). Garey and Johnson have shown that finding the minimum linear arrangement of a graph is NP-hard and the corresponding decision problem is NP-complete [9]. MinLA was first stated by Harper [11]. His aim was to design error-correcting codes with minimal average absolute errors on certain classes

---

* This work is supported by the CONACyT Mexico, the "Contrat Plan Etat Région" project COM (2000-2006) as well as the Franco-Mexican Joint Lab in Computer Science LAFMI (2005-2006).

E. Talbi et al. (Eds.): EA 2005, LNCS 3871, pp. 73–84, 2006.

of graphs. The MinLA problem arises also in other application areas like graph drawing, VLSI layout, software diagram layout and job scheduling [5].

The MinLA problem can be stated formally as follows. Let $G(V, E)$ be a finite undirected graph, where V ($|V| = n$) defines the set of vertices and $E \subseteq V \times V = \{\{i, j\}|i, j \in V\}$ is the set of edges. Given a one-to-one function $\varphi : V \to \{1..n\}$, called a linear arrangement, the total edge length for G with respect to arrangement φ is defined according to the equation 1.

$$LA(G, \varphi) = \sum_{(u,v) \in E} |\varphi(u) - \varphi(v)| \qquad (1)$$

Then the MinLA problem consists in finding an arrangement φ for a given G so that $LA(G, \varphi)$ is minimized.

There exist polynomial time exact algorithms for some special cases of MinLA such as trees, rooted trees, hypercubes, meshes, outerplanar graphs, and others (see [5] for a detailed survey). However, MinLA is NP-hard for general graphs [9] and for bipartite graphs [6]. Therefore, there is a need for heuristics to address this problem in reasonable time. Among the reported algorithms are a) heuristics especially developed for MinLA, such as the binary balanced decomposition tree heuristic (DT) [2], the multi-scale algorithm (MS) [14] and the algebraic multi-grid scheme (AMG) [21]; and b) metaheuristics such as Simulated Annealing [17,18,19] and Genetic Algorithms [20].

This paper aims at developing a powerful Memetic Algorithm (MA) for finding near optimum solutions for the MinLA problem. To achieve this, the new algorithm, called MAMP (standing for Memetic Algorithm for the MinLA Problem), incorporates a highly specialized crossover operator, a fast MinLA heuristic used to create the initial population and a local search operator based on a fine tuned Simulated Annealing algorithm. The performance of MAMP is assessed with a set of 21 benchmark instances taken from the literature. The computational results are reported and compared with previously published ones, showing that our algorithm is able to improve on some previous best results.

The paper is organized as follows. Section 2 reviews some existing solution procedures for the MinLA problem. Then, the different components of our MA are presented in Section 3. Section 4 is dedicated to computational experiments and comparisons with previous results. The last section summarizes the main contributions of this research work.

2   Relevant Existing Procedures

Because of the importance of the MinLA problem, much research has been carried out in developing effective heuristics for it. In this section, we give a brief review of three representative algorithms which were used in our comparisons.

2.1   The SS+SA Heuristic

In 2001 Jordi Petit developed a heuristic for the MinLA problem, called SS+SA [18,19]. It works as follows: First a global solution is obtained by using Spectral

Sequencing (SS), a method originally proposed by Juvan and Mohar, which is based on the computation of the Fiedler vector of G [13]. Then the resulting arrangement is iteratively improved using a SA algorithm previously reported in [17]. It performs local changes based on a special neighborhood distribution, called FlipN, that tends to favor moves with high probability to be accepted.

The SS+SA algorithm proposed by Petit starts at an initial temperature $T_0 = 10$, at each Metropolis round $r = 20n^{3/2}$ moves are generated. Then the current temperature is decremented with the relation $T_k = \alpha T_{k-1}$, with $\alpha = 0.95$ until to reach a final temperature $T_f = 0.2$. The author claims that these parameters were fixed based on some preliminary experiments.

The author makes a computational comparison of the SS procedure, a SA algorithm and the combination of both methods (SS+SA). For this comparison Petit collected a set of 21 benchmark graphs. The test-suite consists of 5 random graphs, 3 "regular" graphs (a hypercube, a mesh, and a binary tree), 3 graphs from finite element discretizations, 5 graphs from VLSI designs, and 5 graphs from graph drawing competitions.

The experiments have shown that for the finite element discretization graphs SS+SA improves the SS and SA solutions by more than 20%, while reducing the running time to a 25% of SA. For the rest of the graphs, SS+SA allways improves the SS solutions and only for two graphs ($c5y$ and $gd96a$) it is unable to improve the SA solution. The running times are usually lower for SS+SA than for SA. The author concludes that the SS+SA heuristic is a valuable improvement over the SS and SA methods.

2.2   The DT+SA Heuristic

Besides Petit's work, Bar-Yehuda *et al.* present in [2] a divide-and-conquer approach to the MinLA problem. Their idea is to divide the vertices into two sets, to recursively arrange each set internally at consecutive locations, and finally to join the two ordered sets, deciding which will be put to the left of the other.

The computed arrangement is specified by a decomposition tree (DT) that describes the recursive partitioning of the subproblems. Each vertex of the tree gives a degree of freedom as to the order in which the two vertex sets are glued together. Thus, the goal of the algorithm is to decide for each vertex of the decomposition tree the order of its two children. The authors propose a dynamic programming algorithm for computing the best possible ordering for a given decomposition tree.

The set of benchmark instances used in [2] is the same proposed by Petit in [17, 18, 19]. They applied their algorithm iteratively, starting each iteration with the result of the previous one. After a few tens of iterations, the algorithm usually yields results within 5-10% of those obtained by Petit's SA, but at a fraction of its running time. They have used these computed arrangements as an initial solution for the SA reported in [17] and slightly better results were obtained.

2.3   The MS Heuristic

In 2002, Koren and Harel present a linear-time algorithm for the MinLA problem, based on the multi-scale (MS) paradigm [14]. MS techniques transform a high-dimensional problem in an iterative fashion into subproblems of increasingly lower dimensions, via a process called coarsening. On the coarsest scale the problem is solved exactly, following which a refinement process starts, whereby the solution is progressively projected back into higher and higher dimensions, updated appropriately at each scale, until the original problem is reproduced and solved.

The algorithm proposed in [14], starts with a preprocessing stage that obtains, rapidly, a reasonable linear arrangement by using spectral sequencing and then improves the result by applying a procedure, that they call *median iteration*, for about 50 sweeps. The median iteration is a randomized algorithm based on a continuous relaxation of the MinLA problem, where vertices are allowed to share the same place, or to be placed on non-integral points.

Then, the MS algorithm starts by refining the arrangement locally. The intention of the refinement is not only to minimize the arrangement cost, but also to improve the quality of the coarsening step that follows. The next step is to coarsen the graph based on restricting consecutive vertex pairs of the current arrangement. The problem is then solved in the restricted solution space, by running all this set of steps (called a *V-cycle*) recursively on the coarse graph. Once a good solution is found in the restricted solution space, the algorithm refines it locally (in the full solution space).

Koren and Harel have also used the set of test instances proposed by Petit. For each graph in this set, they ran their MS algorithm first with a single V-cycle and then with ten. They present these results as well as those obtained during the preprocessing stage (spectral sequencing and median iteration algorithms). The quality of their results after 10 V-cycle iterations is comparable to that of Petit's SA, but the running time is significantly better.

Later in 2004, an improvement to the algorithm proposed by Koren and Harel was presented in [21]. The main difference between these approaches is the coarsening scheme. Koren and Harel use *strict* aggregation, while Safro *et al.* use *weighted* aggregation. In a strict aggregation procedure the nodes of the graph are blocked into small disjoint subsets, called aggregates. By contrast, in the weighted aggregation each node can be divided into fractions, and different fractions belong to different aggregates. Safro *et al.* have shown experimentally that their approach can obtain high quality results in linear time for the MinLA problem and can be considered as one of the best MinLA algorithms known today.

2.4   The Genetic Hillclimbing Algorithm

In [20] a Genetic Hillclimbing (GH) algorithm is proposed. It represents linear arrangements as permutations of vertices and operates as follows: An initial population $|P| = 100$ is created by combining one individual generated with spectral sequencing, 10% of randomly generated individuals and the rest is generated using depth-first and breadth-first search algorithms initialized with a randomly chosen vertex. At each generation $0.5|P|$ pairs of individuals are randomly selected, then a two point crossover with unfeasibility repair is applied with 98% of probability in order to produce two offspring each time. Both resulting offspring are compared with their parents. If offspring has better fitness than one of its parents, then it is inserted in the population else the parent is taken back to the population and the offspring is eliminated. After that, $nlog(n)$ hillclimbing steps are applied to each individual of the population. It allows to obtain locally optimal solutions that will be mutated with probability 15%. The mutation operator consists in applying one random swap. The process is repeated until the number of 20000 generations is reached or when 100 successive generations do not produce a better solution.

For his comparisons the author employs the set of benchmark instances proposed by Petit [17, 18, 19]. Their results show that GH has found slightly better results for 7 instances (over 21 graphs).

3 A New Memetic Algorithm for MinLA

In this section we present a new Memetic algorithm, called MAMP, for solving the MinLA problem. Next all the details of its implementation are presented.

3.1 Search Space, Representation and Fitness Function

The search space $\mathcal{A}$ for the MinLA problem is composed of all possible arrangements from V to $\{1, 2, ..., n\}$. It is easy to see then, that there are $n!$ possible linear arrangements for a graph with n vertices.

In our MA a linear arrangement φ is represented as an array l of n integers, which is indexed by the vertices and whose $i\text{-}th$ value $l[i]$ denotes the label assigned to the vertex i. The fitness of φ is evaluated by using Equation 1.

3.2 The General Procedure

MAMP starts building an initial population P, which is a set of configurations having a fixed constant size $|P|$ (*initPopulation*). Then it performs a series of cycles called generations. At each generation, a predefined number of recombinations (*offspring*) are executed. In each recombination two configurations a and b are chosen randomly from the population (*selectParents*). A recombination operator is then used to produce an offspring c from a and b (*recombineIndividuals*). The local search operator (*localSearch*) is applied to improve c for a fixed number of iterations L and the improved configuration c is inserted in the population. Finally, the population is updated by choosing the best individuals from the pool of parents and children (*UpdatePopulation*). This process repeats until a stop condition is verified, usually when a predefined number of generations (*maxGenerations*) is reached. Note however, that the algorithm may stop before reaching *maxGenerations*, if a better solution is not produced in a predefined number of successive generations (*maxFails*).

3.3   The Initialization Operator

The operator $initPopulation(|P|)$ initiates the population P with $|P|$ configura-
tions. To create a configuration, we use the greedy frontal increase minimization
(FIM) algorithm of McAllister [15], slightly adapted in order to work in a randomized form. The algorithm is based on the following two basic steps: 1) Select a
starting vertex and place it in position 1. 2) For each remaining position 2 through
n, select one of the unplaced vertices for placement in the current position by using
the FIM strategy. It consist in selecting for placement i a vertex that is adjacent
to the fewest vertices in $U_i - F_i$, where $F_i = \{u \in U_i | v \in P_i$ and $(u, v) \in E\}$
denotes the *front* at placement i, P_i represents the set of $i - 1$ vertices placed so
far and U_i the set of currently unplaced vertices.

In order to accomplish this, two measures are defined that enable to know how
highly a vertex $v \in U_i$ is connected to P_i and to U_{i+1}. The measures are defined
respectively as follows: $tl_i(v) = |\{(u, v) \in E | u \in P_i\}|$ and $tr_i(v) = d(v) - tl_i(v)$,
where $d(v)$ denotes the degree of the vertex v. Both measures are used to define a
new selection factor $sf_i(v) = tr_i(v) - tl_i(v)$, which is used at the two-step general
strategy described above as follows: For each placement i in step 2, select $v \in F_i$
with minimum $sf_i(v)$. This algorithm has a linear time complexity with respect
to the number of edges in the graph. This is possible thanks to the use of efficient
data structures that enable to select a vertex with minimum $sf_i(v)$ in constant
time.

Due to the randomness of the greedy algorithm, the configurations in the
initial population are quite different. This point is important for population
based algorithms because a homogeneous population cannot efficiently evolve.

3.4   Selection

Mating selection $(selectParents(P))$ prior to recombination is performed on a
purely random basis without bias to fitter individuals, while selection for survival
$(UpdatePopulation(P))$ is done by choosing the best individuals from the pool
of parents and children. It is done by taking care that each phenotype exists
only once in the new population. Thus, replacement in our algorithm is similar
to the (μ, λ) selection scheme used in [16].

3.5   The Recombination Operator

The main idea of the recombination operator $(recombineIndividuals(a, b))$ is to
generate diversified and potentially promising individuals. To do that, a good
MinLA recombination operator should take into consideration, as much as possible, the individuals' semantic.

In this subsection we present a new recombination operator LGX (*local greedy
crossover*) that is able to preserve certain information contained in both parents,
while some subgraphs are locally improved using a greedy mechanism. The new
LGX operator works in four basic steps:

First, all the labels found at the same vertex in the two parents are assigned
to the corresponding vertex in the offspring. Next, for each labeled vertex in

the offspring a greedy mechanism is applied to find the labels for its adjacent vertices; this procedure tends to minimize the local MinLA contribution of each of these subgraphs. Then, for each unlabeled vertex in the offspring we take, if possible, the label from the same vertex of one of the parents. Finally, the labels for the remaining vertices are randomly assigned. The functioning of the LGX operator is presented in Algorithm 1.

```
recombineIndividuals(a, b)
begin
    // The number of assigned labels in the offspring
    assigned = copyIdenticLabels(a, b, c);
    for each vertex i labeled in c do
    |   assigned += localGreedy(i, c);
    end
    assigned += completeFromParents(a, b, c);
    if assigned < |V| then
    |   completeRandom(c, assigned);
    end
    return The offspring c;
end
```

Algorithm 1. The LGX recombination operator

3.6   The Local Search Operator

The purpose of the local search (LS) operator $localSearch(c, L)$ is to improve a configuration c produced by the recombination operator for a maximum of L iterations before inserting it into the population. In general, any local search method can be used. In our implementation, we have decided to use Simulated Annealing (SA).

In our SA-based LS operator the neighborhood $N(\varphi)$ of an arrangement φ is such that for each $\varphi \in \mathcal{A}$, $\varphi' \in N(\varphi)$ if and only if φ' can be obtained by flipping the labels of any pair of different vertices from φ. We call this flipping operation a *move*. Besides the apparent simplicity of this neighborhood function, the reasons to choose it are: the easiness to perform movements and the low effort necessary to compute incrementally the cost of the new arrangement.

The SA operator starts at an initial temperature $T_0 = 10$, at each Metropolis round $r = 1000$ moves are generated. If the cost of the attempted move decreases then it is accepted. Otherwise, it is accepted with probability $P(\Delta) = e^{-\Delta/T}$ where T is the current temperature and Δ is the increase in cost that would result from that particular move. At the end of each Metropolis round then the current temperature is decremented by a factor of $\alpha = 0.955$. The algorithm stops either if the current temperature reaches $T_f = 0.001$, or when it reaches the predefined maximum of L iterations.

The algorithm memorizes and returns the most recent arrangement φ^* among the best configurations found: after each accepted move, the current configuration φ replaces φ^* if $LA(G, \varphi) \leq LA(G, \varphi^*)$ (and not only if $LA(G, \varphi) < LA(G, \varphi^*)$). The rational to return the last best configuration is that we want

to produce a solution which is as far away as possible from the initial solution
in order to better preserve the diversity in the population.

4 Computational Experiments

In this section, we present a set of experiments accomplished to evaluate the
performance of the MA algorithm presented in Section 3. The algorithms were
coded in C and compiled with *gcc* using the optimization flag -*O3*. They were
run sequentially into a cluster of 10 nodes, each having a Xeon bi-CPU at 2 GHz,
1 GB of RAM and Linux. Due to the non-deterministic nature of the algorithms,
20 independent runs were executed for each of the selected benchmark instances.
When averaged results are reported, they are based on these 20 corresponding
runs.

In all the experiments the following parameters were used for MAMP: a)
population size $|P| = 40$, b) recombinations per generation *offspring* $= 4$, c)
maximal number of local search iterations $L = 150000$, d) maximal number
of generations $maxGenerations = 10000$ and e) maximal number of successive
failed generations $maxFails = 100$.

4.1 Benchmark Instances and Comparison Criteria

The test-suite that we have used in the experiments is the same proposed by
Petit [17] and used later in [2, 14, 20, 21]. It consists of six different families
of graphs: Uniform random (randomA* class), geometric random (randomG*
class), graphs with known optima (trees, hypercubes and meshes), finite element discretizations (3elt, airfoil1 and whitaker3), VLSI design (c*y class) and
graph drawing competitions (gd* class). All of them have 1000 vertices or more,
except for some instances in the gd* class. These instances are available at:
http://www.lsi.upc.es/~jpetit/MinLA/Experiments

The criteria used for evaluating the performance of the algorithms are the
same as those used in the literature: the best total edge length found for each
instance and the CPU time in seconds.

4.2 Comparison Between MAMP and GH

The purpose of the first experiment is to compare our memetic algorithm MAMP
with the previous one of [20] (GH). To enable a fair comparison we have obtained
the GH source code[1]. Then GH and MAMP were compiled and executed in our
hardware and operating system platform 20 times on each benchmark instance.

The parameters for the GH algorithm are those reported in Poranen's work:
a) population with 100 individuals, b) 50 crossovers per generation, c) 98%
crossover rate, d) 15% mutation rate, e) $nlog(n)$ hillclimbing steps, f) a maximum
of 20000 generations and g) at maximum 100 successive failed generations. We
would like to point out that GH employs a population of 100 individuals, while

---

[1] Available at http://www.cs.uta.fi/~tp/optgen/index.html

Table 1. Performance comparison between MAMP and GH

Graph	GH Bc	GH Avg	GH t	MAMP Bc	MAMP Avg	MAMP t	Δ_C
randomA1	878705	883138.2	4079.2	867535	868480.4	918.7	-11170
randomA2	6557701	6564256.4	24010.2	6533999	6536249	3477.4	-23702
randomA3	14253230	14253230	25629.1	14240067	14240757	5221.2	-13163
randomA4	1735414	1735414	10066.2	1719906	1721070.4	1904.1	-15508
randomG4	153470	153470	1924.6	141538	143855	2097.2	-11932
bintree10	3873	3920.6	413.0	3790	3812.8	984.7	-83
hc10	523776	523776	325.8	523776	523776	1152.4	0
mesh33x33	31968	32127.2	1129.9	31917	31979.8	1177.9	-51
3elt	397305	403654.2	41952.7	362209	364403	5758.9	-35096
airfoil1	300656	300656	74023.7	285429	286986.6	5542.4	-15227
whitaker3	1189831	1189831	9006538.7	1167089	1168140.25	15322.4	-22742
c1y	63063	63440.6	783.9	62333	62383.6	651.5	-730
c2y	80453	81914.2	935.4	79017	80998	672.8	-1436
c3y	129775	130789.4	2092.0	123521	123689.4	731.1	-6254
c4y	118270	119277	2796.8	115144	115406	739.4	-3126
c5y	100877	102054.8	1983.7	96952	97219.4	741.5	-3925
gd95c	506	508.4	2.2	506	506.2	1.5	0
gd96a	105947	108714.6	886.7	96253	96384.8	667.9	-9694
gd96b	1416	1417.2	4.5	1416	1416.2	3.3	0
gd96c	519	519.2	2.0	519	520	1.4	0
gd96d	2406	2413.6	10.9	2391	2392	8.1	-15
Average							-8278.8

MAMP has a population size of 40. We have decided to conserve this difference, apparently unfavorable for MAMP, because in a preliminary experiment we have tried to reduce the GH population size to 40, but the results produced by GH were inferior in solution quality.

The results obtained from comparing both algorithms are presented in Table 1. Column 1 shows the name of the graph. Columns 2 to 7 display the best cost (Bc), the average cost (Avg) and the average CPU time (t) in seconds for finding the best solution in each one of the 20 runs of the GH and MAMP algorithms respectively. Last column presents the difference (Δ_C) between the best cost found by MAMP and the best cost produced by GH.

Table 1 shows clearly that MAMP allows us to obtain better results for many classes of graphs with less computing time. We can observe an important improvement in cost in 17 out of 21 instances. For the rest of the instances the results of MAMP equal those produced by GH, but always with less computational effort, thanks to its reduced population size.

4.3   Comparison Between MAMP and the Best Known Results

In the second experiment a performance comparison of our MAMP procedure with the following heuristics was carried out: SS+SA [18, 19], DT+SA [2], AMG [21] and GH [20]. Table 2 presents the detailed computational results

Table 2. Performance comparison between MAMP and several state-of-the-art algorithms

Graph	$|V|$	$|E|$	SS+SA	DT+SA	AMG	GH	MAMP	Δ_C
randomA1	1000	4974	869648	884261	888381	878637	867535	-2113
randomA2	1000	24738	6536540	6576912	6596081	6550292	6533999	-2541
randomA3	1000	49820	14310861	14289214	14303980	14246646	14240067	-6579
randomA4	1000	8177	1721490	1747143	1747822	1735691	1719906	-1584
randomG4	1000	8173	150940	146996	140211	142587	141538	1327
bintree10	1023	1022	4069	3762	3696	3807	3790	94
hc10	1024	5120	523776	523776	523776	523776	523776	0
mesh33x33	1089	2112	31929	33531	31729	32040	31917	188
3elt	4720	13722	363686	363204	357329	383286	362209	4880
airfoil1	4253	12289	285597	289217	272931	306005	285429	12498
whitaker3	9800	28989	1169642	1200374	1144476	1203349	1167089	22613
c1y	828	1749	63145	62333	62262	62562	62333	71
c2y	980	2102	79429	79571	78822	79823	79017	195
c3y	1327	2844	123548	127065	123514	125654	123521	7
c4y	1366	2915	116140	115222	115131	117539	115144	13
c5y	1202	2557	97791	96956	96899	98483	96952	53
gd95c	62	144	509	506	506	506	506	0
gd96a	1096	1676	96366	99944	96249	98388	96253	4
gd96b	111	193	1416	1422	1416	1416	1416	0
gd96c	65	125	519	519	519	519	519	0
gd96d	180	228	2393	2409	2391	2391	2391	0

produced by this experiment. The first three columns in the table indicate the name of the graph, its number of vertices and its number of edges. The rest of the columns indicate the best total edge length found by each of the compared heuristics. These results were taken from their corresponding paper. Finally, last column presents the difference (Δ_C) between the best total edge length found by MAMP and the previous best known solution reported in the literature.

From Table 2, one observes that MAMP is competitive in terms of solution quality. MAMP is able to improve on 4 previous best known solutions and to equal these results in 5 instances. For the other instances, MAMP did not reach the best reported solution, but its results are very close to the best reported (in average 1.009%). Notice that for some instances the improvement is important; leading to a significant decrease of the total edge length (Δ_C up to -6579).

Even if the results obtained by our memetic algorithm are very competitive we observe that MAMP, given that it is a memetic algorithm, consumes considerably more computer time than some heuristics for MinLA such as DT [2], MS [14] and AMG [21].

5   Conclusions

In this paper, a MA designed to compute near optimal solutions for the MinLA problem was presented. This algorithm, called MAMP, is based on the use of a greedy vertex-by-vertex algorithm for generating the initial population of the MA, a fine tuned Simulated Annealing algorithm for finding local optima in the search space, and a highly specialized crossover operator for efficiently explore the space of local optima in order to find the global optimum.

The performance of our MAMP algorithm was assessed through extensive experimentation over a set of well known benchmark instances and compared with four other state-of-the-art algorithms: SS+SA [18, 19], DT+SA [2], AMG [21] and GH [20]. The results obtained by MAMP are superior to those presented by the previous proposed evolutionary approach [20], and permit to improve on some previous best known solutions.

There are some issues for future research. For example, to investigate the behavior of MAMP when it is applied to larger instances, like those proposed by Koren and Harel in [14], in order to study its scalability. Additionally, the performance of MAMP should be more deeply investigated with other parameter settings for population size, operator rates and stopping conditions.

Acknowledgments. The authors would like to thank Andrew J. McAllister who has kindly provided us with his source code. The reviewers of the paper are greatly acknowledged for their constructive comments.

References

1. T. Bäck, U. Hammel, and H. P. Schwefel. Evolutionary computation: Comments on the history and current state. *IEEE Transactions on Evolutionary Computation*, 1(1):3–17, 1997.
2. R. Bar-Yehuda, G. Even, J. Feldman, and S. Naor. Computing an optimal orientation of a balanced decomposition tree for linear arrangement problems. *Journal of Graph Algorithms and Applications*, 5(4):1–27, 1996.
3. D. Corne, M. Dorigo, F. Glover, D. Dasgupta, P. Moscato, R. Poli, and K. V. Price, editors. *New Ideas in Optimization (Part 4: Memetic Algorithms)*. McGraw-Hill, 1999.
4. L. Davis. *Handbook of Genetic Algorithms*. Van Nostrad, New York, 1991.
5. J. Diaz, J. Petit, and M. Serna. A survey of graph layout problems. *ACM Comput. Surv.*, 34(3):313–356, 2002.
6. S. Even and Y. Shiloah. NP-completeness of several arrangement problems. Technical Report CS0043, Computer Science Department, Technion, Israel Institute of Technology, Haifa, Israel, January 1975.
7. B. Freisleben and P. Merz. A genetic local search algorithm for solving symmetric and asymmetric traveling salesman problems. In *Proceedings of the 1996 IEEE International Conference on Evolutionary Computation*, pages 616–621. IEEE Press, 1996.
8. P. Galinier and J. Hao. Hybrid evolutionary algorithms for graph coloring. *Journal of Combinatorial Optimization*, 3(4):379–397, 1999.
9. M. Garey and D. Johnson. *Computers and Intractability: A guide to the Theory of NP-Completeness*. W.H. Freeman and Company, New York, 1979.
10. J. J. Grefenstette. Incorporating problem specific knowledge into genetic algorithms. In L. Davis, editor, *Genetic Algorithms and Simulated Annealing*, pages 42–60, London, 1987. Morgan Kaufmann Publishers.

11. L. Harper. Optimal assignment of numbers to vertices. *Journal of SIAM*, 12(1):131–135, 1964.
12. W. E. Hart, N. Krasnogor, and J. E. Smith, editors. *Recent Advances in Memetic Algorithms and Related Search Technologies*. Springer-Verlag, 2004.
13. M. Juvan and B. Mohar. Optimal linear labelings and eigenvalues of graphs. *Discrete Applied Mathematics*, 36(2):153–168, 1992.
14. Y. Koren and D. Harel. A multi-scale algorithm for the linear arrangement problem. In L. Kucera, editor, *Proceedings of 28th Inter. Workshop on Graph-Theoretic Concepts in Computer Science (WG'02)*, volume 2573 of *LNCS*, pages 293–306. Springer Verlag, 2002.
15. A. J. McAllister. A new heuristic algorithm for the linear arrangement problem. Technical Report TR-99-126a, Faculty of Computer Science, University of New Brunswick, 1999.
16. P. Merz and B. Freisleben. Fitness landscapes, memetic algorithms and greedy operators for graph bi-partitioning. *Evolutionary Computation*, 8(1):61–91, 2000.
17. J. Petit. Approximation heuristics and benchmarkings for the MinLA problem. In *Alex '98 – Building Bridges Between Theory and Applications*, pages 112–128, 1998.
18. J. Petit. *Layout Problems*. PhD thesis, Universitat Politécnica de Catalunya, 2001.
19. J. Petit. Combining spectral sequencing and parallel simulated annealing for the MinLA problem. *Parallel Processing Letters*, 13(1):71–91, 2003.
20. T. Poranen. A genetic hillclimbing algorithm for the optimal linear arrangement problem. Technical report, University of Tampere, Finland, June 2002.
21. I. Safro, D. Ron, and A. Brandt. Graph minimum linear arrangement by multilevel weighted edge contractions. *Journal of Algorithms*, 2004. in press.
22. M. Tomassini. A survey of genetic algorithms. *Annual Reviews of Computational Physics*, III:87–118, 1995.
23. X. Yao, F. Wang, K. Padmanabhan, and S. Salcedo-Sanz. Hybrid evolutionary approaches to terminal assignment in communications networks. In W. E. Hart, N. Krasnogor, and J. E. Smith, editors, *Recent Advances in Memetic Algorithms and Related Search Technologies*, pages 129–160. Springer-Verlag, 2004.

Niching in Evolution Strategies and Its Application to Laser Pulse Shaping

Ofer M. Shir[1], Christian Siedschlag[2], Thomas Bäck[1,3], and Marc J.J. Vrakking[2]

[1] Leiden Institute of Advanced Computer Science,
Universiteit Leiden, Niels Bohrweg 1,
2333 CA Leiden, The Netherlands
[2] FOM-Instituut AMOLF, Kruislaan 407,
1098 SJ Amsterdam, The Netherlands
[3] NuTech Solutions, Martin-Schmeisser-Weg 15,
Dortmund 44227, Germany

Abstract. Evolutionary Algorithms (EAs), popular search methods for optimization problems, are known for successful and fast location of single optimal solutions. However, many complex search problems require the location and maintenance of multiple solutions. *Niching methods*, the extension of EAs to address this issue, have been investigated up to date mainly within the field of Genetic Algorithms (GAs), and their applications were limited to low-dimensional search problems.

In this paper we present in detail the background for *niching methods* within Evolution Strategies (ES), and discuss two ES niching methods, which have been introduced recently and have been tested only for theoretical functions. We describe the application of those ES niching methods to a challenging real-life high-dimensional optimization problem, namely *Femtosecond Laser Pulse Shaping*. The methods are shown to be robust and to achieve satisfying results for the given problem.

1   Introduction

Evolutionary Algorithms (EAs) have the tendency to converge quickly into a single solution [1], which means that all the individuals of the artificial population evolve to become nearly identical. Given a problem with multiple solutions, the traditional evolutionary algorithms will locate a single solution. *Niching methods* aim to maintain the diversity of certain properties within the population, and by that allow parallel convergence within those subpopulations into multiple good solutions for the given problem. Up to date, *niching methods* have been studied mainly within the field of Genetic Algorithms (GAs). The research in this direction has yielded various successful methods which have been shown to find multiple solutions efficiently [1]. In the context of real-valued multivariable function optimization, Evolution Strategies (ES) are the most commonly used technique, and some would argue the most natural environment among all the branches of EAs. This is simply due to their straightforward encoding, as well as to their successful performance in this domain in comparison to other

E. Talbi et al. (Eds.): EA 2005, LNCS 3871, pp. 85–96, 2006.

methods. The higher the dimensionality of the function is, the more obvious becomes the advantage of ES with respect to GAs. Two ES niching methods have been proposed only recently [2, 3], and have been applied successfully to high-dimensional theoretical functions. The purpose of this joint study of Physicists and Computer Scientists, is to apply those niching methods to an 80-dimensional real-life optimization problem.

This paper is organized as follows. Section 2 presents the basis for ES-niching, which is followed by the description of two proposed algorithms. Section 3 describes the application of the given methods to a real-life Physics optimization problem, and presents the experimental results which were obtained. In section 4 we draw our conclusions and give a summary of this study.

2 Niching in Evolution Strategies

2.1 The Motivation: ES Diversity

The promotion of *diversity* in the traditional GA had been originally the main motivation for the development of niching methods, as was deeply investigated by Mahfoud [1]. In this section we give a brief review of ES diversity, with respect to the tools given by Mahfoud, and by that supply the motivation for niching within ES.

We consider three main effects which cause the standard ES to lose diversity: *selective pressure, operator disruption* and *random genetic drift*.

Selective Pressure. The standard ES [4] has a strictly deterministic, rank-based approach, to selection. In the two traditional strategies, (μ, λ) and $(\mu + \lambda)$, an approach of deterministically selecting the best individuals (out of the appropriate set - the next generation or the union of the two generations, respectively) is applied, which intuitively implies high *selective pressure*. Due to the crucial role of the selection operator within the evolution process, its impact within the ES field has been widely investigated. It should be noted that the term selective pressure is occasionally associated with the ratio $\frac{\lambda}{\mu}$. Furthermore, Goldberg and Deb introduced the important concept of *takeover time* [5], which gives a quantitative description of selective pressure with respect only to the selection operator:

Definition 1. The takeover time τ^* *is the minimal number of generations until repeated application of the selection operator yields a uniform population filled with copies of the best individual.*

The selective pressure has been further investigated by Bäck [6], who analyzed all the ES selection mechanisms also with respect to takeover times, and showed that under the traditional values of the standard ES the takeover times for the two standard selection mechanisms as well as for tournament selection are very short. This implies high *selective pressures*.

Operator Disruption. In the standard ES the mutation operator typically has a small effect, which means "staying in the neighbourhood". In that sense, the mutation operator can be regarded in the standard ES as an operator with negligible disruption. The recombination operator has a bigger effect though. In the standard ES, where *discrete* and *intermediate recombination operators* are in use [4], the disruptive nature is also highly intuitive - modifying a coordinate of the decision parameters to be optimized, not in a local manner (averaging or taking a value from a different individual), has the potential to shift the offspring not in a negligible way.

Random Genetic Drift. *Genetic drift* is a stochastic process in which the diversity is lost in finite populations [7]. A distribution of genetic properties is transferred to the next generation in a limited manner, due to the finite number of offspring. As a result the distribution will approach an equilibrium distribution. In small populations this process can occur fast and become significant. Since small population sizes are used in the standard ES, the effect of *random genetic drift* occurs and causes the loss of diversity within the population. In multimodal functions, it was shown that the effect of *genetic drift* in ES causes a convergence to an equilibrium distribution around a single attractor [8].

ES Diversity: Conclusions. The standard ES is exposed to several strong effects which interrupt the formation and maintenance of multiple solutions and push the evolution process towards a rapid convergence into a single solution.

2.2   ES Dynamic Niching

The ES dynamic niching algorithm [2] was introduced recently as the first niching method within the Evolution Strategies framework. The inspiration for this algorithm was given by various niching algorithms of the GAs field, and in particular by the *fitness sharing* [9] and *crowding* [10] concepts, as well as by the *dynamic niche sharing* method [11].

The basic idea of the algorithm is to *dynamically identify the various fitness-peaks of every generation* that define the niches, *classify all the individuals into those niches*, and apply *a mating restriction scheme which allows competitive mating only within the niches* (every niche can produce a defined number of offspring, following a *fixed mating resources* concept). Additionally, a fixed number of random individuals is generated independently, in order to take part in the peak identification of the next generation. Furthermore, the unique selection mechanism replaces individuals from each niche only with individuals from the same niche - an idea which originates from the *crowding* method. Finally, we imitate the niche formation technique of the *dynamic niche sharing method*.

Distance Metric. Given that the *individual space* (the decision parameters space) is of dimension n, the distance is calculated using the *euclidean distance* in the n-dimensional space. Given the individuals $\boldsymbol{x}_i = [x_{i,1}, x_{i,2}, \ldots, x_{i,n}]$ and $\boldsymbol{x}_j = [x_{j,1}, x_{j,2}, \ldots, x_{j,n}]$ the distance $d_{i,j}$ is calculated as:

$$d_{i,j} = \sqrt{\sum_{k=1}^{n}(x_{i,k} - x_{j,k})^2} \tag{1}$$

Assumptions. The algorithm holds two assumptions:

1. The expected/desired number of peaks, q, is given or can be estimated.
2. All peaks are at least in distance 2ρ from each other, where ρ is the fixed radius of every niche.

Although those assumptions could be considered to be rather strong, they are applicable to many cases, and are also held by most of the GAs' niching methods. It is important to remark that the formulas for determining the value of the so-called niche radius ρ, to be given shortly, depend on q, the number of peaks of the target function.

The Algorithm. Given a population of individuals, a standard ES *mutation operator* is applied as the first step:

$$\boldsymbol{x}' = \boldsymbol{x} + \boldsymbol{z} \tag{2}$$

where $\boldsymbol{z}$ is a vector of random variables with a joint-normal distribution:

$$\boldsymbol{z} \sim N(\boldsymbol{0}, \boldsymbol{\Sigma}) : \Phi(\boldsymbol{z}) = \sqrt{\frac{\det \boldsymbol{\Sigma}}{(2\pi)^n}} \cdot \exp\left(-\frac{1}{2} \cdot \boldsymbol{z}^T \cdot \boldsymbol{\Sigma} \cdot \boldsymbol{z}\right) \tag{3}$$

A single step size is used per an individual, so our distribution is based on a covariance matrix proportionate to the identity matrix:

$$\boldsymbol{\Sigma} = \sigma \cdot \mathbf{I} \tag{4}$$

The adaptation of the step size is done according to the traditional standard-ES:

$$\sigma' = \sigma \cdot \exp\left(\tau' \cdot N_1(0,1) + \tau \cdot N_2(0,1)\right) \tag{5}$$

where $N_1(0,1)$ and $N_2(0,1)$ denote independent random variables, and τ and τ' are the traditional constants taken from Bäck [4].

After evaluating the fitness of the individuals, the fitness-peaks identification takes place - a greedy approach is applied in identifying the dynamic peaks of each generation, using the *dynamic peak identification algorithm* (DPI), which was introduced by Miller and Shaw [11], with the distance metric given earlier. The method is given as algorithm 1. By having the estimated niche radius ρ, it is straightforward to classify all the individuals of the population into those peaks and populate those niches. At this point the *mating phase* begins, which is a closed competitive mating session within every niche. Each niche gets fixed

Algorithm 1. Greedy Dynamic Peak Identification (DPI)

```
input:     Pop - array of population members
           N - population size
           q - number of peaks to identify
           ρ - niche radius.

Sort Pop in decreasing fitness order
i := 1
NumPeaks := 0
DPS := ∅ (Dynamic Peak Set)
loop until NumPeaks = q or i = N + 1
   if Pop[i] is not within ρ of peak in DPS
       DPS := DPS ∪ {Pop[i]}
       NumPeaks := NumPeaks + 1
   endif
   i := i + 1
endloop

output:   Dynamic Peak Set
```

mating resources (number of individuals in the next generation), i.e. independent of the fitness value of its peak. In this manner we prevent the best niche to take over the population's resources and flood the next generation with its offspring. This is also meant to prevent a *genetic drift* into a single distribution. In particular, a uniform distribution of the resources to q niches is considered:

$$\tilde{\mu} = \frac{\mu}{q} \qquad \tilde{\lambda} = \frac{\lambda}{q} \tag{6}$$

meaning that each niche has $\tilde{\mu}$ parents and produces $\tilde{\lambda}$ offspring in every generation. The *selection mechanism* in the algorithm can be considered as a combination of the two traditional ES strategies, (μ, λ) and $(\mu + \lambda)$. $\tilde{\lambda}$ individuals are produced within every niche in the following way - the first parent is chosen via *tournament selection*, where the second parent is the best individual in the niche which is different than the first parent (this is known as the *line breeding mechanism*). In case that the niche contains only one individual, the second parent will be the best individual of another niche (aiming by that to explore the search space). Given those $\tilde{\lambda}$ pairs of parents, the standard-ES *recombination operator* is applied: *intermediate recombination* for the strategy parameters and *discrete recombination* for the decision parameters [4]. The $\tilde{\mu}$ parents of the next generation are selected as follows: the best η of the $\tilde{\lambda}$ offspring along with the best $\delta = \tilde{\mu} - \eta$ individuals of the current niche (the latter group proceeds to the next generation without recombination). If the niche does not have δ individuals, new randomly-generated individuals will be added on that niche's resources. At this point, additional $\omega = \tilde{\mu}$ uniformly distributed random individuals are added to the whole population, and take part in the next round of the *dynamic peak identification algorithm*.

The algorithm is summarized in the pseudo-code given as algorithm 2.

The Niche Radius ρ. The original formula for ρ for *phenotypic sharing* in GAs was derived by Deb and Goldberg [5]. By following the trivial analogy and

---

Algorithm 2. The ES Dynamic Niching Algorithm: A Generation Loop

---
```
Apply Mutation on the population
Evaluate fitness of population and Sort
Compute the Dynamic Peak Set using the DPI (Algo-1)
for every niche i = 1..q produce the next generation:
    Generate λ̃ offspring as follows:
        Choose 1st parent via Tour-Selec. of the niche
        Choose the best indiv. of that niche as the 2nd parent
        Apply standard recombination
    Select the best η of the λ̃ offspring and the best
    μ̃ − η indiv. of the niche to form the next gen.
endfor
Generate additional ω = μ̃ random indiv.,
Join all q niches, to yield the new population
```
---

considering the decision parameters as the decoded parameter space of the GA, the same formula can be applied, using the metric introduced earlier. Given q, the number of peaks in the solution space, every niche is considered to be surrounded by an n-dimensional hypersphere with radius ρ which occupies $\frac{1}{q}$ of the entire volume of the space. The volume of the hypersphere which contains the entire space is

$$V = cr^n \tag{7}$$

where c is a constant, given explicitly by:

$$c = \frac{\pi^{\frac{n}{2}}}{\Gamma(\frac{n}{2}+1)}, \qquad \Gamma(n) = \int_0^\infty x^{n-1}\exp(-x)dx \tag{8}$$

Given lower and upper boundary values $x_{k,min}$, $x_{k,max}$ of each coordinate in the decision parameters space, r is defined as follows:

$$r = \frac{1}{2}\sqrt{\sum_{k=1}^n (x_{k,max} - x_{k,min})^2} \tag{9}$$

If we divide the volume into q parts, we may write

$$c\rho^n = \frac{1}{q}cr^n \tag{10}$$

which yields

$$\rho = \frac{r}{\sqrt[n]{q}} \tag{11}$$

2.3 Dynamic Niching with Covariance Matrix Adaptation ES

The dynamic niching with CMA-ES algorithm was introduced recently [3] as the successor of *the ES dynamic niching algorithm*. We provide here a short introduction of the CMA-ES method, followed by a description of the algorithm.

The CMA-ES: A Brief Overview. The *covariance matrix adaptation evolution strategy* [12], is a specific variant of ES that has been successful for treating correlations among object variables. This method tackles the critical element of Evolution Strategies, the adaptation of the mutation parameters. We shall provide here only a short description of the principal elements of the $(1, \lambda)$-CMA-ES.

The fundamental property of this method is the exploitation of information obtained from previous successful mutation operations. Given an initial search point $\mathbf{x}^0$, λ offspring are sampled from it by applying the mutation operator. The best search point out of those λ offspring is chosen to become the parent of the next generation. The action of the *mutation operator* for generating new samples of search points in generation $g + 1$ is defined as follows:

$$\boldsymbol{x}^{g+1} = \boldsymbol{x}^g + \delta \cdot \mathbf{B} \cdot \boldsymbol{z} \tag{12}$$

where δ is the global step size, which is adaptive with respect to the optimization process, and z is a vector of random variables drawn from the *multivariate normal distribution*. The matrix $\mathbf{B}$, the crucial element of this process, is composed of the eigenvectors of the covariance matrix with the appropriate scaling of the eigenvalues - defining the distribution of a sequence of successful mutation points. It is initialized as the *unity matrix* and is updated according to cumulative data from the evolution process itself. We omit most of the details and refer the reader to Hansen and Ostermeier [12].

Dynamic Niching with CMA. The algorithm uses the skeleton of *the ES dynamic niching algorithm* but changes the evolutionary core mechanism from the standard ES to the CMA-ES, and in particular to the $(1, \lambda)$-CMA. A brief description of the algorithm follows.

Given q, the estimated/expected number of peaks, $q + 1$ "CMA-sets" are initialized, where a CMA-set as is defined as the collection of all the dynamic variables of the CMA algorithm which uniquely define the search at a given point of time. Such dynamic variables are the current search point (the decision parameters to be optimized), the covariance matrix, the step size, as well as other auxiliary parameters. At every point in time the algorithm stores exactly $q + 1$ CMA-sets, which are associated with $q+1$ search points: q for the peaks and 1 for the "non-peaks domain". The $(q+1)^{th}$ CMA-set is associated with an individual which is randomly re-generated in every generation in order to explore the search space and produce candidates for niche formation. Until stopping criteria are met, the following procedure takes place. Each search point generates in every generation λ samples (offspring) based on its evolving sampling distribution - its step size as well as the covariance matrix. After the fitness evaluation of the new $\lambda \cdot (q+1)$ individuals, the classification into niches of the entire population is done using the DPI, introduced earlier as algorithm 1. The peaks of the *dynamic peak set*, given as output of the DPI, are chosen to become the new search points, and their CMA-sets are inherited from their parents (which are uniquely defined, due to the lack of *recombination*) and updated according to the CMA method. The dynamic peak set may contain up to q peaks, and in case there are less than q individuals in that set, the rest of the search points are randomly re-generated.

Algorithm 3. Dynamic Niching with $(1, \lambda)$-CMA-ES: A Single Generation Loop

---

```
for all i = 1..q + 1 search points
    Generate λ samples based on the CMA distribution of i
endfor
Evaluate Fitness of the population.
Compute the Dynamic Peak Set of the λ · (q + 1) individuals using the DPI
for every given peak of the dynamic-peak-set do:
    Set peak as a search point of the next generation
    Inherit the CMA-set and update it respectively
endfor
if N_dps =size of dynamic-peak-set < q
    Generate q − N_dps new search points, reset CMA-sets
endif
Reset the (q + 1)^th search point
```

In any case, the $(q+1)^{th}$ search point is randomly generated at this stage, as the representative of the "non-peaks" domain, as explained earlier. This concludes a single generation loop.

This algorithm holds the same assumptions as *the ES dynamic niching algorithm*. It uses the same distance metric, as well as the niche radius calculations. The algorithm is summarized as algorithm 3.

3   The Application: Femtosecond Laser Pulse Shaping

3.1   General

To investigate and, more importantly, to control the motion of atoms or molecules by irradiating them with laser light, one has to provide laser pulses with durations on the same time scale as the motion of the particles. Recent technological development has made lasers with pulse lengths on the order of femtoseconds (1 fs=$10^{-15}s$) routinely available. Moreover, the time profile of these pulses can be shaped to a great extent. By applying a self-learning loop using an evolutionary mechanism, the interaction between the system under study and the laser field can be steered, and optimal pulse shapes for a given optimization target can be found. In our work, the role of the experimental feedback in the self-learning loop is played by numerical simulations. The target function we aimed to optimize was the *alignment* of an ensemble of molecules after the interaction with a shaped laser pulse. There is currently a great interest in the atomic and molecular physics community to align molecules with laser pulses, since dealing with an aligned sample of molecules simplifies the interpretation of experimental data. The alignment's quantity is defined as the expectation value of the *cosine-squared* of the angle of the molecular axis with respect to the laser polarization axis (i.e. success-rate or fitness are given as real values between 0 and 1). To calculate the time-dependent alignment, the Schrödinger's equation for the angular degrees of freedom of a model diatomic molecule under the influence of the shaped laser field is solved. Explicitly, the time-dependent profile of the pulse in our simulations has been given by

$$f(t) = \int_{-\infty}^{\infty} A(\omega) \exp(i\phi(\omega)) \exp(i\omega t) \, d\omega, \tag{13}$$

where $A(\omega)$ is a Gaussian window function describing the contribution of different frequencies to the pulse and $\phi(\omega)$, the *phase function*, equips these frequencies with different complex phases. Hence, by changing $\phi(\omega)$, the temporal structure of $f(t)$ can be altered. In a real life pulse shaping experiment, $A(\omega)$ is fixed and $\phi(\omega)$ is used to control the shape of the pulses. We have used the same approach in our numerical simulations, i.e. the search space is in the frequency domain while the fitness evaluations is performed in the time domain. To this end, we interpolated $\phi(\omega)$ at n frequencies ω_n; the n values $\phi(\omega_n)$ are our decision parameters to be optimized. In order to achieve a good trade-off between high resolution and optimization efficiency, the value of $n = 80$ turned to be a good compromise.

3.2   The Application of Niching to the Problem

Aiming to apply niching, we were required to define an appropriate *distance metric*. We should note at this point that the function entering the simulation was actually $f^2(t)$, the *time-dependent laser intensity*. Hence, the outcome of the calculations was invariant under the transformation $\tilde{\phi}(\omega) = \phi(\omega) + \phi_0$. Furthermore, adding a linear term to the phase function (i.e. $\tilde{\phi}(\omega) = \phi(\omega) + c \cdot \omega$) simply shifts the whole pulse with respect to the time origin and therefore has also no observable effect. This had to be taken into account when defining a distance metric between two individuals $\phi(\omega)$ and $\phi'(\omega)$, as it is clear that using the straightforward approach would not achieve the goal: due to the fact that $\phi(\omega)$ is invariant under the specified transformations, calculating the distance between two feasible solutions $\phi(\omega)$, $\tilde{\phi}(\omega)$ would not guarantee that the derived pulses $f(t)$, $\tilde{f}(t)$ respectively will have a different profile. Our proposed solution, concluded from the specified *invariance properties*, was to calculate the distance in the **second-derivative space** of $\phi(\omega)$. Explicitly, given that the discretization is to n function values, the distance between $\phi_i(\omega)$, $\phi_j(\omega)$ is given by:

$$d_{i,j} = \sqrt{\sum_{k=1}^{n} \left(\left(\frac{d^2\phi_i(\omega)}{d\omega^2} \right)_k - \left(\frac{d^2\phi_j(\omega)}{d\omega^2} \right)_k \right)^2} \tag{14}$$

3.3   Experimental Results

Experimental Setup. We provide here a few technical details concerning our experiments. Every fitness evaluation takes approximately 35 seconds. Taking this into account, our experimental setup was set to a minimal configuration:

- The CMA-ES based niching method was set to $(1, 10)$ core strategy.
- The parameters of the Standard-ES based method were set to $\{\mu = 5, \lambda = 10, \eta = 5, \omega = 0\}$.

In our experiments we have mostly aimed for a fixed number of solutions, $q = 3$.

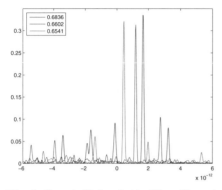

Fig. 1. Run A: Pulses in the Time Domain

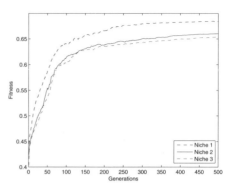

Fig. 2. Evolution of Run A

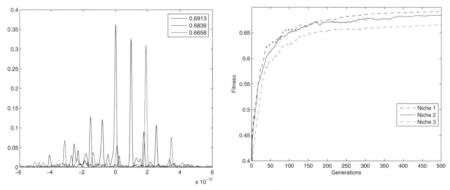

Fig. 3. Run B: Pulses in the Time Domain

Fig. 4. Evolution of Run B

Results. The fitness gets a real value in the interval $[0, 1]$, subject to maximization, as was introduced earlier. A random feasible solution gets on average a value of 0.333 (isotropic $3D$ space), and the best result known to us up to date is around 0.7.

The results of our experiments will be discussed at several levels:

1. Our definition of the distance metric for this problem has been proven to be successful. The obtained pulses in the time domain had indeed different characteristics, and in particular their shapes differed in a satisfying manner. Illustrative examples are given in figures 1 and 3.
2. The CMA-ES based niching method has achieved better alignment results in comparison with the Standard-ES based method: the best niche's fitness was always higher with CMA. Moreover, the CMA niching method achieved the higher result known to us, obtained with any other optimization method that we have used so far (fitness of 0.7). For both methods, the 2nd and 3rd niches also obtained good results, usually very close to the result of the best niche. Plots with typical simulation runs are given as figures 2 and 4.
3. The obtained pulses have interesting profiles from the *Physics* point of view. Some of the profiles were obtained for the first time by the niching

methods. It seems that the *niching pressure*, which has been introduced by the algorithms, is responsible for the generation of those unique pulse profiles. It does so by forcing the population to look for several attractors, and in our case managed to push it to some new attractors which were not obtained by other optimization methods.

4. Due to the cost of a single fitness evaluation, the number of simulations was limited, and we do not provide a statistical analysis of the results.

4   Conclusions

We have applied two ES niching methods, which have been tested so far only on theoretical functions, to a real-life challenging problem, namely *Femtosecond Laser Pulse Shaping*. The application was successful at several levels. We have managed to tackle the problem of defining the distance metric for the niching algorithms. The simulations themselves yielded highly satisfying results, with respect to fitness values and to uniqueness. This study has been successful from the Physics as well as from the Computer Science point of views.

Acknowledgments

This work is part of the research programme of the 'Stichting voor Fundamenteel Onderzoek de Materie (FOM)', which is financially supported by the 'Nederlandse Organisatie voor Wetenschappelijk Onderzoek (NWO)'.

References

[1] Mahfoud, S.: Niching Methods for Genetic Algorithms. PhD thesis, University of Illinois at Urbana Champaign (1995)

[2] Shir, O.M., Bäck, T.: Niching in evolution strategies. In: Proceedings of the Genetic and Evolutionary Computation Conference, GECCO-2005, New York, NY, USA, ACM Press (2005)

[3] Shir, O.M., Bäck, T.: Dynamic niching in evolution strategies with covariance matrix adaptation. In: Proceedings of the 2005 Congress on Evolutionary Computation CEC-2005, Piscataway, NJ, USA, IEEE Press (2005)

[4] Bäck, T.: Evolutionary algorithms in theory and practice. Oxford University Press, New York, NY, USA (1996)

[5] Deb, K., Goldberg, D.E.: An investigation of niche and species formation in genetic function optimization. In: Proceedings of the third international conference on Genetic algorithms, San Francisco, CA, USA, Morgan Kaufmann Publishers Inc. (1989) 42–50

[6] Bäck, T.: Selective pressure in evolutionary algorithms: A characterization of selection mechanisms. In Michalewicz, Z., Schaffer, J.D., Schwefel, H.P., Fogel, D.B., Kitano, H., eds.: Proc. First IEEE Conf. Evolutionary Computation (ICEC'94), Orlando FL. Volume 1., Piscataway, NJ, USA, IEEE Press (1994) 57–62

[7] Kimura, M.: The neutral theory of molecular evolution. Cambridge University Press, Cambridge (1983)

[8] Schönemann, L., Emmerich, M., Preuss, M.: On the extiction of sub-populations on multimodal landscapes. In: Proc. of the Int'l Conf. on Bioinspired optimization Methods and their Applications, BIOMA 2004, Jožef Stefan Institute, Slovenia (2004) 31–40

[9] Holland, J.H.: Adaptation in Natural and Artificial Systems: An Introductory Analysis with Applications to Biology, Control and Artificial Intelligence. MIT Press, Cambridge, MA, USA (1992)

[10] Jong, K.A.D.: An analysis of the behavior of a class of genetic adaptive systems. PhD thesis (1975)

[11] Miller, B., Shaw, M.: Genetic algorithms with dynamic niche sharing for multimodal function optimization. In: Proceedings of the 1996 IEEE International Conference on Evolutionary Computation (ICEC'96), New York, NY, USA (1996)

[12] Hansen, N., Ostermeier, A.: Completely derandomized self-adaptation in evolution strategies. Evolutionary Computation **9** (2001) 159–195

A Modified Genetic Algorithm for the Beam Angle Optimization Problem in Intensity-Modulated Radiotherapy Planning

Yongjie Li[1], Dezhong Yao[1], Jiancheng Zheng[2], and Jonathan Yao[2]

[1] School of Life Science and Technology,
University of Electronic Science and Technology of China,
Chengdu 610054, China
Liyj999@yahoo.com, Dyao@uestc.edu.cn
[2] Topslane Inc,
Pleasant Hill, CA 94523, USA
{Zhengjc, Jonathanyao}@topslane.com.cn

Abstract. In this paper, a modified genetic algorithm (GA) is proposed to improve the efficiency of the beam angle optimization (BAO) problem in intensity-modulated radiotherapy (IMRT). Two modifications are made to GA in this study: (1) a new operation named *sorting operation* is introduced to sort the gene in each chromosome before the crossover operation, and (2) expert knowledge about tumor treatment is employed to guide the GA evolution. Two types of expert knowledge are employed, i.e., beam orientation constraints and beam configuration templates. The user-defined knowledge is used to reduce the search space and guide the optimization process. The sorting operation is introduced to inherently improve the evolution performance for the specified ABO problem. The beam angles are selected using GA, and the intensity maps of the corresponding beams are optimized using a conjugate gradient (CG) method. The comparisons of the preliminary optimization results on a clinical prostate case show that the proposed optimization algorithm can slightly or heavily improve the computation efficiency.

1 Introduction

Intensity-modulated radiotherapy (IMRT) is a powerful technology to potentially improve the therapeutic ratio by using modulated beams from multiple spatial directions to irradiate the tumors. The conventional IMRT planning starts with the selection of suitable beam angles, followed by an optimization of beam intensity maps using inverse optimization method under the guidance of a objective function [1] [2]. The set of such beam directions should be chosen such that the plan with this beam combination could produce highly three-dimensional conformal dose distributions to the target, while sparing those organ-at-risks (OARs) and normal tissues as much as possible.

Beam angle selection is an important but also challenging issue for IMRT planning because of the inherent complexity of the problem, mainly the large search

E. Talbi et al. (Eds.): EA 2005, LNCS 3871, pp. 97–106, 2006.

space and the coupling between the beam configuration and the intensity maps of the beams [3] [4]. A mass of studies have demonstrated that the selection of suitable beam angles is most valuable for a plan with a small number of beams (<5) [1], and is also clinically meaningful for plans with large number of beams (>9) in some complicated cases, where the tumor volume surrounds a critical organ, or is surrounded by multiple critical organs [3] [5].

At present, the selection of beam angles is generally based upon the experience of the human planner in the clinical practice. Several trial-and-error attempts are normally needed in order to find a group of acceptable beam angles, mainly because of the facts that beam directions are case dependant and they are coupled with the intensity profiles of the incident beams, which result in the less straightforwardness for selection, compared to the conventional conformal radiotherapy (CRT) [3]. To date, extensive efforts have been made by many researchers to facilitate the technique of computer-assisted beam angle selection for IMRT planning [3~9]. Though there are fruitful improvements achieved and the function of computer-aided automatic selection of beam angles for IMRT planning is provided in some of the newest commercial treatment planning systems (TPS), it still can not act as a routine planning tool in clinical practice because of the limitation of the associated intrinsic extensive computation time.

To further improve the performance of the optimization, two issues are the directions for the ongoing studies: (1) optimization algorithms themselves, and (2) the external intervention or guidance to the optimization process. As for the first issue, we introduce a new operation, named *sorting operation*, to GA to sort the gene in each chromosome before the selection operation in order to enhance the quality of the children after the crossover operation. As for the second issue, we employ the expert knowledge about tumor treatment accumulated by the oncologists and physicists over time during their clinical practice to guide the optimization process.

The remainder of the paper is organized as follows. Section 2 describes in details the modifications to the standard GA, as well as the objective function and the fitness value defined for the GA-based optimization solution. In Section 3 we give a clinical prostate tumor case to demonstrate the validity and performance of the proposed algorithm. Finally, some discussion and conclusions and the directions for future works are briefly presented in Section 4.

2 Materials and Methods

In order to simplify the optimization and decrease the computation burden, the beam angle selection and the beam intensity map optimization in BAO are normally separated into two iterative steps [3] [4] [7]. In this paper, the beam angles are selected using a modified GA, and the intensity maps of the selected beams are optimized using a conjugate gradient (CG) method. This section will detailedly describe the sorting operation and the strategy of combining expert knowledge with GA, as well as the objective function and the fitness value.

2.1   Genetic Algorithm with Sorting Operation

The beams in this study are restricted to the coplanar ones, and the beam angles are specified according to the International Electrotechnical Commission (IEC) convention. The search space covering the total 360° gantry rotational angles are discretized into equally spaced directions with a given angle step, such as 5° or 10°. These discrete angles are called trial beam angles and they compose of the search space of beam angles of the BAO problem.

In this study, a one-dimensional integer-coding scheme is adopted, in which the combination of beam angles is represented by a chromosome (an individual) with a length of user-specified beam number of the plan, and each gene in the chromosome represents a trial beam angle. For example, the individual *parent 1* shown in Fig. 1 demonstrates a five-beam plan with angles of 10°, 80°, 120°, 200° and 250°. The genes in one chromosome are required to be different with each other, which means that there should be no two beams with same angles in one treatment plan.

The standard GA consists of three genetic operations: selection, crossover and mutation (Fig. 1). Parent individuals with higher fitness are selected into the next generation with a higher probability according to a simple strategy of proportional fitness assignment. To any two randomly selected parent individuals (angle sets), a crossover operation will be applied according to a specified crossover probability, normally 0.5~0.95. Then a mutation operation to the two children angle sets will be done according to a mutation probability, normally 0.001~0.02.

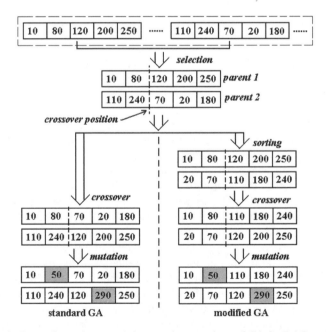

Fig. 1. The coding scheme and the genetic operations of GA for BAO problem

In essence, BAO is a combinational optimization problem, in which specified number of beams are to be selected among a beam candidate pool. That is to say, the genes (i.e. the beam angles) are order-independent. For example, there is no clinical difference between the individual (10°, 80°, 120°, 200°, 250°) and (80°, 120°, 10°, 200°, 250°). These two individuals represent a same treatment plan, and mathematically, they have the equal fitness value (to be described in *Section 2.3*). The irradiation order of the beams will be clinically determined by the planner. However, the order of the genes in a chromosome has meaningful impact on the evolution performance, especially on the crossover operation, which is clearly demonstrated by Fig. 1. For the parents *parent 1* and *parent 2*, two children (10°, 50°, 70°, 20°, 180°) and (110°, 240°, 120°, 290°, 250°) would be produced by standard GA (sGA) after the crossover operation and mutation operation. These two children are not clinically preferable because the separation between most of the neighboring beams is so small that it is hard to produce acceptable dose distribution. On the contrary, the modified GA (mGA) would generate two children (10°, 50°, 110°, 180°, 240°) and (20°, 70°, 120°, 290°, 250°) when a sorting operation is applied to the two parent individuals before the crossover operation. These two children have higher possibility to produce superior dose distributions because the beams are approximately uniformly distributed in the whole 360° gantry angle space. Such strategy has been proved valid by most of the manual plans designed by those experienced oncologists and physical therapists. In summary, the introduced sorting operation is used to avoid the beams to be distributed in a small incidence range, and consequently, to potentially improve the optimization efficiency.

2.2 Expert Knowledge Guided Genetic Algorithm

There are two types of expert knowledge about individual treatment used in our optimization method, both of which are defined by the planner through a graphical user interface (GUI): (1) beam orientation constraints, which define the orientation scopes through which no beam can pass, and (2) one or more groups of beam configuration templates that are the most possible beam angles suitable to the current treatment site (each group is a plan containing several beams). The first type of knowledge is used to define the search space by reducing the defined constraint scopes from the whole space with 360°, which may largely shorten the optimization time by reducing the search space. The left of the total 360° are divided into discrete angles with an angle increment, such as 5° or 10°. The second type of the knowledge is used (1) to initialize some of the individuals in the first generation of GA (the left individuals are initialized randomly), and (2) to replace the worst individual in each new generation. The scheme of expert knowledge guided GA is shown in Fig. 2.

It should be pointed out that, no more than a quarter of the total beam configurations (individuals) in the first generation of GA are allowed to be initialized with the expert templates, in order to avoid that the expert knowledge dominates the GA operations at the beginning of the optimization. If there are plan templates remained after the initialization operation, they will be used to replace the worst individual in each new generation, until no template remains.

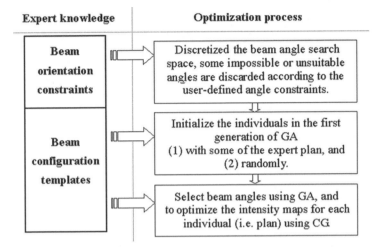

Fig. 2. The scheme of expert knowledge guided GA for beam angle optimization

2.3 Objective Function and Fitness Value

For each new individual (i.e. a new plan), a CG method is employed to optimized the corresponding beam intensity maps [2] [7], and then the dose distributions calculated using these optimized intensity maps are used to calculate the fitness value for evaluation of the individual. The optimization aims to minimize the dose difference between the prescribed and the calculated dose distributions, which can be mathematically described by the following objective function

$$F_{obj}(\vec{x}) = \alpha \cdot F_{OAR}(\vec{x}) + \beta \cdot F_{PTV}(\vec{x}) \tag{1}$$

$$F_{OAR}(\vec{x}) = \sum_{i=1}^{Noar} \sum_{j=1}^{NT_i} \delta \cdot w_j \cdot \left(d_j(\vec{x}) - p_j\right)^2 \tag{2}$$

$$F_{PTV}(\vec{x}) = \sum_{j=1}^{NT_{PTV}} \delta \cdot w_j \cdot \left(d_j(\vec{x}) - p_j\right)^2 \tag{3}$$

$$d_j(\vec{x}) = \sum_{m=1}^{Nray} a_{jm} \cdot \vec{x}_m \tag{4}$$

Where $\vec{x} = (x_1, x_2, \cdots, x_{N_B})$ is the beam set, N_B is the specified number of the beam in a treatment plan. $F_{obj}(\vec{x})$ is the value of objective function of the beam set $\vec{x}$, $F_{OAR}(\vec{x})$ is the part associated with all the OARs, and $F_{PTV}(\vec{x})$ is the part associated with the target. N_{OAR} is the total number of the OARs, NT_i is the point number in the ith OARs, NT_{PTV} is the point number in the target, $\delta = 1$ when point dose in the volume breaks the constraints, else $\delta = 0$, w_j is the weight of jth point, d_j is the calculated dose of the jth point in the volume, p_j is the prescribed dose of the jth point in the volume. α and β are the regularizing factors that balance the importance between the target and the OARs. All of the selected beams in $\vec{x}$ are divided into rays (also called pencil beamlets), N_{ray} is the total number of the ray. a_{jm} is the dose deposited to the jth point from the mth ray with a unit weight. $\vec{x}_m$ is the intensity of the mth ray.

The quality of each individual is evaluated by a fitness value, and the purpose of optimization is to find the individual (plan) with maximum fitness. The fitness value is calculated by

$$Fitness(\vec{s}) = F_{max} - Fobj(\vec{s}), \quad \vec{s} = \left(s_1, s_2, \cdots, s_{N_{angle}}\right) \qquad (5)$$

Where F_{max} is a rough estimation of the maximum value of the objective function, which makes sure that all the fitness values are positive, a requirement of the selection operation. $\vec{s}$ is a group of angles to be selected, and N_{angle} is the number of the beam angles of the plan. Both F_{max} and $Fobj(\vec{s})$ are calculated using Eq. (1) ~ (4).

The whole optimization is terminated when no better plan can be found in the specified number of successive generations of GA, and the individual with the highest fitness in the last generation will be regarded as the optimal set of beam angles. The details of beam angle optimization without using expert knowledge can be found in our previously published paper [7].

3   Results

A clinical case with prostate tumor (planning tumor volume, PTV) shown in Fig. 3 is optimized using the proposed method. There are four organ-at-risks (OARs) needed to be considered during the irradiation: rectum, bladder, left and right femur head. The sizes and relative positions of the volumes change substantially from slice to slice, and on the most of the slices the contours of the rectum and bladder are overlapped with the tumor. Seven 6MV coplanar photon beams are used to irradiate the tumor.

The selection of parameters in the GA, such as population size, crossover probability and mutation probability, is an important issue for the optimization performance of GA. Though some theoretical studies have been made for the determination of these parameters [10], all these three parameters are mostly empirically selected in engineering applications [11] [12]. The population size of GA is empirically set to the double of the total number of angle candidates [11]. For example, if there are five beams to be selected, the population size can be set to 10 or a little more. As for this seven-beam plan of the clinical case, the population size is set to 20. The other two parameters, the crossover probability and mutation probability, are empirically set to 0.9 and 0.01, respectively. These parameters have been experimentally proved suitable for the beam angle optimization problem, though the optimization perform-ance would be better by fine-tuning of these parameters.

First, the optimization results are compared between the GA with and without the sorting operation. Then, the results are compared between the optimization with and without the expert knowledge. For our new method, two beam orientation constraints are defined ((a) and (b) in Fig. 3), and a plan configuration candidate with beam an-gles of 0°, 50°, 100°, 150°, 210°, 260° and 310° is defined as an expert knowledge, shown as the dotted white straight lines in Fig. 3. This plan candidate has become an informal standard for the prostate case in the clinical IMRT practice in some institu-tions and oncology centers.

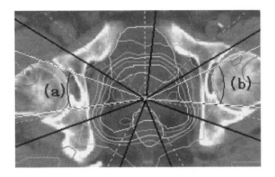

Fig. 3. The clinical prostate case and the dose distribution of the optimized plan. The arcs (a) and (b) are the two beam orientation constraint. The dotted white straight lines are the angles of a beam configuration template. The solid black straight lines are the optimized beam angles.

Table 1. The comparison of mean computation time among different algorithm

Algorithms	Mean computation time
standard GA (sGA)	45 min 26 sec
sGA + sorting operation	32 min 18 sec
sGA + sorting operation + expert knowledge	27 min 43 sec

For a convincing comparison, all the optimization tasks are run ten times. The op-timization will be terminated if the generation number is reached to 200, or there is no better individual found in 20 successive generations.

Just as expected, all the runs find the same optimal beam angles: 10°, 60°, 110°, 155°, 200°, 250° and 300°, shown as the thick black straight lines in Fig. 3. The mean computation time is listed in Table 1. About 45 min 26 sec are taken by the standard GA (i.e., neither sorting operation nor external knowledge is used), but the computa-tion time is reduced to 32 min 18 sec when the sorting operation is applied to GA, and 27 min 43 sec are used when both the sorting operation and the defined knowledge is incorporated into the optimization progress.

The fitness value versus generation number curves for one run of each algorithm are shown in Fig. 4. From the figure we can clearly find that, the convergence is meaningfully obtained by combining the sorting operation into the optimization, and is further improved by utilizing the expert knowledge.

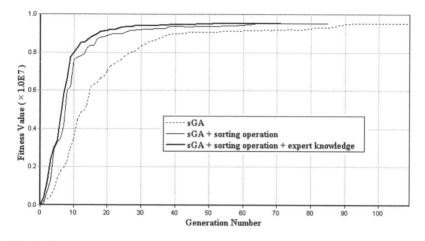

Fig. 4. The fitness value versus generation number curve for different algorithms. The fitness values are the currently best values among all the individuals.

4 Discussion and Conclusions

In this paper, a modified GA (mGA) was developed for beam angle optimization in IMRT planning. In mGA, a sorting operation was introduced to avoid the good information in chromosomes being destroyed by the crossover operation. The beam angles are selected with GA guided by the user-defined expert knowledge. For each new plan selected by GA, the corresponding beam intensity maps are fast optimized using CG. The calculated dose distributions are used to calculate the fitness value in order to evaluate the plan. A clinical prostate tumor case is employed to test the performance of the proposed algorithm. The comparison of the optimization time shows that the optimization efficiency is improved by the proposed mGA.

One could question that why the full search process does not operate exclusively on such sorted vectors (individuals). Mutation disturbs the ordering. In fact, the proposed algorithm is equivalent to always using normalized (sorted) individuals.

The optimization of beam angles for IMRT planning is an important but also a difficult thing because of the extensive computation. Many efforts are needed to be made before the automatic selection of beam angles becomes a routine tool for IMRT planning. By fully and easily making use of the plentiful expert knowledge accumulated by the oncologists and physicists over time, the presented technique is hoped to be more feasible and practicable for routine IMRT planning. The optimization time will be slightly or heavily shortened, and the optimized beam angles are better, at least not worse than that of not utilizing expert knowledge. The degree of the improvement depends on the quantity and quality of the prior knowledge provided by the planner.

In fact, the value of the crossover operation is still controversial. It maintains the diversity of the population. On the other hand, however, it brings the risks to destroy the good information about the solution, which is the partial reason that GA converges relatively slow at the later stage of the optimization process. De Jong *et al* (1997) had noted that crossover might not perform especially well on functions featuring high modality [13]. The proposed sorting operation aims to improve the convergence, and the limited results show that it works well. It should be emphasized that the sorting operation is problem-dependent, and perhaps not suitable for other engineering problems.

The idea and the implementation of GA is simple, however, it is not a piece of cake for GA to solve a specified engineering optimization problem. Now it is a trend to explore some novel schemes to incorporate the expert knowledge into the optimization algorithms. The presented algorithm has just provided a preliminary frame for the combination of expert knowledge with the GA. We are currently working on the building of an easily accessed knowledge database and on the more valid scheme for the guiding of the genetic evolution with plan template. Also, the influence of the quality and quantity of the expert knowledge on the performance of GA are being studied in-depth. For example, if some bad knowledge is provided for a specified case, an insight research is needed to evaluate their influence on the genetic process of GA.

Acknowledgments

This work is supported by a grant from the 973 Project of China (Grant No. 2003CB716106), a grant from NSFC of China (Grant No. 90208003), a grant from the Doctor Training Fund of the Ministry of Education (MOE) of China, and a grant from TRAPOYT of China. The work is also partially supported by Topslane Inc. The authors would like to thank Wenyan Chen and Yu Wu of Topslane for their helpful discussions and assistance.

References

1. Webb S.: Intensity-modulated Radiation Therapy. Bristol and Philadelphia, Institute of Physics Publishing (2000)
2. Spirou S. V., Chui C. S.: A gradient inverse planning algorithm with dose-volume constraints. Med. Phys. 25 (1998) 321–333
3. Pugachev A., Boyer A. L., Xing L.: Beam orientation optimization in intensity-modulated radiation treatment planning. Med. Phys. 27 (2000) 1238–1245
4. Hou Q., Wang J., Chen Y., Galvin J. M.: Beam orientation optimization for IMRT by a hybrid method of genetic algorithm and the simulated dynamics. Med. Phys. 30 (2003) 2360–2376
5. Gaede S., Wong E. and Rasmussen H.: An algorithm for systematic selection of beam directions for IMRT. Med. Phys. 31 (2004) 376–388
6. Djajaputra D., Wu Q., Wu Y. Mohan R.: Algorithm and performance of a clinical IMRT beam-angle optimization system. Phy. Med. Biol. 48 (2003) 3191–3212
7. Li Y., Yao J., Yao D.: Automatic beam angle selection in IMRT planning using genetic algorithm. Phy. Med. Biol. 49 (2004) 1915–1932.
8. Souza W. D., Meyer R. R., Shi L.: Selection of beam orientations in intensity-modulated radiation therapy using single-beam indices and integer programming. Phy. Med. Biol. 49 (2004) 3465–3481

9. Wang X., Zhang X., Dong L., Liu H., Wu Q., Mohan R.: Development of methods for beam angle optimization for IMRT using an accelerated exhaustive search strategy. Int. J. Radiat. Oncol. Boil. Phys. 60 (2004) 1325–1337
10. Goldberg D. E.: Genetic Algorithms in Search, Optimization, and Machine Learning. Addison-Wesley, Reading, Massachusetts (1989)
11. Yu Y. and Schell M. C.: A genetic algorithm for the optimization prostate implants. Med. Phys. 23 (1996) 2085–2091
12. Wu X. and Zhu Y.: A mixed-encoding genetic algorithm with beam constraint for conformal radiotherapy treatment planning. Med. Phys. 27 (2000) 2508–2516
13. De Jong K., Potter M. and Spears W.: Using problem generators to explore the effects of epistasis. Proceedings of the Seventh International Conference on Genetic Algorithms (1997) 338–345

On a Property Analysis of Representations for Spanning Tree Problems

Sang-Moon Soak[1], David Corne[2], and Byung-Ha Ahn[1]

[1] Dept. of Mechatronics, Gwangju Institute of Science and Technology, South Korea
{soakbong, bayhay}@gist.ac.kr
[2] Dept. of Computer Science, University of Exeter, Exeter EX4 4QJ, UK
D.W.Corne@exeter.ac.uk

Abstract. This paper investigates on some properties of encodings of evolutionary algorithms for spanning tree based problems. Although debate continues on how and why evolutionary algorithms work, many researchers have observed that an EA is likely to perform well when its encoding and operators exhibit locality, heritability and diversity. To analyze these properties of various encodings, we use two kinds of analytical methods; static analysis and dynamic analysis and use the Optimum Communication Spanning Tree (OCST) problem as a test problem. We show it through these analysis that the encoding with extremely high locality and heritability may lose the diversity in population. And we show that EA using Edge Window Decoder (EWD) has high locality and high heritability but nevertheless it preserves high diversity for generations.

1 Introduction

For a long time, many researchers have proposed various analytical methods to reveal the basic principle of encodings in EAs. Manderick et al. [6] used correlation coefficients for the fitness values of solutions before and after operators are applied. Sendhoff et al. [17] proposed the concept of "causality" to analyze the locality of EAs. Gottlieb et al. [1],[4],[9] proposed "mutation innovation", "crossover innovation" and "crossover loss" to emphasize the importance of locality and heritability. Merz et.al [7], Reeves et.al [11] and Watson et.al [19] used the fitness landscape analysis. Besides those, many literatures have dealt with methods for analyzing the properties of encodings [5],[12],[15]. In this paper, we concentrate on the analysis of locality, heritability and diversity of encodings based on Gottlieb et al.'s study and the fitness landscape analysis.

A difficulty of the population-based optimization is that once the search has narrowed near the previous optimal solution, the diversity in the population may not be enough for the search to get out of there and proceed towards the new optimal solution. Especially, if an evolutionary algorithm has very high locality or very high heritability as well, it may suffer from much serious problem. (Often, in these cases, diversity preserving mechanisms were used for avoiding these problems [9],[10].)

E. Talbi et al. (Eds.): EA 2005, LNCS 3871, pp. 107–118, 2006.

In this paper we show it through empirical tests that locality, heritability and diversity are in conflict with each other. In other word, if an encoding has extremely high locality and extremely high heritability, it may lose the diversity in population after offsprings are created and lead the search toward the narrow space (exploitation) because offsprings generated by operators will be very similar to their parents. Therefore, as generation goes, it will be deprived of the ability of exploration. But, note that high diversity does not imply the loss of locality and heritability. However, to obtain a good performance of evolutionary algorithms the harmony of these properties is needed.

For empirical tests, we compare five encodings, the Prüfer encoding [3], the network random key encoding (NetKey) [12], the link and node bias encoding (LNB) [8], the edge set encoding (Edge Set With Heuristic (ESWH) and Edge Set Without Heuristic (ESWOH) : ESs) [9] and the edge-window-decoder encoding (EWD) [18]. These encodings have been applied very successfully to spanning tree based problems like optimum communication spanning tree problem, degree constrained minimum spanning tree problem and quadratic spanning tree problem. For more details about each encoding, refer to the references.

This paper is organized as follows. The optimum communication spanning tree problem is described in Section 2. Section 3 presents the analysis of encodings. We make some concluding remarks in Section 4.

2   Optimum Communication Spanning Tree Problem: OCST

We perform an empirical analysis with OCST problem, which is one of the well-known NP-hard constrained spanning tree problems.

Consider an undirected complete graph $G = (V, E)$, where $V = \{1, 2, ..., N\}$ is the set of N nodes and $E = \{1, 2, ..., M\}$ is the set of M edges with given distance (or cost). Generally, the MST is to find the minimal cost spanning tree. In the case of the OCST, there are also "communication requirements" associated with each pair of nodes, specified by $R(i, j)$. E.g. these may represent the number of expected daily telephone calls between two cities. For any spanning tree T of G, the communication cost between two cities i and j is defined to be the communications requirement multiplied by the distance between the two cities on T, and the communication cost of T itself is the total communication cost summed over all pairs of nodes.

The goal is to construct a spanning tree with minimum communication cost. That is to find a spanning tree T such that formula (1) is minimized, where $d_T(i, j)$ is the sum of the distance of edges along the route between i and j on T.

$$Min \left[\sum_{i,j \in V} R(i, j) d_T(i, j) \right] \tag{1}$$

3 Analysis of Encodings

The properties, locality, heritability and diversity, of an encoding in evolutionary algorithms are the core factors for the effective search toward an optimal or
near optimal solution. Though debate still continues on, many researchers have
observed that an EA is likely to perform well when its encoding and operators
exhibit these properties [8],[10]. Therefore, we want to analyze the difference
among various encodings. To analyze this, we use the locality [1],[4], the heritability [5] and the fitness landscape analysis [7],[11],[19].

3.1 Metrics

In order to analyze the properties of an encoding, suitable metrics have to be
defined.

First of all, there are two search spaces in computational space of evolutionary
algorithms; the genotypic search space and the phenotypic search space. Most
of the genetic operators work on the genotypes and the movement of genotypes
on the genotypic search space by the genetic operators results in the change of
corresponding phenotypes on the phenotypic search space. Finally, it makes the
fitness value of corresponding solutions be changed. Therefore, the genotypic
distance have to be defined preferentially. But since the genotypic distance is
dependent on the encoding used, it must have universality.

Since the majority of the research follow the concept of evolutionary biology [16] when defining the genotypic distance, the genotypic distance is generally
defined as follow;

– *The genotypic distance* is the smallest number of individual mutations required for the inter-conversion of two genotypes.

On the other hand, "the phenotypic distance" and "the fitness distance" are
independent on the encoding used, but they are dependent on the problem used.
So, these two metrics should be defined as the problem.

Next, we define "phenotypic distance (d_p)" and "fitness distance (d_f)" based
on the OCST problem which is used as the test problem in this paper.

– *The phenotypic distance* is the total number of different edges between two
  phenotypes (spanning trees). Therefore, the phenotypic distance is *the Hamming distance*.

$$d_p(T_i, T_j) = \frac{1}{2} \sum_{u,v \in V} |E_{uv}^i - E_{uv}^j| \tag{2}$$

  where E_{uv}^i is 1 if an edge (u, v) exists in a tree T_i, otherwise 0.

– *The fitness distance* is the difference between the fitness values of two phenotypes (spanning trees).

$$d_f(T_i, T_j) = |f(T_i) - f(T_j)| \tag{3}$$

3.2   Locality

The locality can be defined as how well neighboring genotypes correspond to neighboring phenotypes [1], [4], [15]. Therefore, the locality of representation is high if small changes in the genotype result in small changes in the corresponding phenotype. In this context, it is appropriate to measure the locality of encodings using the mutation operator instead of the crossover operator, because the mutation operator is usually responsible for small steps in the phenotypic space, hence for gradual changes which we want to analysis.

Gottlieb and Eckert [1], [4] introduced *the mutation innovation* to measuring the locality. Mutation operators work in the genotype, but their effect can only be analyzed in the corresponding phenotype, which involves structural information of candidate solutions. So, the effect of mutation can be measured by the distance between the involved phenotypes. Therefore, the mutation innovation (MI) is equal to the phenotypic distance (d_p) but, only difference is to be compared between parent and its mutant.

$$MI = d_p(x, x^m) \tag{4}$$

where x and x^m indicate parent and its mutant respectively.

To analyze the locality of each encoding, we generated 1,000 random initial solutions in compliance with the used encoding, applied only a mutation operator to each encoding and performed the experiment on the selected benchmark instances (Palmer24 and Berry35U) and random generated instances ($N = 10 \sim 100$).

In this experiment, the reciprocal exchange mutation is used for Prüfer, NetKey and EWD, the random perturbation mutation is used for LNB and the specialized mutation operator is used for ESs [10]. If two genes with the identical gene value are selected when the reciprocal exchange mutation is applied to an encoding, it never generates a different offspring from the parent. So, in this case two genes with different gene values are selected again.

Table 1 shows the locality comparison among encodings and here a mutation was applied once to each encoding. In case of ESWH and ESWOH, all solutions had $MI = 1$ at all instances. The reason is for their specialized mutation operator; each mutation process changes exactly one edge on the genotype, and for the genotype and the phenotype are the same; non-redundant encoding. On the other hand, the others are a kind of redundant encodings except the Prüfer encoding. So, although a mutation operator is applied to encodings, sometimes it does not cause the change at the phenotype (the redundancy) or the different genotypes can be mapped to the same phenotype (the heuristic bias).

$P(MI = 0)$ represents these things. NetKey and LNB show higher frequency than Prüfer and EWD. It relates to *the degree of redundancy* and *the heuristic bias of encodings*. In case of NetKey, exactly two genes are exchanged by the mutation (the reciprocal exchange mutation) and it results in the change of sorting order at exactly two genes. So, if the selected genes are the genes which are not selected for the previous phenotype, it never makes a difference between the phenotypes. Therefore, $P(MI = 0)$ will be increased as the size of network is increased because only $N-1$ edges among the total edges $N(N-1)/2$ are selected

Table 1. Comparison of locality on Palmer24, Berry35U and random generated Instances (Rand10 ∼ 100), based on randomly generating 1,000 genotypes and applying mutation once to each

Palmer24	Prufer	LNB	NetKey	ESWH	ESWOH	EWD	Berry35U	Prufer	LNB	NetKey	ESWH	ESWOH	EWD
$P(MI=0)(\%)$	0.00	80.5	80.3	0.00	0.00	4.10		0.00	93.5	87.3	0.00	0.00	0.00
$E(MI\|MI>0)$	4.49	7.87	1.51	1.00	1.00	2.42		5.08	33.0	1.49	1.00	1.00	11.70
$Max(MI)$	12	22	2	1	1	7		17	33	3	1	1	18
$\sigma(MI\|MI>0)$	2.01	5.67	0.50	0.00	0.00	1.02		2.64	0.00	0.51	0.00	0.00	2.24
Rand10							**Rand20**						
$P(MI=0)(\%)$	0.00	55.7	61.9	0.00	0.00	70		0.00	56.3	77.7	0.00	0.00	4.7
$E(MI\|MI>0)$	3.20	2.43	1.43	1.00	1.00	2.10		4.30	3.11	1.45	1.00	1.00	2.39
$Max(MI)$	6	8	2	1	1	5		10	11	2	1	1	6
$\sigma(MI\|MI>0)$	1.12	1.71	0.49	0.00	0.00	0.89		1.65	2.05	0.49	0.00	0.00	0.96
Rand30							**Rand40**						
$P(MI=0)(\%)$	0.00	61.1	84.0	0.00	0.00	3.90		0.00	64.8	87.2	0.00	0.00	2.60
$E(MI\|MI>0)$	4.97	4.03	1.60	1.00	1.00	2.48		5.66	4.67	1.56	1.00	1.00	2.58
$Max(MI)$	14	22	2	1	1	9		19	18	2	1	1	7
$\sigma(MI\|MI>0)$	2.39	3.46	0.49	0.00	0.00	1.01		3.19	3.47	0.49	0.00	0.00	1.00
Rand50							**Rand60**						
$P(MI=0)(\%)$	0.10	64.5	89.7	0.00	0.00	3.80		0.00	66.7	91.6	0.00	0.00	2.80
$E(MI\|MI>0)$	6.29	4.49	1.51	1.00	1.00	2.59		7.20	5.40	1.57	1.00	1.00	2.65
$Max(MI)$	21	23	2	1	1	6		27	23	3	1	1	8
$\sigma(MI\|MI>0)$	3.87	3.91	0.50	0.00	0.00	0.98		4.86	4.44	0.54	0.00	0.00	1.00
Rand70							**Rand80**						
$P(MI=0)(\%)$	0.00	68.1	91.0	0.00	0.00	2.40		0.00	70.1	90.1	0.00	0.00	2.80
$E(MI\|MI>0)$	7.38	5.53	1.58	1.00	1.00	2.63		8.02	5.92	1.48	1.00	1.00	2.65
$Max(MI)$	31	22	2	1	1	7		29	23	3	1	1	9
$\sigma(MI\|MI>0)$	5.56	4.22	0.49	0.00	0.00	1.01		5.94	4.52	0.54	0.00	0.00	1.01
Rand90							**Rand100**						
$P(MI=0)(\%)$	0.00	71.1	92.2	0.00	0.00	1.70		0.00	69.4	90.5	0.00	0.00	1.50
$E(MI\|MI>0)$	8.59	5.62	1.46	1.00	1.00	2.71		9.49	6.03	1.45	1.00	1.00	2.69
$Max(MI)$	36	23	3	1	1	6		38	23	3	1	1	5
$\sigma(MI\|MI>0)$	6.88	4.41	0.55	0.00	0.00	1.01		7.99	4.84	0.57	0.00	0.00	0.98

to generate a spanning tree. The test results show that. And LNB has also the same redundancy as NetKey in terms of the length of the encoding, but in addition to that it has a strong heuristic bias in the context of having a preference toward a specific spanning tree [2]. Therefore, mutants over 80% at Palmer 24, over 93% at Berry35U and over avg. 70% at the random generated instances are the same as their parents. Especially, in Berry35U instance LNB shows much higher $P(MI=0)$ value (93.5%) in comparison to those of the other encodings. That is for the strong heuristic bias of LNB using Prim's algorithm for sorting all edges with the modified cost matrix and for the instance's data set; all edge distances are the same (In this case LNB can only generate a star tree [2]). In all random instances, also LNB and NetKey show relatively higher values than the others. In case of EWD, even though it is a redundant representation, it has much lower redundancy comparing to NetKey and LNB. So, it exhibits relatively much lower probability at $P(MI=0)$. The ESs and Prüfer show that all offsprings were different from their parents ($P(MI=0)=0$). As mentioned above, the reason is that the specialized mutation operator of ESs exactly changes one edge to a different edge which is not included in the tree and in case of the Prüfer encoding the mutation exchanges exactly two different genes. So, while $P(MI=0)$ was 0%, $P(MI=1)$ was 100%.

In addition, this table shows three other indicators of locality, $E(MI \mid MI > 0)$, $\sigma(MI \mid MI > 0)$ and $Max(MI)$. $E(MI \mid MI > 0)$ represents the expected mutation innovation in the case that some phenotypic property has actually been affected. So, high values represent low degree of locality. Especially, ESs can be seen as a very ideal case because single change in genotypes exactly causes 1 distance in phenotype. EWD shows low locality in comparison with NetKey and

ESs but high locality in comparison with Prüfer and LNB. But, when considering the redundancy of encodings, it is very difficult to distinguish which encoding has better locality between EWD and NetKey. In case of $\sigma(MI \mid MI > 0)$, also this table shows the similar results. NetKey, ESs and EWD are much stable than Prüfer and LNB. In Berry35U, LNB shows $E(MI \mid MI > 0) = 33$ and $\sigma(MI \mid MI > 0) = 0$. The reason is for LNB implies a strong heuristic bias. And the maximum number of edges modified ($Max(MI)$) does not exceed 38 at Prüfer, 33 at LNB, 3 at NetKey, 1 at ESs and 9 at ESW.

In figure. 1, the upper two figures indicate the frequency of solutions with the identical phenotypic distance (d_p) when a mutation is applied to each encoding once. EWD and Prüfer show that the solutions with various phenotypic distance are generated by a mutation. That means the exploration ability of EWD and Prüfer encoding. On the other hand, the other encodings exhibit their exploitation ability.

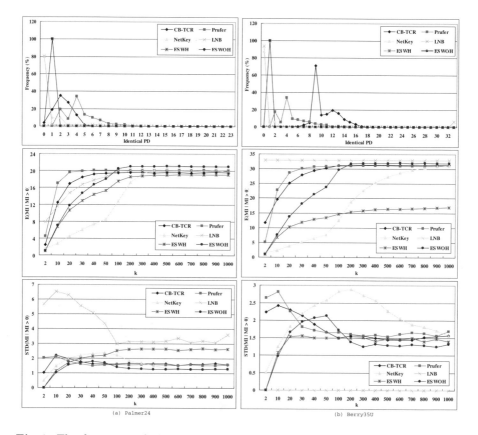

Fig. 1. The frequency of identical locality and the value of $E(MI \mid MI > 0)$ and $\sigma(MI \mid MI > 0)$ according to generation. In here, $E(MI \mid MI > 0)$ and $\sigma(MI \mid MI > 0)$ represent values obtained by the phenotypic distance between the original solution and its mutant generated after k generation ($P_m=100$).

The other figures show the value of $E(MI \mid MI > 0)$ and $\sigma(MI \mid MI > 0)$ according to generation. $E(MI \mid MI > 0)$ and $\sigma(MI \mid MI > 0)$ represent the values obtained by the phenotypic distance between the original solution and its mutant generated after k mutation. As k increases, it shows a significant difference within 200 generation. NetKey and ESWH exhibit lower mean values than the other encodings. That indicates high locality of the two encodings. But NetKey is very unpredictable at Berry35U and ESWH indicates slightly higher standard deviation values (STD) at Palmer24. On the other hand, even though LNB shows relatively high mean values and especially at Berry35U instance the mean values were all 33. As mentioned above, it is for strong bias toward a specific tree structure - a star tree. Moreover, at Palmer24 LNB is very unstable and unpredictable. Prüfer exhibits relatively high mean values -low locality- at both instances but $\sigma(MI \mid MI > 0)$ is very predictable at palmer24 instance.

Although ESWH and ESWOH start at the same $E(MI \mid MI > 0)$ value at the beginning of generation, the difference between ESWH and ESWOH becomes large because of the heuristic bias of ESWH. EWD starts slightly high mean value but finally the mean value becomes very similar to other encodings' mean value.

3.3   Heritability

The locality is a feature of the interaction between a coding and mutation operator. On the other hand, the heritability is a feature of the interaction between a coding and crossover operator. Julstrom [5] defined the heritability as the number of edges in the offspring's spanning tree that appeared in neither parent's tree. We define the heritability as a similar way.

— *The heritability* is the number of edges in the offspring's spanning tree that appeared in either parent's tree.

Table 2. Comparison of Average Heritability based on randomly generating 1,000 genotypes and applying crossover once to each

Heritability	Prufer	LNB	NetKey	ESWH	ESWOH	EWD
Palmer24	15.72	19.08	18.79	23.00	23.00	16.77
Berry35U	22.27	26.63	27.90	34.00	34.00	17.77
Rand10	6.92	7.80	7.67	9.00	9.00	7.46
Rand20	13.24	16.20	15.56	19.00	19.00	15.80
Rand30	19.39	24.11	23.57	29.00	29.00	24.34
Rand40	25.90	32.83	31.72	39.00	39.00	33.27
Rand50	31.60	40.05	39.96	49.00	49.00	41.67
Rand60	37.91	47.71	47.82	59.00	59.00	51.02
Rand70	43.94	57.01	56.17	69.00	69.00	60.04
Rand80	49.96	64.78	64.32	79.00	79.00	67.93
Rand90	55.57	73.46	72.73	89.00	89.00	77.51
Rand100	62.49	81.14	80.81	99.00	99.00	86.17

$$d_h(P_i, P_j, O) = |(P_i \cup P_j) \cap O| \tag{5}$$

where P and O represent parent and offspring.

Each encoding uses different crossover operators considering which crossover operator can give better performance for the considering encoding [18]. So, Prüfer uses two-point crossover, LNB one-point crossover, NetKey uniform crossover, ESs their specialized crossover and EWD adjacent node crossover.

Table. 2 exhibits the average heritability of each encoding and the high d_h values imply the high heritability.

ESs show very ideal case at all instances because of their specialized crossover operator. All of the generated offsprings are created by their parents' edges. However, they show the highest locality and heritability. In this empirical com-

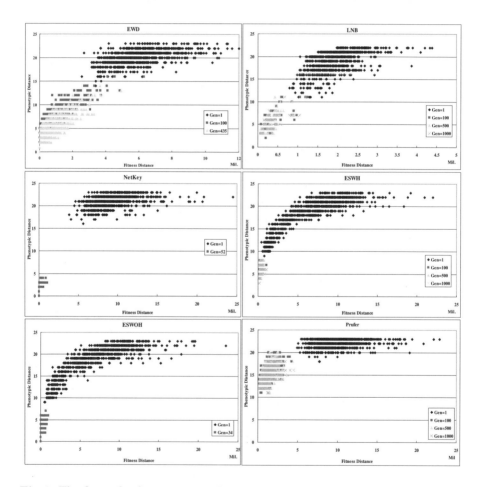

Fig. 2. The fitness landscape. 1,000 solutions are generated using each representation methods and genetic operators are applied to a representation (Palmer24).

parison, EWD exhibits higher heritability than the other encodings except ESs and Prüfer encoding exhibits the lowest heritability as the locality comparison.

3.4 Diversity

– If various different solutions coexist in population, the representation has high diversity. Otherwise, it has low diversity.

We analyzed the diversity of encodings at two instances (palmer24 and berry35U) and in this experiment all genetic operators were applied to encodings as following [18].

Figure. 2 and figure. 3 show the relation between phenotypic distance and fitness distance to optimum solution (palmer24) or the best known solution

Fig. 3. The fitness landscape. 1,000 solutions are generated using each representation methods and genetic operators are applied to a representation (Berry35U).

(berry35U) according to generation. Here, different shape points represent the distribution of solutions scattered at each generation. Note that EWD keeps preserving the diversity of solution over generations at both instances and has high diversity in population. On the other hand, the other encodings dramatically lose the diversity of solution.

We conclude that the reason why the other encodings lose the diversity of population is for the strong heuristic bias or extremely high locality. If an encoding has extremely high locality, after operators are applied it keeps generating very similar offsprings or nearly the same offsprings, and finally the population will be filled by similar offsprings very fast as generation goes. As a result, the search space which the encoding explores will narrow and then it will become to lose the balance between "exploration" and "exploitation". For example, at the locality comparison NetKey and ESs exhibited higher locality. But, if considering the diversity distribution, the solutions are distributed in a very limited space.

Observing the distribution of solutions, we can also estimate the difficulty of problems. In Palmer24 instance, the solution distribution of each encoding shows the positive correlation [15] between phenotypic distance and fitness distance. It means that an algorithm which guides toward solutions with small fitness value can easily find the optimum solution. On the other hand, in case of Berry35U, the solution distribution of each encoding shows no correlation between phenotypic distance and fitness distance. For example, ESWH and ESWOH use the specialized initialization operator, which is derived from Kruskal's algorithm and the operator prefers to shorter edges. So, the initial solution distributions of ESs are scattered along x-axis and even though they find solutions with less fitness distance, the phenotypic distance is still large. As a result, the figure shows that if an algorithm guides toward a better fitness solution, it may fall into a local optima, and preserving the diversity of solution can give a help to escape the local optima.

4   Conclusions

We investigated the locality, heritability and diversity of encodings of evolutionary algorithms for spanning tree based problems and performed empirical tests on the optimum communication spanning tree problem.

Generally, the Prüfer encoding has low locality so that it did not give good performance in several literatures. We could also confirm low locality and low heritability of the Prüfer encoding. And it is known that if an encoding has high locality and high heritability, the evolutionary algorithm will give good performance. But, in our experiment, we showed that if an encoding has extremely high locality and heritability like ESs, it can lose the diversity of population. So, some researchers used the diversity preservation strategy [9], [10] to avoid this problem. However, it can be a good strategy. LNB and Netkey showed high locality and high heritability because of the heuristic bias or the redundancy of the encoding. But, these encodings also showed a feature of the diversity loss.

On the other hand, EWD showed slightly low locality and high heritability and the highest diversity in all test instances. And EWD showed a feature which it is independent on the property of problem. That shows EWD can be applied various spanning tree based problems and may give good performance.

As a result, EWD and ESs seem to be good encodings for the OCST, and potentially other spanning tree based problems.

References

1. C. Eckert and J. Gottlieb, Direct Representation and Variation Operators for the Fixed Charge Transportation Problem, PPSN VII, LNCS, vol. 2439, (2002) 77–87.
2. T. Gaube and F. Rothlauf, The Link and Node Biased Encoding Revisited: Bias and Adjustment of Parameters, EvoWorkshop 2001, LNCS Vol.2037, (2001) 1–10.
3. M. Gen and R. Chen, Genetic Algorithms and Engineering Design, Wiley, (1997). Also see (for Prüfer encoding):
http://www.ads.tuwien.ac.at/publications/bib/pdf/gottlieb-01.pdf.
4. J. Gottlieb and C. Eckert, A Comparision of Two Representations for the Fixed Charge Transportation Problem, PPSN VI, LNCS, vol. 1917, (2000) 345–354.
5. B.A. Julstrom, The Blob Code: A Better String Coding of Spanning Trees for Evolutionary Search, in Genetic and Evolutionary Computation Conference Workshop Program. Morgan Kaufmann, (2001) 256–261.
6. B. Manderick, M. de Weger, and P. Spiessens, The genetic algorithm and the structure of the fitness landscape, Proceedings of the 4th International Conference on Genetic Algorithms, (1991) 143–150.
7. P. Merz and B. Freisleben, Fitness Landscapes, Memetic Algorithms, and Greedy Operators for Graph Bipartitioning, Evolutionary Computation, vol. 8, no. 1, (2000) 61–91.
8. C.C. Palmer and A. Kershenbaum, An Approach to a Problem in Network Design Using Genetic Algorithms, Networks, Vol. 26, (1995) 151–163.
9. G.R. Raidl, Empirical Analysis of Locality, Heritability and Heuristic Bias in Evolutionary Algorithms: A Case Study for the Multidimensional Knapsack Problem, Evolutionary Computation Journal, MIT Press, 13(4), to appear in 2005.
10. G.R. Raidl and B.A. Julstrom : Edge-Sets: An Effective Evolutionary Coding of Spanning Trees, IEEE Transactions on Evolutionary Computation, 7(3),pp. 225-239, 2003.
11. C.R. Reeves and T. Yamada, Genetic algorithms, path relinking, and the flowshop sequencing problem, Evolutionary Computation, vol. 6, pp. 45–60.
12. F. Rothlauf, Locality, Distance Distortion, and Binary Representations of Integers, Working Papers, July (2003).
13. F. Rothlauf, D.E. Goldberg and A. Heinzl, Network Random Keys - A Tree Network Representation Scheme for Genetic and Evolutionary Algorithms, Evolutionary Computation, Vol. 10 (1), (2002) 75–97.
14. F. Rothlauf, J. Gerstacker and A. Heinzl, On the Optimal Communication Spanning Tree Problem, Working Papers in Information Systems, University of Mannheim, (2003)
15. F. Rothlauf, On the Locality of Representations, Working Paper in Information Systems, University of Mannheim, (2003)
16. P. Schuter, Artificial Life and Molecular Evolutionary Biology, In F. Moran et al. (Eds.), Advances in Artificial Life, Springer, (1995) pp. 3–19.

17. B. Sendhoff, M. Kreutz and W.V. Seelen, A condition for the genotype-phenotype mapping: Causalty, Proceedings of the Seventh International Conference on Genetic Algorithms, Morgan Kauffman, 1997.
18. S.M, Soak, D. Corne and B.H. Ahn, The Edge-Window-Decoder Representation for Tree-Based Problems, submitted to IEEE Transaction on Evolutionary Computation (2004).
19. J.P. Watson, L. Barbulescu, L.D. Whitley and A.E. Howe, "Constrasting Structured and Random Permutation Flow-Shop Scheduling Problems: Search-Space Topology and Algorithm Performance," http://www.cs.colostate.edu/ genitor/Pubs.html.

A Cooperative Multilevel Tabu Search Algorithm for the Covering Design Problem

Chaoying Dai, (Ben) Pak Ching Li, and Michel Toulouse

Department of Computer Science, University of Manitoba
{chaoying, lipakc, toulouse}@cs.umanitoba.ca

Abstract. This work describes an adaptation of multilevel search to the covering design problem. The search engine is a tabu search algorithm which explores several levels of overlapping search spaces of a $t - (v, k, \lambda)$ covering design problem. Tabu search finds "good" approximations of covering designs in each search space. Blocks from those approximate solutions are transferred to other levels, redefining the corresponding search spaces. The dynamics of cooperation among levels tends to re-group good approximate solutions into small search spaces. Tabu search has been quite effective at finding re-combinations of blocks in small search spaces which provide successful search directions in larger search spaces.

Keywords: Multilevel algorithms, Covering design problem, Tabu search meta-heuristic.

1 Introduction

A $t - (v, k, \lambda)$ *covering design* is a pair (X, B), where X is a set of size v, called *points* and B is a collection of k-subsets of X, called *blocks*, such that every t-subset of X is contained in at least λ blocks of B. Let $C_\lambda(v, k, t)$ denote the minimum number of blocks in any $t - (v, k, \lambda)$ covering design. A $t - (v, k, \lambda)$ covering design is *optimal* if it has $C_\lambda(v, k, t)$ blocks [12]. The *covering design problem* is the problem of determining the value of $C_\lambda(v, k, t)$. The covering design problem has applications in lottery design, data compression and error-trapping decoding [5].

The value $C_\lambda(v, k, t)$ can be determined using an exact search algorithm. Unfortunately, such algorithms are ineffective for all but a few set of parameters, due to the effects of combinatorial explosion. Therefore, search heuristics may be a viable option for improving upper bounds on $C_\lambda(v, k, t)$.

In this paper, we introduce a cooperative multilevel search heuristic method to improve upper bounds on $C_\lambda(v, k, t)$. Assume we are looking for a $t - (v, k, \lambda)$ covering design of b blocks. Let $\binom{X}{k}$ be the set of all k-subsets in X and let $\mathcal{S}_0 = \{S \subset \binom{X}{k} | |S| = b\}$ be the solution space for $t - (v, k, \lambda)$. Assume $S_1, S_2, \ldots, S_l$ are subsets of $\binom{X}{k}$ such that $S_l \subset S_{l-1} \subset \cdots \subset S_1 \subset S_0 = \binom{X}{k}$. Each subset S_i defines a search space $\mathcal{S}_i$ on $t - (v, k, \lambda)$ in the same way as $\mathcal{S}_0$. A tabu search algorithm explores independently each search space to seek sets of b blocks that

E. Talbi et al. (Eds.): EA 2005, LNCS 3871, pp. 119–130, 2006.

cover as many t-subsets as possible. The cooperative multilevel search strategy consists of substituting some blocks of the smallest search spaces by sets of good blocks discovered by tabu search. This exchange of blocks eventually brings combinations of good blocks in small search spaces where tabu search is quite effective finding covering for large number of t-subsets. We have tested our algorithm on covering design problems with tight gaps between lower and upper bounds [8, 9]. Those are the most difficult upper bounds to improve. We were able to find known upper bounds for all the problem instances tested and found new upper bounds for several of them.

The subsequent sections of this paper are organized as follows. Section 2 provides background information on the covering design problem. In section 3, we describe the multilevel paradigm. Section 4 summarizes the implementation of our cooperative multilevel algorithm. Section 5 reports experimental results and we conclude in Section 6.

2   Background

In this section we provide a short background on covering designs and search heuristics for covering designs.

2.1   Covering Designs

The study of covering designs began around the end of the 1930's. Turán (see [5]) was one of the first researchers to study covering designs. Since then, many researchers have studied covering designs from various directions. One such direction is the determination of $C_\lambda(v, k, t)$ by means of computer programs. Because the exact value of $C_\lambda(v, k, t)$ has been computed only for small set of values for v, k, t and λ, most research on covering designs has focused on determining the upper and lower bounds for $C_\lambda(v, k, t)$. In this section, we briefly describe some important results about the lower bounds and upper bounds for $C_\lambda(v, k, t)$.

The Schönheim lower bound $(L_\lambda(v, k, t))$ [19] provides a lower bound for $C_\lambda(v, k, t)$ given by:

$$L_\lambda(v, k, t) := \left\lceil \frac{v}{k} \left\lceil \frac{v-1}{k-1} \cdots \left\lceil \frac{v-t+1}{k-t+1} \lambda \right\rceil \cdots \right\rceil \right\rceil \leq C_\lambda(v, k, t).$$

This bound is a very good general lower bound for $C_\lambda(v, k, t)$. For many values of v, k, and t where $C_\lambda(v, k, t)$ is known, $L_\lambda(v, k, t)$ attains the value $C_\lambda(v, k, t)$ [13].

In 1963, Erdős and Hanani [7] conjectured that for fixed values of t and k, where $t < k$.

$$\lim_{v \to \infty} \frac{C_1(v, k, t)\binom{k}{t}}{\binom{v}{t}} = 1.$$

This result was shown to be true in 1985 by Rödl [18], using probabilistic methods. This result implies that $C_1(v, k, t) = (1 + o(1))\frac{\binom{v}{t}}{\binom{k}{t}}$.

Various techniques have been used to construct covering designs [9]. One of the earliest constructions involved using finite geometries to construct covering designs. For example, it has been found that the hyperplanes of the affine geometry $AG(t,q)$ form an optimal (q^t, q^{t-1}, t) covering design with $\frac{q^{t+1}-q}{q-1}$ blocks. Another common approach is to use recursive techniques for constructing covering designs. That is, using smaller covering designs to construct larger covering designs [14]. For example, if S_1 is a $t - (v - 1, k, \lambda)$ covering design and S_2 is a $(t-1) - (v-1, k-1, \lambda)$ covering design, then a $t - (v, k, \lambda)$ covering design can be constructed by taking all blocks from S_2 with adding a new point v to all of these blocks and including all blocks from S_1.

Exact search methods have also been used to construct covering designs. Bate [2] developed a backtracking algorithm to exhaustively search for generalized covering designs to determine $C_1(v, k, t)$. In 2003, Margot [11] used integer programming techniques, branch-and-cut and isomorphism rejection to design an algorithm for computing $C_1(v, k, t)$. However, such algorithms are effective for only a few set of parameters.

2.2 Search Heuristics for Covering Designs

Search heuristic methods are used to search for a $t - (v, k, \lambda)$ covering design which is smaller than the best known upper bound for $C_\lambda(v, k, t)$. These methods have worked well for small values of v, k, λ [15, 16]. For $\lambda = 1$, the covering design problem can be modeled as a combinatorial optimization problem in $\binom{v}{k}$ Boolean decision variables, one for each k-subset. A feasible solution is a Boolean vector where at most b variables are set to 1, where b is the size of the covering design we are looking for. The cost function optimized by the search heuristic is the number of t-subsets not covered at least one time by the current solution (a set B of b blocks). More precisely, let $\binom{X}{t}$ be the set of t-subsets and let $cover_y$ be the number of times the t-subset $y \in \binom{X}{t}$ is covered by the b blocks in B. Let $notcover_y = \max\{0, \lambda - cover_y\}$ denote the number of times the b blocks in B fails to cover the t-subset y. The cost of solution B is given by

$$cost(B) = \sum_{y \in \binom{X}{t}} notcover_y.$$

When $cost(B) = 0$, then all t-subsets are covered at least λ times, meaning we have discovered a $t - (v, k, \lambda)$ covering design with b blocks.

A natural mapping function to define neighborhoods for covering design problems consists of choosing m points among the k points of a block and replace these by m other points from the $v - k$ points not belonging to this block. Such a move can replace $1 \leq m \leq \min(k, v - k)$ points belonging to a same block. The neighborhood $\mathcal{N}(B)$ of solution B is a subset of $\mathcal{S}_0$ such that $I \in \mathcal{N}(B)$ if I has $b - 1$ identical blocks with B and one block which differs by exactly m points. The size of the neighborhood $\mathcal{N}(B)$ is given by $|\mathcal{N}(B)| = b \times \binom{k}{m} \times \binom{v-k}{m}$. Since the size of neighborhoods based on swapping points increases rapidly in terms of m, typical move based heuristics for covering designs are based on neighborhood where $m = 1$.

3   The Multilevel Paradigm

Multilevel approaches have first been proposed in the field of numerical approximation [3]. Based on the original problem domain discretization, coarser discretizations (levels) are recursively constructed by increasing the grid spacing in comparison with the latest generated grid. In *nested iteration*, the simplest multilevel scheme [4], starting with the coarser grid, an approximation is computed and then interpolated on the next grid, which is less coarsened. The approximation is then refined using an iterative solver. The latest refined approximation is used as initial point of the relaxation in the original problem domain discretization. The nested iteration scheme helps improve convergence in the original domain discretization by providing a good initial point to the relaxation method. In the *V-cycle scheme*, a first approximation is computed on the original grid spacing. The residual error associated to the approximation is *projected* on the next coarser grid where the system of linear equations is solved for this residual error. One V-cycle consists of projections upward from less coarsened grids toward coarser grids. Then interpolations from coarser grids toward less coarsened grids refine the approximation. Projections change the problem definition by solving for a new residual error at each level. They also help to improve convergence of iterative solvers by focusing on the oscillatory component of the error function at each level.

The multilevel approach has been adapted recently to combinatorial optimization problems in combination with search algorithms. The basic framework of multilevel search is the following: Let $\mathcal{A}$ denote a given combinatorial optimization problem and A_0 a problem instance of $\mathcal{A}$. During the **coarsening phase**, a succession $A_1, \ldots, A_l$ of increasingly smaller problem instances of $\mathcal{A}$ is generated by reducing the number of decision variables in comparison with the definition of problem instance A_0. During the **initial search phase**, a feasible solution s_l is computed for the smallest problem instance A_l. During the **refinement phase**, the feasible solution s_l is used to *interpolate* values for the decision variables of problem instance A_{l-1}. This setting of decision variables in A_{l-1} is used as initial solution for a search algorithm which explores the search space of A_{l-1}. The optimization of the cost function for A_{l-1} using a search algorithm also improves (refines) the feasible solution s_l obtained from the problem instance A_l. The refinement phase consists of interpolating and refining feasible solutions until the values of the decision variables of s_0 can be interpolated from a feasible solution of problem instance A_1. This last interpolation provides for an initial solution for a search algorithm to optimize the cost function problem instance A_0.

The coarsening phase is critical to multilevel algorithms. It reduces the size of the problem instance and, more importantly, it determines which regions of the solution space will be explored during the initial search and refinement phases. Coarsening strategies were first proposed in the context of applications of the multilevel paradigm to the graph partitioning problem [1, 10]. These strategies are based on clustering decision variables. Starting from the original graph instance $G_0 = (V_0, E_0)$, pairs of adjacent vertices x, y are selected randomly and merged together to become a single vertex xy in the coarsened graph $G_1 = (V_1, E_1)$. Edge $\{x, y\} \in G_0$ is removed, edges $\{u, x\}$ or $\{u, y\}$ in G_0 are replaced by edge $\{u, xy\}$ in G_1. The coarsening of G_0 yields a graph G_1 where $|V_1| \approx |\frac{V_0}{2}|$ and $E_1 \subset E_0$.

Let $GPP(G_x)$ be the graph partitioning problem for the graph instance G_x and $\mathcal{S}_x$ the solution space of $GPP(G_x)$. The number of decision variables in $GPP(G_1)$ is about half of $GPP(G_0)$. Nonetheless, a feasible solution of $GPP(G_1)$ can be interpolated in the solution space of $GPP(G_0)$ by expanding vertex $xy \in V_1$ into vertices $x, y \in V_0$ and placing x, y in the same partition as xy. Consequently, $\mathcal{S}_1 \subset \mathcal{S}_0$, i.e., any feasible solution to $GPP(G_1)$ is also a feasible solution to $GPP(G_0)$. The coarsening phase recursively applies the above coarsening strategy to the latest coarsened graph and outputs a succession of increasingly coarsened graph $G_1, G_2, \cdots, G_l$ which satisfy the condition $\mathcal{S}_l \subset \mathcal{S}_{l-1} \subset \cdots \subset \mathcal{S}_1 \subset \mathcal{S}_0$.

For several combinatorial optimization problems, coarsening by clustering decision variables is hardly applicable. In [6], the authors proposed a coarsening strategy by fixing the state of some decision variables. A decision variable is fixed if its value cannot be changed by the solution process. Let $x_1, x_2, \ldots, x_n$ be the set of decision variables of a problem instance A_0. By fixing some decision variables of A_0, a new problem instance A_1 is defined where $\mathcal{S}_1 \subset \mathcal{S}_0$. Any solution to A_1 can be trivially interpolated in the solution space of A_0. Fixing recursively the state of some decision variables has the effect of coarsening the original problem instance A_0 into problem instances with fewer decision variables. Furthermore, the strict inclusion condition $\mathcal{S}_l \subset \mathcal{S}_{l-1} \subset \cdots \subset \mathcal{S}_1 \subset \mathcal{S}_0$ is also satisfied.

In most multilevel algorithms applied to combinatorial optimization problems, the refinement phase is reminiscent of the nested iteration scheme in multigrid approximation. Recently, a multi-cycle refinement phase has been proposed [6, 17] in the context of parallel cooperative search algorithms. In multi-cycle refinement, projection operators transformed the problem instance searched at each level by changing its coarsening. Modification to the coarsening of levels define new regions of the solution space that can be explored by search heuristics. This allows for a new sequence of interpolations and searches, closing one cycle. There are several possible variations in the multi-cycle refinement phase, we propose a new one in this paper.

4   Multilevel Tabu Search Algorithm for the Covering Design Problem

In this section, we introduce the design of our multilevel algorithm for the covering design problem. We describe our strategy to coarsen covering design problem instances as well as the projection operator and re-coarsening strategies applied to transform subsequently the initial coarsening. Next, we describe the tabu search algorithm which is used to explore the search space defined by each level of coarsening. Finally, we describe a variation of the multi-cycle refinement phase adapted for the covering design problem.

4.1 The Coarsening Phase

Our coarsening procedure defines search spaces (levels) by fixing recursively subsets of decision variables. The decision variables for the covering design problem correspond to the $\binom{v}{k}$ blocks of a given problem instance. Each block is assigned exclusively a level during the coarsening phase. An integer array (*multilevel*) of dimension $\binom{v}{k}$ expresses this assignment. Entry j of the array takes a value i in the range 0 to l to indicate that the block j is assigned to level i. At the initialization, all the $\binom{v}{k}$ blocks are assigned to level 0. Then, through random selection, blocks assigned to level 0 are re-assigned to level 1. This procedure is repeated for each level i, re-assigning randomly blocks from level $i-1$ to level i.

The number of blocks assigns to each level is decided by the *coarsening factor* cf. The value of this coarsening factor is a function of the total number of blocks $\binom{v}{k}$, the number of levels $l+1$ and $|L_l|$ the number of blocks required at top level l. It is computed as follows:

$$cf = \frac{\binom{v}{k} - (|L_l| \times (l+1))}{(l+1) \times \frac{l}{2}}.$$

The value cf expresses the difference between the number of blocks assigned to two adjacent levels i and $i+1$. Assume L_i represents the set of blocks assigned to level i. The number of blocks $|L_i|$ that must be assigned to level i is given by the following formula:

$$|L_i| = |L_l| + (l - i) \times cf$$

Therefore, the number of blocks $|L_{l-1}|$ at level $l-1$ is $|L_{l-1}| = |L_l| + cf$, the number of blocks at level $l-2$ is $|L_{l-2}| = |L_l| + 2 \times cf$, etc. The number of blocks at level 0 is $|L_0| = |L_l| + (l \times cf)$. The sum of blocks assigned to all the levels must be $\sum_{i=0}^{l} |L_i| = \binom{v}{k}$.

A block is considered to be fixed for level i if it is assigned to a level lower than i. Therefore, the number of decision variables at level i is $S_i = \sum_{k=i}^{l} |L_k|$, the set of blocks assigned to levels greater or equal to i. The search space at level i is constituted by all the possible combinations of b blocks in the set $S_i = L_i \cup L_{i+1} \cup \cdots \cup L_l$. The search space $\mathcal{S}_0$ at level 0 corresponds to all combinations of b blocks in the set $\binom{X}{k} = \cup_{i=0}^{l} L_i$. Note that the set of decision variables $S_i = L_i \cup L_{i+1} \cup \cdots \cup L_l$ of level i is a strict subset of $S_{i-1} = L_{i-1} \cup L_i \cup \cdots \cup L_l$ of level $i-1$. Consequently, the strict inclusion condition $\mathcal{S}_l \subset \mathcal{S}_{l-1} \subset \cdots \subset \mathcal{S}_1 \subset \mathcal{S}_0$ among search spaces is satisfied by this coarsening procedure.

4.2 Projection and Re-coarsening

The projection operator copies from level i to level l the blocks B of the best covering design approximation at level i. This operator is implemented by re-assigning blocks in B to level l, as shown in the **while** loop of Fig. 1.

According to our coarsening procedure, each block in B is assigned to a level greater or equal to i. Line 1 obtains the current level assignment of block s from

projection(B)
   while $(B \neq \emptyset)$ **do**
1. $s = s \in B; \ B = B \setminus s; \ j = multilevel[s];$
    if $(j \neq l)$ **then**
2. $multilevel[s] = l;$
3. $u =$ randomly select a free block assigned to level $l; \ multilevel[u] = j;$

Fig. 1. The projection procedure

the array *multilevel* (the array which stores the assignment of each block to a specific level). Line 2 re-assigns to level l those blocks of B not already assigned to level l.

Each time a block s is re-assigned by the projection operator from level j to level l, it removes one block from level j and adds one block to level l. Given the way levels are defined in our coarsening procedure, assigning a new block s to level l is equivalent to adding the decision variable s to sets $S_{j+1}, S_{j+2}, \ldots, S_l$ such that $S_{j+1} = S_{j+1} \cup s, \ S_{j+2} = S_{j+2} \cup s, \ \ldots, \ S_l = S_l \cup s$. The operation of line 2 is in fact a re-coarsening of levels $j + 1$ to l, modifying the search space of all these levels. In order to keep the number of blocks constant at each level, line 3 re-assigns a block from level l to level j. Line 3 changes the coarsening of levels $j + 1$ to l: $S_{j+1} = S_{j+1} \setminus u, S_{j+2} = S_{j+2} \setminus u, \ldots, S_l = S_l \setminus u$.

Re-coarsening is designed to re-focus the search space of each level toward better regions of the solution space. To achieve this purpose, the re-coarsening of each level is biased by the cost function through the projection of the best solutions to level l. In order to have a chance to influence the multilevel search, a block entering the search space of level i through projection must not exit this level before performing a search of the corresponding level. To enforce this condition, the blocks of solutions that have been projected to level l must stay assigned to level l for the duration of a search. The *free* blocks in line 3 of the projection procedure are blocks that do not belong to any of the best solutions recently projected to level l. In this manner, blocks that seem to contribute to find good solutions are kept at level l. Blocks at level l are re-combined together by the tabu search procedure or re-combined in the same manner with blocks from any of the other levels. On the other hand, blocks not belonging to any of the current best solutions are sent back to a lower level j through the last operation of line 3, they then become excluded from combining with blocks belonging to levels $j + 1$ to l.

4.3   The Tabu Search Procedure

The search space of each level is explored using a tabu search procedure. This tabu search procedure uses two tabu lists. A first tabu list prohibits moves that undo swaps of blocks $x \rightarrow y$ by entering in the tabu list the move $y \rightarrow x$. A second tabu list disallows a block from leaving the current solution B for a certain number of tabu iterations after entering B. The size of the tabu list varies randomly in a pre-defined range for each call to the tabu procedure. We found that variations in the length of the tabu lists is helpful to diversify the exploration of search spaces when projection fails to re-coarsen some of the levels. The termination criterion for this tabu search procedure is a pre-defined number of iterations without improving the best known solution. Our tabu search procedure is described in Fig. 2.

tabu_search(*initial_solution*)
  best = *initial_solution*; B = *initial_solution*;
  while (termination criterion not satisfied) **do**
    $B = V \in \mathcal{N}(B) \wedge V$ not tabu; (V is the best solution in the neighborhood of B
            and V is not in any of the two tabu lists)
    update tabu lists;
    if $(cost(B) \leq cost(best))$ **then**
      best = B;
    return *best*;

Fig. 2. The tabu search procedure

4.4   The Multi-cycle Refinement Phase

This multilevel algorithm is based on a multi-cycle refinement phase. Refinement cycles are divided into two categories: interpolation cycles and search cycles.

Interpolation Cycles. Interpolation cycles are initiated at level 0 as described in Fig. 3. An interpolation operation at level $i \neq l$ uses the best solution of level $i+1$ to restart the tabu search procedure at level i (line 1). At level l, the search is restarted from the current best solution at level l (line 3). An interpolation cycle ends by a restart of the search at level 0 using the current best solution at level l (line 4). Each time the tabu search procedure has completed the search initiated from the interpolated solution, the best solution in the search sequence is projected to level l (lines 2 and 4). In an interpolation cycle, information move downward through the interpolation operations and upward through the projection operations performed at levels 0 to $l - 1$.

Interpolation_cycle()
       for $(i = 0; i \leq l - 1; i + +)$ **do**
1. $best_i = $ tabu_search$(best_{i+1})$;
2. $projection(best_i)$;
       $B_l = $ best solution of level l from the previous cycle;
3. $best_l = $ tabu_search(B_l);
4. $best_0_tmp = $ tabu_search$(best_l)$; $projection(best_0\_tmp)$;
       if $(cost(best_0\_tmp) \leq cost(best_0))$ **then** $best_0 = best_0\_tmp$;

Fig. 3. The interpolation cycle

Search Cycle. Search cycles run a tabu search procedure at each level, starting at level l toward level 0. Search cycles have a dual purpose. One is to discover improving solutions once re-coarsening has modified the search space of each level. The second purpose is to diversify the exploration of the solution space $\mathcal{S}_0$.

During a search cycle, the tabu search procedure at level i starts with the current best solution at this level. If the search fails to improve the current best solution, the exploration of the search space is then restricted to blocks in L_i, the blocks assigned to level i (line 3 in Fig. 4). In the search space defined uniquely by blocks of L_i, tabu search cannot access the blocks of the best solutions, which are assigned to level l. Constrained to blocks in L_i, the search usually enters a sequence of uphill moves where blocks enter B that would not have been included if all candidate neighbors had been considered. Then, search is re-opened to the whole search space of level i (line 5). The last search sequence at level i is initiated from the last solution visited in the restricted search space (line 6). This is usually a poor solution, consequently, the last search sequence is a sequence of downhill moves, replacing blocks in B with other blocks improving the cost of B. The solution that is projected at the end of the search at level i may or may not have a better cost than the best solution in the previous cycle. However, because of the uphill and downhill search moves, level i is likely to project to level l a more diversified set of blocks than if the search has been performed uniquely in the search space of level i. This speed-up the re-coarsening of each level, which in turn diversifies the exploration of the solution space S_0.

Search_cycle()
    for $(i = l; i \geq 0; i - -)$ **do**
        $B_i = best_i$; ($best_i$ is the current best solution at level i)
1.   search space = any combination of b blocks in $L_i \cup L_{i+1} \cup \cdots \cup L_l$;
        $best_i = tabu\_search(best_i)$;
2.   **if** $(cost(best_i) \geq cost(B_i))$ **then**
3.   search space = any combination of b blocks in L_i;
4. $best_i =$ last solution of $tabu\_search(best_i)$;
5.   search space = any combination of b blocks in $L_i \cup L_{i+1} \cup \cdots \cup L_l$;
6. $best_i = tabu\_search(best_i)$;
        if $(i \neq l)$ **then** $projection(best_i)$;

Fig. 4. Search procedure for refinement cycles

The Initial Search Phase. To compute the initial state of the multi-cycle refinement phase, we run a pre-defined number of p search cycles (the value of parameter is determined empirically). In the first search cycle, each tabu search procedure is started from a randomly generated solution. The random initial solution at level i is computed by selecting b blocks in the set of blocks $S_i = L_i \cup L_{i+1} \cup \cdots \cup L_l$, as described in Fig. 5. Searches in the first cycle from initial random solutions are likely to generate significant re-coarsenings at all levels above level 0. The **for** loop of line 3 launch a sequence of $p - 1$ search cycles in order to explore the new search spaces created by the re-coarsenings.

Refinement Phase. The entire multi-cycle refinement sequence is summarized in Fig. 6. Beyond the initial search phase, the refinement phase is decomposed into sequences of p cycles: first cycle is an interpolation cycle and it is followed by $p - 1$ cycles.

Initialization_sequence()
1.   **for** $(i = l; i \leq 0; i--)$ **do**
        $B_i = \emptyset$;
        for $(j = 1; j \leq b; j++)$ **do**
            $block =$ a randomly selected block in S_i;
            $B_i = B_i \cup block$;
        $best_i = tabu\_search(B_i)$;
2.   **if** $(i \neq l)$ **then** $projection(best_i)$;
3.   **for** $(j = 2; j \leq p; j++)$ **do**
        $Search\_cycle(j)$;

Fig. 5. The initial search phase

Multi-cycle_refinement_phase()
    $Initialization\_sequence()$;
    while (not found solution or number of cycles smaller than limit) **do**
        $Interpolation\_cycle()$;
        for $(j = 1; j \leq p-1; j++)$ **do**
            $Search\_cycle()$

Fig. 6. Multi-cycle refinement phase

5   Experimentation

Several tests have been performed during the development and validation phases of this algorithm, we report the results in Table 1 below. The column "$t - (v, k, \lambda)$" describes the parameters of the covering design problem while the column "# of runs" reports how many time we have run our algorithm on each problem. A large number of runs (such as 50 for 3-$(14,5,1)$) indicates that the corresponding problem has been used as a test problem during the development phase. The column "b" indicates the size of the covering design we have tested. All values of b are one block less than the best known upper bounds, except for some of the problems for which we have been able to improve the best known upper bounds. (Our tests are based on the best known covering design upper bounds as published on the web site [8] in Spring 2005). The columns "$Cost$" reports, for all runs, the solution with the smallest number of t-subsets not covered. For example, a cost of 2 indicates the best set of b blocks failed to cover 2 t-subsets. A cost of 0 indicates that we have improved the best known upper bound. In this case, on the corresponding row under columns b, we report the previous best known upper bound in () beside our new upper bound.

For runs where new upper bounds have been found, the total number of cycles executed varies between 7 and 175. For these runs, the range in computational time varies between 20 minutes to 15 hours on a 500 MHz sequential computer (many factors impact the computational time requirements of a cycle, among them the size of the covering design parameters). A run is aborted once 1000 cycles have been executed without discovering a new upper bound. For the tests reported in Table 1, the computational time requirements vary between 1 hour up to 168 hours (1 week) for runs that didn't improve the best known upper

Table 1. Experimental results

$t-(v,k,\lambda)$	Cost	b	# of runs	$t-(v,k,\lambda)$	Cost	b	# of runs
3-(12,5,1)	2	28	5	3-(13,5,1)	1	33	10
3-(14,5,1)	1	42	50	3-(15,5,1)	2	55	5
3-(16,5,1)	2	64	5	3-(17,6,1)	1	43	50
3-(19,6,1)	5	62	3	3-(20,6,1)	30	71	8
4-(13,6,1)	2	65	5	4-(14,6,1)	8	79	5
4-(15,6,1)	32	116	5	4-(14,7,1)	2	43	5
4-(15,7,1)	7	56	5	4-(16,7,1)	1	75	2
4-(17,7,1)	53	98	3	4-(17,8,1)	4	53	5
4-(16,9,1)	6	25	5	4-(17,10,1)	5	22	5
5-(11,6,1)	6	99	3	5-(12,7,1)	5	58	2
5-(13,8,1)	1	42	8	5-(14,7,1)	10	137	2
5-(14,8,1)	6	54	4	5-(15,8,1)	7	88	8
5-(16,9,1)	0	61(62)	6	5-(16,10,1)	0	36(37)	8
6-(13,8,1)	7	99	2	6-(14,9,1)	0	72(75)	8
6-(15,9,1)	1	99	3	6-(15,10,1)	0	53(55)	7
6-(16,10,1)	4	76	8	6-(16,11,1)	11	43	8
6-(17,12,1)	31	35	8	7-(13,9,1)	7	78	8
7-(14,10,1)	0	56(57)	8	5-(17,10,1)	2	48	7

bound. Finally, in terms of comparison, we have ran extensive tests against simulated annealing [15] for all the problems reported in Table 1, none was able to improve the best known upper bound.

6   Conclusion

The general strategy of cooperative multilevel algorithms is to solve several problems and use the solutions to define a new set of problems. This paper has described an exploratory application of this approach to covering designs. Blocks of successful approximate solutions discovered by a tabu search procedure are substituted to some blocks of an existing problem description, yielding a new problem definition. The key observation here is that the new problem definition hold at its core a successful combination of blocks. By making the problem small enough such that it holds only successful combination of blocks, we create conditions to obtain successful search directions from the re-combinations of blocks in the smaller problem. Furthermore, under the strict inclusion condition, blocks of the smaller problem are included in the definition of all the other problems. This provide for individual blocks to be tested inside good combinations of blocks, which often provides small increments in the definition of new successful combination of blocks. Overall, this multilevel strategy has already delivered interesting numerical results and seems to hold the potential to deliver more for covering designs and other problems in the field of combinatorial designs.

References

1. S.T. Barnard and H.D. Simon. A Fast Multilevel Implementation of Recursive Spectral Bisection for Partitioning Unstructured Problems. *Concurrency: Partice & Experience*, 6(2):111–117, 1994.
2. J.A. Bate. *A Generalized Covering Problem*. PhD thesis, University of Manitoba, 1978.
3. A. Brandt. Multi-level adaptive solutions to boundary value problems. *Mathematics of Computation*, 31:333–390, 1977.
4. W.L. Briggs, V.E. Henson, and S.F. McCormick. *A Multigrid Tutorial*. SIAM, 1999.
5. C.J. Colbourn and J.H. Dinitz, editors. *The CRC Handbook of Combinatorial Designs*. CRC Press, 1996.
6. T.G. Crainic, Y. Li, and M. Toulouse. A Simple Cooperative Multilevel Algorithm for the Capacitated Multicommodity Network Design. *Computer & Operations Research*, Accepted for publication.
7. P. Erdős and H. Hanani. On a limit theorem in combinatorial analysis. *Publicationes Mathematicae Debrecen*, 10:10–13, 1963.
8. C.J. Gordon. Web site of covering bounds. *http://www.ccrwest.org/cover.html*.
9. C.J. Gordon, O. Patashnik, and G. Kuperberg. New constructions for covering designs. *Journal of Combinatorial Designs*, 3(4):269–284, 1995.
10. B. Hendrickson and R. Leland. The Chaco User's Guide: Version 2.0. Report SAND95-2344, Sandia National Laboratories, 1995.
11. F. Margot. Small covering designs by branch-and-cut. *Mathematical Programming*, 94:207–220, 2003.
12. W. H. Mills and R. C. Mullin. Coverings and packings. In *Contemporary Design Theory: A Collection of Surveys*, pages 371–399. Wiley-Interscience Series in Discrete Mathematics and Optimization, 1992.
13. W.H. Mills. Covering designs I: coverings by a small number of subsets. *Ars Combinatoria*, 8:199–315, August 1979.
14. K. J. Nurmela. Constructing combinatorial designs by local search. Technical report, Helsinki University of Technology, November 1993.
15. K. J. Nurmela and P. R. J. Östergård. Constructing covering designs by simulated annealing. Technical report, Helsinki University of Technology, January 1993.
16. K. J. Nurmela and P. R. J. Östergård. New coverings of t-sets with (t+1)-sets. *Journal of Combinatorial Designs*, 7:217–226, 1999.
17. M. Ouyang, M. Toulouse, K. Thulasiraman, F. Glover, and J.S. Deogun. Multilevel Cooperative Search for the Circuit/Hypergraph Partitioning Problem. *IEEE Transactions on Computer-Aided Design*, 21(6):685–693, 2002.
18. V. Rödl. On a packing and covering problem. *European Journal of Combinatorics*, 5:69–78, 1985.
19. J. Schönheim. On coverings. *Pacific Journal of Mathematics*, 14:1405–1411, 1964.

Enhancements of NSGA II and Its Application to the Vehicle Routing Problem with Route Balancing

Nicolas Jozefowiez[1], Frédéric Semet[2], and El-Ghazali Talbi[1]

[1] Université des Sciences et Technologies de Lille,
Laboratoire d'Informatique Fondamentale de Lille,
59655 Villeneuve d'Ascq Cedex, France
{jozef, talbi}@lifl.fr
[2] Université de Valenciennes et du Hainaut-Cambrésis, Laboratoire d'Automatique,
de Mécanique et d'Informatique industrielles et Humaines,
59313 Valenciennes Cedex 9, France
frederic.semet@univ-valenciennes.fr

Abstract. In this paper, we address a bi-objective vehicle routing problem in which the total length of routes is minimized as well as the balance of routes, *i.e.* the difference between the maximal route length and the minimal route length. For this problem, we propose an implementation of the standard multi-objective evolutionary algorithm NSGA II. To improve its efficiency, two mechanisms have been added. First, a parallelization of NSGA II by means of an island model is proposed. Second, an elitist diversification mechanism is adapted to be used with NSGA II. Our method is tested on standard benchmarks for the vehicle routing problem. The contribution of the introduced mechanisms is evaluated by different performance metrics. All the experimentations indicate a strict improvement of the generated Pareto set.

1 Introduction

This paper investigates the use of two variants of NSGA II to solve a bi-objective vehicle routing problem. The elementary version of the vehicle routing problem is the capacitated vehicle routing problem (CVRP). It can be modeled as a problem on a complete graph where the vertices are associated to a unique depot and to m customers. Each customer must be served a quantity q_i of goods ($i = 1, \ldots, m$) from the unique depot. To deliver these goods, vehicles are available. With each vehicle is associated a maximal amount Q of goods it can transport. A solution of the CVRP is a collection of routes where each customer is visited only once and the total demand for each route is at most Q. With each arc (i, j) is associated the distance between vertex i and vertex j. The CVRP aims to determine a minimal total length solution. It has been proved NP-hard [1] and solution methods range from exact methods to specific heuristics, and meta-heuristic approaches [2].

E. Talbi et al. (Eds.): EA 2005, LNCS 3871, pp. 131–142, 2006.

Table 1. Objective values for the best found solutions by Taburoute and by Prins' GA

	Taburoute		Prins' GA	
Instance	Distance	Balance	Distance	Balance
E51-05e	524.61	20.07	524.61	20.07
E76-10e	835.32	78.10	835.26	91.08
E101-08e	826.14	97.88	826.14	97.88
E151-12c	1031.17	98.24	1031.63	100.34
E200-17c	1311.35	106.70	1300.23	82.31
E121-07c	1042.11	146.67	1042.11	146.67
E101-10c	819.56	93.43	819.56	93.43

Another natural objective to consider in addition to the minimization of the total length is the balance of the routes. Route balancing can be expressed in several ways. In [3], the authors balance the time needed for each trip. It is computed as the sum of the differences between each route length and the shortest route length. Route balancing is also an objective in [4] which addresses a three objective multi-period vehicle routing problem. In this paper the balance is measured by the standard deviation and the load of a route consists in the number of visited customers. In [5], the minimization of the time spent on a bus, which has some common points with the route balancing, is considered. In [6], the authors take into account 8 objectives in the context of a real-life VRP faced by a Belgian transportation firm. One of them is identical to our second objective; i.e. the minimization of the difference between the maximal route length and the minimal route length.

In this paper, we address a variant of the CVRP: the vehicle routing problem with route balancing (VRPRB). The following two objectives are considered:

1. Minimization of the distance traveled by the vehicles.
2. Minimization of the difference between the longest route length and the shortest route length.

In Table 1, the seven CVRP benchmarks proposed by Christofides and Eilon [7], and Christofides and al. [8], are considered. Following the naming scheme used in Toth and Vigo [2], the name of each instance has the form $Ei - jk$. E means that the distance metric is Euclidean. i is the number of vertices including the depot vertex. j is the number of available vehicles. k is a character which identifies the paper where the distance data are provided. $k = e$ refers to Christofides and Eilon [7], $k = c$ to Christofides et al. [8]. For each instance, we report both objective values associated with the best solutions obtained using Taburoute [9] and Prins' GA [10]. These methods, which can be regarded as some of the best algorithms for the CVRP, do not take into account the route balancing objective. This clearly appears in Table 1 where the best solutions are of poor quality regarding the additional objective.

Our solution to generate the Pareto set is based on the standard multi-objective evolutionary algorithm (MOEA) NSGA II proposed by Deb et al. [11]. Our choice of meta-heuristics is motivated by the difficulty of solving the problem with exact approaches. Since a Pareto set has to be generated, a population based method like NSGA II seems well-fitted. To improve the results of NSGA II on the VRPRB, we propose a parallelization of the problem. To obtain well-diversified approximations of the Pareto set, we have adpated the elitist diversification mechanism initially proposed in [12, 13] for NSGA II.

The paper is organized as follows. Section 2 presents our implementation of NSGA II for the VRPRB and its parallelization into an island model. In section 3, we specify the adaptation of the elitist diversification mechanism for NSGA II. In section 4, we assess the efficiency of the new mechanisms on a set of standard benchmarks. Conclusions are drawn in section 5.

2   NSGA II for the Vehicle Routing Problem with Route Balancing

We first describe the general framework of NSGA II in subsection 2.1. Then, the recombination phase (*i.e.* STEP 4) is given in the subsection 2.2 since it is the only step which needs to be adapted for the VRPRB. Finally, an improvement of NSGA II by means of an island model is proposed in subsection 2.3.

2.1   NSGA II

NSGA II can be described as follows. Its population R_t, where t is the number of the current generation, is divided into two subpopulations P_t and Q_t. The sizes of P_t and Q_t are equal to N and, therefore, the size of R_t is $2N$. The subpopulation P_t corresponds to the parents and Q_t to the offspring. The four main steps of NSGA II are presented below without going into the details of the mechanisms used such as the ranking and the crowding distance. It is sufficient to recall that a solution i has two fitnesses according to the current population: a rank r_i which represents its quality in terms of convergence toward the optimal Pareto set, and a crowding distance d_i which corresponds to its quality in terms of diversification. The lower the rank and the crowding distance are, the better the solution is. For additional details about NSGA II, the reader is refered to [11]. At generation t, the different steps are:

STEP 1. Combine the parent and offspring populations to create $R_t = P_t \cup Q_t$. Compute the ranks and crowding distances of the solutions in R_t. Sort the solution according to their ranks in an increasing order. Identify the fronts $\mathcal{F}_i$, $i = 1, \ldots, r$, where i represents a rank.
STEP 2. Create a new population $P_{t+1} = \emptyset$. Set $i = 1$. While $|P_{t+1}| + |\mathcal{F}_i| < N$, do $P_{t+1} = P_{t+1} \cup \mathcal{F}_i$ and $i = i + 1$.
STEP 3. Sort the solutions of $\mathcal{F}_i$ according to their crowding distance in a decreasing order. The $(N - |P_{t+1}|)$ first solutions of $\mathcal{F}_i$ (*i.e.* the most diversified solutions) are included to P_{t+1}.
STEP 4. Create Q_{t+1} from P_{t+1}.

The solution provided by NSGA II is the set of solutions not dominated in the final population R. However, experiments have shown that the size of

134   N. Jozefowiez, F. Semet, and E.-G. Talbi

Algorithm 1. recombination_phase(P, Q: POPULATION)

$Q \leftarrow \emptyset$
for $i \leftarrow 1, \ldots, N$ **do**
    $pa_1 \leftarrow tournament(P \cup C)$
    $pa_2 \leftarrow tournament(P \cup C)$
    if $rand() < 0.5$ **then**
        $s \leftarrow RBX(pa_1, pa_2)$
    else
        $s \leftarrow SPLIT(pa_1, pa_2)$
    end if
    if $rand() < 0.4$ **then**
        $s \leftarrow or\_opt(s)$
    end if
    $2opt\_local\_search(s)$
    $Q \leftarrow Q \cup \{s\}$
end for

the potentially Pareto optimal solution set can be very large for the VRPRB. Therefore, we have added an archive to NSGA II whose only purpose is to save the potentially Pareto optimal solutions identified during the search. It prevents such solutions to be lost due to the stochastic behavior of the algorithm and the limited size of the population.

2.2   The Recombination Phase

The recombination phase is described in Algorithm 1. The tournament operator is the binary tournament as described by Deb *et al.*. Two solutions are randomly selected and the solution with the best rank is kept. To break the tie, the solution with the greatest crowding distance is selected. The crossover operators are the route based crossover (RBX) [14] and the SPLIT crossover [12, 13] inspired by Prins' genetic algorithm [10]. When a solution is created, a 2-opt local search is applied on each route in order to avoid artificially balanced solutions [12, 13].

2.3   Parallelization

To improve the results obtained by NSGA II, we have implemented it in an island model. The model is built as follows: each island corresponds to one instantiation of NSGA II with its own population. The communication network is a ring, and therefore each island has two neighbors. One island sends information to its neighbors regularly in terms of generations. When the generation corresponds to a communication phase, which is performed instead of recombination (STEP 4). Due to the fact that the communication network is a ring, an island receives information at the same time it sends information. The computations of a given island do not begin again until it has received the information from its two neighbors.

The communication phase runs as follows. An island sends to its two neighbors the $\frac{N}{2}$ best solutions from its population (*i.e.* the $\frac{N}{2}$ first solutions, according

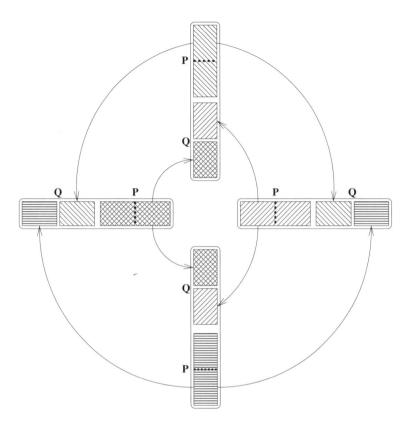

Fig. 1. Extension of NSGA II into an island model

to the ranking and crowding distance sort, of the population after the selection phase (STEP 1 to STEP 3). Therefore, an island receives $\frac{N}{2}$ solutions twice. These solutions replace those from Q_t since they would have been lost in the case of a standard recombination phase. Figure 1 illustrates the communications in the case of four islands.

3 Using the Elitist Diversification Mechanism in NSGA II

In this section, we propose the enhancement of NSGA II by means of a diversification mechanism called the elitist diversification mechanism initially proposed in [12, 13]. First, the mechanism is presented. Then, the general parallel model is described as well as its use in the case of NSGA II.

3.1 The Elitist Diversification Mechanism

In the elitist diversification, additional archives are considered. They contain the potentially optimal Pareto solutions (PPS) when one objective is maximized

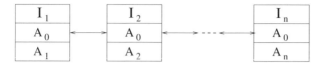

Fig. 2. The basic co-operative model - the toric structure is not shown in order not to obfuscate the figure

instead of being minimized. It may be noted that we suppose that every objective is to be minimized. Let $S(A)$ be the subset of solutions of the decision space found by an algorithm A, and k the index of the objective function component which is maximized. To define new archives, the dominance operator $\prec_k$ is introduced:

$$\forall y, z \in S(A), y \prec_k z \Longleftrightarrow (\forall i \in \{1 \ldots n\} \setminus \{k\}, f_i(y) \leq f_i(z))$$
$$\wedge (f_k(y) \geq f_k(z))$$
$$\wedge ((\exists i \in \{1 \ldots n\} \setminus \{k\}, f_i(y) < f_i(z))$$
$$\vee (f_k(y) > f_k(z)))$$

Then, we have $A_k = \{s \in S(A) | \forall s' \in S(A), s' \nprec_k s\}$, with $k = 1, \ldots, n$, the archive of PPS associated with the maximization of the kth objective component instead of the minimization. We denote $\prec_0$ the classical dominance operator *i.e.* a solution x is said to dominate a solution y if x is not worse than y on every objective and there is at least one objective where x is strictly better than y.

Like in the elitism strategy, solutions from the new archives are included into the population of the MOEA at each generation. The role of these solutions is to attract the population to unexplored areas, and so to avoid the premature convergence to a specific area of the objective space. Indeed, using solutions from these archives ensures that an exploration is done while favorising one objective. Preliminary experiments point out that the improvement is less important when all archives are embedded in the same MOEA. This leads us to distribute the archives among several searches resulting in a co-operative model. In the general case with n objectives, the co-operative model is composed of n islands denoted I_k. Each island I_k has two types of archive: A_0 and A_k. At each *Migration$_t$* generation, I_k sends its A_0 archive to its two neighbors I_{k-1} and I_{k+1}. The communication topology is toric, therefore k is computed modulo n. This co-operative model and its communication topology consist in the model described in Figure 2.

3.2 Parallel Extension of the Elitist Diversification Mechanism

The co-operative model described previously formed the elementary brick of a more general island model used to favor the convergence and diversification tasks (see Figure 3). This parallelization is not used in order to speed up the search but to search a larger part of the solution space in a given time. Since every island will be executed at the same time, it will take the same computational time as a single island while the number of solutions created will be multiplied by the

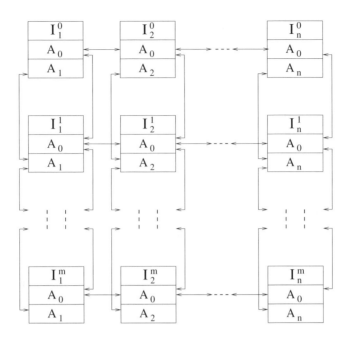

Fig. 3. The complete co-operative model - the toric structure is not shown in order not to obfuscate the figure

number of islands. An island is denoted I_j^i. It means it belongs to the ith brick and its additional archive is of A_j type. The island I_j^i sends its A_0 archive to all its neighbors: I_{j-1}^i, I_{j+1}^i, I_j^{i-1}, and I_j^{i+1}. It only communicates its A_j archive to I_j^{i-1} and I_j^{i+1}. Since the communication topology between and within the bricks is toric, the indexes are computed modulo n.

3.3   Inclusion of the Elitist Diversification Mechanism in NSGA II

The goal is to add the management of the additional archives in NSGA II. It must be noted that NSGA II initially used no archive and the main population plays the role of the A_0 archive, *i.e.* it saves the non-dominated solutions found during the search. Therefore, each island of the parallel model described before corresponds to one instantiation of NSGA II to which one additional archive has been added. This archive is used during the recombination phase: k individuals are chosen among those belonging to the additional archive and form the set C_t. Then, the recombination phase is the same as the standard one except that the tournament used to select the parents is modified as follows. Two solutions are selected randomly in $P_t \cup C_t$. If pa_1 or pa_2 belongs to C_t, the solution from C_t wins the tournament. If both solutions belong to C_t, one is chosen randomly. Finally, if both solutions come from P_t, the standard binary tournament of NSGA II is applied. The additional archive is updated after each recombination phase ; we try to include the solutions generated during the phase.

The exchange strategy between the islands is the same as the one used in 2.3. However, since an island has four neighbors in this model, it communicates only the $\frac{N}{4}$ best solutions from its population after the selection phase. Therefore, an island receives four times $\frac{N}{4}$ solutions which replace those from Q_t. However, there are two special cases. First, in the bi-objective case, an elementary brick is formed of only two genetic algorithms. Then, an island receives twice the $\frac{N}{4}$ best solutions from the other genetic algorithm of the brick. It is not relevant and, in this case, the strategy is modified for the algorithms from a same brick to exchange $\frac{N}{2}$ solutions between them. The same difficulty occurs when there are only two elementary bricks and can be solved in a similar way.

4 Computational Results

4.1 Protocol

NSGA II for the VRPRB, the parallel model pNSGA II, the variant with elitist diversification NSGAED, and the parallel variant with elitist diversification pNS-GAED have been coded in C. MPI has been used for the parallel aspect of the implementation. Experiments have been realised on an IBM RS6000/SP equipped with Power4 1.1 Ghz processors.

Evaluations have been made on the benchmark by Christofides et al. [7] for the capacitated vehicle routing problem. Each instance has been solved 10 times by each method.

The parameterization of the methods has been set experimentally. For the population of NSGA II, N has been fixed to 128. NSGA II and pNSGA II stopped after 100000 generations while NSGAED and pNSGAED stopped after 50000. Thus, we insure that each process generates the same number of solutions. For the elitist diversification, 15 solutions were used from each archive.

As suggested in [15], the S metric [16] was used. $S(A)$ gives the size of the area dominated by the approximation generated by A. The values of the objectives were normalized according to the reference point used in the S metric.

4.2 Contribution of the Parallelization

We have tested the contribution of the parallelization scheme when 1, 4, 8, and 16 processors were used. Table 2 reports the mean values and the standard deviations of the S metric for the different cases. As it can be expected, the results are improved with the number of processors used. However, the impact of more than 4 processors is less significant than the difference between the sequential version and the one with 4 processors. According to the behavior of the standard deviation, it seems that increasing in the number of processors contributes to improve the robustness of the method.

The impact of communications on computational times have also been assessed. The average computational times in seconds according to the number of processors are reported in Table 3. It seems that communication times do not play a significant role.

Table 2. Mean values and standard deviations of the S metric for NSGA II according to the number of used processors

Instance		1 proc.	4 proc.	8 proc.	16 proc.
E51-05e	Mean	0.511232	0.527863	0.527733	0.530235
	standard deviation	0.006132	0.001838	0.004329	0.001987
E76-10e	Mean	0.414035	0.420253	0.425498	0.426979
	standard deviation	0.002988	0.001714	0.002892	0.002052
E101-08e	Mean	0.570935	0.576901	0.577431	0.579026
	standard deviation	0.001779	0.001638	0.000724	0.000418
E151-12c	Mean	0.618357	0.631726	0.634581	0.637956
	standard deviation	0.006315	0.001637	0.003460	0.001426
E200-17c	Mean	0.607886	0.628112	0.632612	0.639964
	standard deviation	0.014343	0.005537	0.008276	0.002474
E121-07c	Mean	0.516538	0.526248	0.527154	0.527934
	standard deviation	0.007145	0.001480	0.001405	0.000637
E101-10c	Mean	0.584904	0.620338	0.627408	0.629321
	standard deviation	0.018182	0.004675	0.003061	0.002398

Table 3. Average computation times of NSGA II according to the number of processors

Instance	E51-05e	E76-10e	E101-08e	E151-12c	E200-17c	E121-07c	E101-10c
4 proc.	993.4	1453.4	2451.3	4082.1	4996.3	4615.1	2640.1
8 proc.	937.8	1300.8	2406.1	3621.9	4463.7	4791.1	2425.3
16 proc.	1080.7	1329.0	2289.6	3794.5	4677.7	5171.1	2451.6

Table 4. Mean values and standard deviation of the S metric for NSGA II without and with the elitist diversification mecanism

		1 processor		8 processors	
Instance		NSGA II	NSGAED	pNSGA II	pNSGAED
E51-05e	Mean	0.511232	0.521232	0.527733	0.529467
	standard deviation	0.006132	0.004139	0.004329	0.001282
E76-10e	Mean	0.414035	0.415599	0.425498	0.425809
	standard deviation	0.002988	0.003651	0.002892	0.002992
E101-08e	Mean	0.570935	0.573612	0.577431	0.577501
	standard deviation	0.001779	0.001800	0.000724	0.001430
E151-12c	Mean	0.618357	0.619450	0.634581	0.635170
	standard deviation	0.006315	0.007012	0.003460	0.003016
E200-17c	Mean	0.607886	0.617594	0.632612	0.643165
	standard deviation	0.014343	0.006185	0.008276	0.004848
E121-07c	Mean	0.516538	0.518553	0.527154	0.527442
	standard deviation	0.007145	0.007998	0.001405	0.000478
E101-10c	Mean	0.584904	0.602430	0.627408	0.629226
	standard deviation	0.018182	0.020408	0.003061	0.003343

Table 5. Best solutions found for each objective with the associated values of the other objective for the different implementations of NSGA II

Instance	NSGA II								NSGAED			
	1 proc.		4 proc.		8 proc.		16 proc.		1 proc.		8 proc.	
E51-05e	524.61	20.07	524.61	20.07	524.61	20.07	524.61	20.07	524.61	20.07	524.61	20.07
	0.51	694.79	0.17	648.80	0.23	643.89	0.18	779.88	0.48	1093.31	0.32	645.29
	44.8		44.1		46.5		50.4		45.8		47.6	
E76-10 e	835.32	78.10	842.06	67.55	835.32	78.10	835.32	78.10	835.89	85.60	835.32	78.10
	1.01	1268.02	1.10	1117.29	0.64	997.97	0.48	1153.25	0.85	1110.78	0.33	1380.01
	109.7		124.7		108.8		135.2		115.9		127.00	
E101-08e	827.39	67.55	827.39	67.55	827.39	67.55	827.39	67.55	827.39	67.55	827.29	67.55
	0.64	1596.59	0.09	1695.55	0.35	984.76	0.27	1132.28	0.43	1129.10	0.10	1610.86
	129.4		161.1		175.6		185.8		130.2		171.9	
E151-12c	1047.02	87.10	1044.69	91.14	1043.83	77.89	1032.95	97.57	1043.71	87.67	1034.91	109.06
	1.22	2290.53	0.56	1443.31	0.51	1382.28	0.44	1551.50	0.66	2523.00	0.21	1523.94
	204.1		310.10		332.9		408.5		200.1		322.1	
E200-17c	1358.08	68.77	1349.48	79.14	1332.82	87.50	1328.44	96.61	1353.17	87.84	1327.89	97.20
	3.04	1838.32	2.17	3006.26	1.85	1976.10	0.86	3490.14	2.18	3436.43	0.58	3746.57
	206.6		316.4		373.6		479.7		192.1		354.8	
E121-07c	1043.78	146.67	1042.11	146.67	1042.11	146.67	1042.11	146.67	1043.11	147.27	1042.11	146.67
	0.15	2258.21	0.06	2314.61	0.06	2348.74	0.07	2296.43	0.06	2285.22	0.07	1649.76
	441.5		847.40		903.1		921.1		452.1		943.2	
E101-10c	819.56	93.43	819.56	93.43	819.56	93.43	819.56	93.43	819.56	93.43	819.56	94.43
	1.16	2035.80	0.77	2107.14	0.65	1425.43	0.21	1324.53	1.10	1303.67	0.35	1664.80
	269.1		441.4		487.7		499.6		278.6		490.4	

4.3   Contribution of the Elitist Diversification Mechanism

We have evaluated the performance of NSGAED, pNSGAED with 8 bricks compared to the performance of NSGA II and pNSGA II with 8 processors. The mean values and the standard deviation of the S metric are reported in Table 4.

It appears that the elitist diversification is always able to improve the results of NSGA II when only one processor is used. The improvement is more important for large instances such as E200-17c. We have also evaluated the contribution when eight processors are used. The contribution is less important on the smallest instances since the parallelization without the elitist diversification was already able to improve the results significantly. However, the contribution is still important on the largest instances.

4.4   Global Efficiency of NSGA II for the Vehicle Routing Problem with Route Balancing

Optimal Pareto sets are not known for the VRPRB. Therefore, we have compared the results of our MOEA with the best-known values on the length objective and with the evident lower bound that is 0 for the balance objective. We have also reported the number of potentially Pareto optimal solutions in Table 5 as follows: for each entry, the first line corresponds to the best found length with its associated balance, the second line to the best found balance with its associated length, and the third line to the average number of solutions in the approximations. It appears that the elitist diversification is able to improve the results toward the best-known values for the total length objective. Since the best balance is very close to 0, we may assume that very well-balanced solutions are obtained.

5   Conclusions

In this paper, we have described an implementation of NSGA II for a bi-objective vehicle routing problem, called the vehicle routing problem with route balancing, where both the minimization of the total length and the balance of the routes, *i.e.* the minimization of the difference between the longest route length and the shortest route length, have to be optimized. Two enhancements of NSGA II have been proposed. The first one is the parallelization of NSGA II by means of an island model. The second one is the use of the elitist diversification mechanism, which aims to improve the diversification in NSGA II. Their contributions were evaluated on a set of standard benchmarks with standard metrics. The positive impact of both mechanisms has been observed through computational experiments. Since optimal Pareto sets remain unknown for the problem, the fact that the values found for the total length objective are close to the best-known ones, and that the best values for the route balancing objective are quite small tends to indicate that our generated approximations are of good quality.

References

1. Lenstra, J.K., Rinnooy Kan, A.H.G.: Complexity of vehicle routing and scheduling problem. Networks **11** (1981) 221–227
2. Toth, P., Vigo, D., eds.: The vehicle routing problem. Volume 9 of SIAM Monographs on Discrete Mathematics and Applications. SIAM (2001)
3. Lee, T., Ueng, J.: A study of vehicle routing problems with load-balancing. International Journal of Physical Distribution and Logistics Management **29** (1999) 646–658
4. Ribeiro, R., Lourenço, H.R.: A multi-objective model for a multi period distribution management problem. In: MIC'2001. (2001) 97–102
5. Corberan, A., Fernandez, E., Laguna, M., Marti, R.: Heuristic solutions to the problem of routing school buses with multiple objectives. Journal of the Operational Research Society **53** (2002) 427–435
6. El-Sherbeny, N.: Resolution of a vehicle routing problem with multi-objective simulated annealing method. PhD thesis, Faculté Polytechnique de Mons (2001)
7. Christofides, N., Eilon, S.: An algorithm for the vehicle dispatching problem. Operational Research Quarterly **20** (1969) 309–318
8. Christofides, N., Mingozzi, A., Toth, P., Sandi, C., eds.: 11. In: Combinatorial Optimization. John Wiley, Chichester (1979)
9. Gendreau, M., Hertz, A., Laporte, G.: A tabu search heuristic for the vehicle routing problem. Management Science **40** (1994) 1276–1290
10. Prins, C.: A simple and effective evolutionary algorithm for the vehicle routing problem. Computers and Operations Research (2004) (Article in Press).
11. Deb, K., Agrawal, S., Pratab, A., Meyunivan, T.: A fast elitist non-dominated sorting genetic algorithm for multi-objective optimization: NSGA-II. IEEE Transactions on Evolutionary Computing (2002) 182–197
12. Jozefowiez, N., Semet, F., Talbi, E.G.: Parallel and hybrid models for multi-objective optimization: Application to the vehicle routing problem. In Guervos, J.M., et al., eds.: PPSN VII. Volume 2439 of Lecture Notes in Computer Science., Springer-Verlag (2002) 271–280
13. Jozefowiez, N.: Modélisation et résolution approchées de problèmes de tournées multi-objectif. PhD thesis, Laboratoire d'Informatique Fondamentale de Lille, Université des Sciences et Technologies de Lille, Villeneuve d'Ascq, France (2004)
14. Potvin, J.Y., Bengio, S.: The vehicle routing problem with time windows part ii: genetic search. INFORMS Journal on Computing 8 **8** (1996)
15. Knowles, J., Corne, D.: On metrics for comparing nondominated sets. In: Congress on Evolutionary Computation (CEC'2002). Volume 1., IEEE Service Center (2002) 711–726
16. Zitzler, E.: Evolutionary algorithm for multiobjective optimization: Methods and applications. PhD thesis, Swiss Federal Institute of Technology (ETH), Zurich, Switzerland (1999)

The Importance of Scalability When Comparing Dynamic Weighted Aggregation and Pareto Front Techniques

Grzegorz Drzadzewski and Mark Wineberg

Computing and Information Science, University of Guelph,
Guelph, Ontario, N1G 2W1, Canada
gdrzadze@uoguelph.ca,
wineberg@cis.uoguelph.ca

Abstract. The performance of the Dynamic Weight Aggregation system as applied to a Genetic Algorithm (DWAGA) and NSGA-II are evaluated and compared against each other. The algorithms are run on 11 two-objective test functions, and 2 three-objective test functions to observe the scalability of the two systems. It is discovered that, while the NSGA-II performs better on most of the two-objective test functions, the DWAGA can outperform the NSGA-II on the three-objective problems. We hypothesize that the DWAGA's archive helps keep the searching population size down since it does not have to both search and store the Pareto front simultaneously, thus improving both the computation time and the quality of the front.

1 Introduction

At the present moment in the field of evolutionary computation there is very little research being conducted to investigate the behaviour of newly developed work by researchers other than the original creator. This is a major deficiency in the field. In most other disciplines of science the important aspect is the repeatability of experiments and confirmation of results by other independent research teams. In this paper we are performing an un-bias study reproducing the newly developed Dynamic Weight Aggregation Evolutionary Strategy (DWAES) algorithm and comparing it to a popular Pareto front style algorithm (NSGA-II).

There are two main approaches to evolutionary multi-objective optimization: weighted aggregation approaches and Pareto-based approaches.

The weighted aggregation approaches are easier to implement and understand, as well as being the first of the Evolutionary Multi-Objective Optimization (EMOO) algorithms created. However, recently they have been deemed flawed since they only produce a single solution along the Pareto front, and in many circumstances cannot find particular solutions along the front, no matter what weightings are used. Consequently Pareto-based approach has risen in popularity and now dominates the literature. This group of algorithms work by dividing its population into dominated and non-dominated solutions [1], where a non-dominated solution is one where no other solution is better than it across every objective. These groups of algorithms have often been analysed and compared with each other.

E. Talbi et al. (Eds.): EA 2005, LNCS 3871, pp. 143–154, 2006.

Recently, a modification to the simplistic weighted aggregation approach was proposed: the Dynamic Weight Aggregation (DWA) system [2], [3]. This system, while based on the weighted aggregation approach, was designed to overcome the two shortcomings mentioned above[1]. Experimentation on traditional EMOO problems seemed to verify the technique. However, the DWA system was never directly compared to the Pareto based EMOO methods.

In this paper we compare the DWA method as applied to the GA against a Pareto base EMOO system, the NSGA-II, to see if the DWA produces solutions of as high quality (as close to the Pareto front and covering the front as evenly).

2 The Two Systems

2.1 Non-dominated Sorting GA

One of the most popular of the Pareto-based approaches is the NSGA-II algorithm, which is an enhancement of the original non-dominated sorting GA (NSGA) proposed by Srinivas and Deb in 1994 [4]. The NSGA algorithm first sorts the solutions by fronts: each subset of the population that is not dominated by any other member of population is separated from those that are, with this definition recursively applied as each front is removed from the population. From this sorted population, standard reproduction techniques are applied using the front levels as fitness.

The NSGA-II uses a new non-dominated sorting approach, which is more efficient than the original method [5]. The old sorting algorithm used in NSGA has a complexity of $O(mN^3)$. The NSGA-II algorithm has improved the performance of the sort so it now has a complexity of $O(mN^2)$, where m is the number of objectives and N is the population size – this improves the execution time significantly. The NSGA-II also incorporates elitism and has a parameter-less diversity preservation mechanism.

2.2 The Dynamic Weighted Aggregation Systems

The conventional weighted aggregation (CWA) approach, which is a simple weighted sum of the different objective fitness values into a single fitness value, while being the simplest approach to Evolutionary Multi-Objective Optimization (and the first utilized), has been severely criticized on account of two main weaknesses [1]: First, the conventional weighted aggregation can provide only one Pareto solution from one run of optimization. Second, it has been shown that weighted aggregation is unable to deal with multi-objective optimization problems with a concave Pareto front.

Recently a new dynamic weight aggregation algorithm was proposed with the claim that it has eliminated the two problems associated with the conventional approach [2], [3]. The idea behind the algorithm is that "if the weights for the different objectives are changing during optimization, the optimizer will go through all points on the Pareto front. If the found non-dominated solutions are archived, the whole Pareto front can be achieved"[3]. This works for both the convex and concave

---

[1] Similar dynamic weighting techniques have also been used in non-evolutionary search methods such as Pareto Simulated Annealing [10], and Multi-Objective Tabu Search [11].

Pareto fronts. A theory for why the CWA algorithm does not work on concave Pareto front is provided in [2], which states that the CWA can only converge to a Pareto-optimal solution if the Pareto solution corresponding to the given weight combination is stable. Since all points on convex Pareto front are stable CWA has no trouble with it, but it is unable to reach points on the concave Pareto front. DWA algorithm on the other hand is able to go through all the points on the concave and convex Pareto front.

Using the CWA approach, a total fitness value for the chromosome is computed from the multiple fitness functions by performing a weighted sum

$$f(c) = w_1 f_1(c) + w_2 f_2(c) = w_1 f_1(c) + (1 - w_1) f_2(c) \tag{1}$$

where w_1 and w_2 are constant weights (which must sum to 1).

In the DWA, the constant weights are changed to time varying weights, $w_1(t)$ and $w_2(t)$, where t is 'time' measured in generations. The equations used in [2] for the two dynamic weights are:

$$w_1(t) = |\sin(2\pi t / T)| \tag{2}$$

and

$$w_2(t) = 1.0 - w_1(t) \tag{3}$$

where T is the period, a user defined parameter that controls how rapidly the weights cycle from 0 to 1 and back again.

In the case of a three objective problem, the weights are computed similarly, except that now there is rotation about two axes instead of just one and the weights are determined based on variables α and β.

$$
\begin{aligned}
w_1(\alpha) &= |\sin(2\pi\alpha)| \\
w_2(\alpha, \beta) &= (1 - w_1(\alpha)) |\sin(2\pi\beta)| \\
w_3(\alpha, \beta) &= 1 - w_1(\alpha) - w_2(\alpha, \beta),
\end{aligned}
\tag{4}
$$

where $0 \le \alpha, \beta \le \pi/2$.

Since the fitness function changes from generation to generation, it becomes important to store good solutions found in each generation. These good solutions are stored in the *archive*. A solution is added to the archive if it is not Pareto-dominated by any member of the archive. If a new solution Pareto-dominates members of the archive then all the dominated solutions are removed from it while the new solution is added.

3 Experimental Design

3.1 Algorithms and Parameters

To compare NSGA-II with the DWA system, it is important to isolate the various features of the two systems. This is both to assure a fair comparison, and to prevent extraneous factors from obscuring the underlying differences or similarities. Consequently, we chose to keep the underlying evolutionary algorithms the same for both systems. This means that all the parameters, with the exception of any system specific parameters, are set in common.

Table 1. Parameter Settings

Common parameter

	2 obj	3 obj		2 obj	3 obj
Population	100	{600,800}	Length	10	{14,16}
Generations	150	{900,1200}	Tournament Sel Pres.	0.9	.9
Prob. of cross over	0.8	0.8	Uniform xover prob.	0.4	.4
Mutation Rate	0.1	{0.071,0.0625}	Alphabet Size	100	100

DWAGA only parameters

	2 obj	3 obj		2 objectives	3 obj
# of 90° rotations	2	n/a	Gen	{150, 250, 600, 900, 2250}	n/a
Archive Size	100	1000			

To accomplish this uniformity for comparison we had to choose which evolutionary algorithm to base the two systems on. The NSGA, as its name implies, was designed to work on top of a Genetic Algorithm. The DWA, on the other hand, was originally written for an Evolutionary Strategy system. Since the DWA is just a modification of the fitness weights, which can be trivially used for either ES or GA, we chose to implement a Dynamic Weighted Aggregation Genetic Algorithm (DWAGA) to compare against the NSGA-II system.

3.1.1 Two Objective Problems

The performance of DWAGA was examined using 5 different period values (T). The values for the period length varied all the way from 200 to 7500 depending on the test function. It was discovered that DWAGA worked best when the period was set to a value that makes the number of 90° rotations equal to 2 (using equation 2).

Using 150 generations the DWAGA with a period of 600 will perform one 90° rotation; a period of 300 will result in two 90° rotations, 200 results in 3 rotations, 150 in 4 rotations, and 120 in 5 rotations.

When testing we discovered that our implementation of the DWAGA was, in general, faster than the NSGA-II. Therefore the DWAGA could perform more generations and improve the solutions that it had obtained and still finish at the same time as the NSGA-II. Consequently we ran DWAGA for a varying number of generations, making sure that the time equaled that of the NSGA-II.

The details for parameter values used for two objective problems can be found in table 1.

3.1.2 Three Objective Problems

When dealing with 3 objective problems we have to vary both α and β for the DWAGA system. Consequently, there are two periods for the 3-objective DWAGA system, with β cycling through its settings for every setting of α. Instead of complicating maters with two user-defined parameter both periods are set to be inversely proportional to the number of generations. Also the DWAGA system only goes through one 90° rotation for both α and β instead of 180°.

The details for parameter values used for three objective problems can again be found in table 1. All experiments are repeated 30 times for statistical accuracy.

Table 2. Function definitions for two tri-objective functions used in the test suite

F12	F13
$f_1 = x_1^2 + (x_2 - 1)^2$	$f_1 = 0.5(x_1^2 + x_2^2) + \sin(x_1^2 + x_2^2)$
$f_2 = x_1^2 + (x_2 + 1)^2 + 1$	$f_2 = \dfrac{(3x_1 - 2x_2 + 4)^2}{8} + \dfrac{(x_1 - x_2 + 1)^2}{27} + 15$
$f_3 = (x_1 - 1)^2 + x_2^2 + 2$	$f_3 = \dfrac{1}{x_1^2 + x_2^2 + 1} - 1.1\exp(-x_1^2 - x_2^2)$
where $-2 \le x_1, x_2 \le 2$	where $-3 \le x_1, x_2 \le 3$

3.1.3 Test Functions

Since we are reconstructing the experiments of the creators of DWA we are comparing the NSGA-II algorithm against the DWAGA on the same five test functions that were used by them in [3], which we similarly label F1 to F5 (three of them F2, F3 and F5 were also used in [6] and called T1 to T3).

In addition we are using an extra six multi-objective test functions that are used in test suites [6] and [7]. F6 – F8 corresponds to F3 – F5 as found in [7] and F9 to F11 corresponds to T4 – T6 as found in [6].

We then tried two tri-objective functions to see how the two algorithms scale, see Table 2 for function definitions.

3.1.4 Performance Measures

The performance of the EMOO systems is evaluated by examining the following measures as suggested by [8]: the spacing, diversity, coverage and execution time of the respective systems. Again, all measurement statistics are based on 30 repetitions.

Spacing is a measure of how evenly the solutions are spaced on the Pareto front. Each distance between neighbouring solutions is compared against the average of the distance between neighbours. If all solutions are evenly spaced, the measure will read 0, the more non-uniform the distribution along the Pareto front, the higher the number. The formula for Spacing is:

$$S = \sqrt{\frac{1}{n-1}\sum_{i=1}^{n-1}(d_i - \bar{d})^2} \qquad (5)$$

where d_i is the distance between two neighbouring solutions and $\bar{d}$ is the average distance between neighbours.

In the case of three objective problems the Pareto front is a plane instead of a line. As a result the distance there is measured between a solution and its closest neighbour.

Diversity is similar to Spacing, but instead of being based on the L_2-norm (associated with the Euclidean distance) it is based on the L_1-norm (associated with the Hamming distance). Also, Diversity is designed to take into account the full range of the Pareto front. With Spacing, the system could produce solutions that are evenly spaced but only cover a small section of the Pareto front, yet produce the same result as a system that evenly covers the entire Pareto front. Diversity compensates for this effect.

$$Diversity = \frac{d_f + d_l + \sum_{i=1}^{N-1} |d_i - \bar{d}|}{d_f + d_l + (N-1)\bar{d}} \tag{6}$$

Here d_f and d_l are the distances between the end points of the found Pareto front and the (known) extreme solutions of the true Pareto front. N is the size of the solution set.

In the case of three objective problems the corners of the Pareto front plane are taken as the extreme solutions.

Coverage of Two Sets: this measure compares the size of the Pareto front from one of the optimization techniques with the size of the Pareto front formed from the combined fronts of each of the two techniques.

$$Coverage\_1(\alpha) = \#(A \cap C) / \#(C) \tag{7}$$

$$Coverage\_2(\alpha) = \#(A \cap C) / \#(A) \tag{8}$$

$$Coverage\_3(\alpha, \beta) = (\#(A \cap C) - \#(A \cap B \cap C)) / \#(C) \tag{9}$$

where A is a Parcto front found by algorithm α, B is a Pareto front found by algorithm β, and C is a Pareto front formed when combining Pareto fronts A and B. Coverage_1(α) is the percentage of the combined Pareto front discovered by algorithm α and Coverage_2(α) is the percentage of the Pareto front discovered by α that is used in the combined Pareto front. Coverage_3(α, β) is the percentage of the combined Pareto front discovered by algorithm α that was not discovered by algorithm β.

Execution time: the time it took on the computer that executed the two algorithms. Both programs were written in Java and run on AMD Athlon XP 1800, with a CPU Clock speed of 1150Mhz and with 512MB of RAM DDR of memory.

Through experimentation it was discovered that the coverage-of-two-sets measurement was the most important measurement; often by itself it was informative enough to determine which algorithm is better. When the Coverage measurement did not indicate a clear winner, the diversity measurement was a good way of breaking the tie and determining the winner. When the diversity measurement did not indicate a clear winner, the spacing measurement was used to break the tie.

Finally, for statistical accuracy, all experiments have been run 30 times each for each setting, i.e. all statistics are based on 30 repetitions.

4   Results

4.1   Results When NSGA-II Is Victorious

The Coverage measurements indicate that for all these test functions the combined Pareto front consists entirely of the solutions found by NSGA-II algorithm (see Table 3). This clearly shows that DWA is inferior for these test functions. Since the performance difference on the coverage measurements between these two methods is so drastic, further measurements on diversity and spacing are not necessary.

Table 3. Coverage_1 and Coverage_2 measurements for the NSGA-II and DWAGA algorithms

	Coverage_1(DWA)				Coverage_1(NSGA)			
	avg	std	Conf Interval		avg	std	Conf Interval	
f2	0.0%	0.0%	0.0%	0.0%	100.0%	0.0%	100.0%	100.0%
f3	0.0%	0.0%	0.0%	0.0%	100.0%	0.0%	100.0%	100.0%
f4	0.1%	0.3%	-0.2%	0.3%	99.9%	0.3%	99.7%	100.2%
f5	0.0%	0.0%	0.0%	0.0%	100.0%	0.0%	100.0%	100.0%
f9	0.0%	0.0%	0.0%	0.0%	100.0%	0.0%	100.0%	100.0%
f10	7.7%	19.8%	-7.5%	22.9%	92.3%	19.8%	77.1%	107.5%
f11	0.0%	0.0%	0.0%	0.0%	100.0%	0.0%	100.0%	100.0%

	Coverage_2(DWA)				Coverage_2(NSGA)			
	avg	std	Conf Interval		avg	std	Conf Interval	
f2	0.0%	0.0%	0.0%	0.0%	100.0%	0.0%	100.0%	100.0%
f3	0.0%	0.0%	0.0%	0.0%	100.0%	0.0%	100.0%	100.0%
f4	0.1%	0.5%	-0.3%	0.5%	100.0%	0.0%	100.0%	100.0%
f5	0.0%	0.0%	0.0%	0.0%	100.0%	0.0%	100.0%	100.0%
f9	0.0%	0.0%	0.0%	0.0%	100.0%	0.0%	100.0%	100.0%
f10	11.2%	27.3%	-9.9%	32.2%	97.0%	16.3%	84.5%	109.6%
f11	0.0%	0.0%	0.0%	0.0%	100.0%	0.0%	100.0%	100.0%

avg = average   std= standard deviation   Conf Interval= Confidence Interval

4.2 Results When NSGA-II Is Challenged on Two Objective Problems

When the DWAGA and NSGA-II algorithms were tested on functions f1, f6, f7, and f8 it was observed that the NSGA-II was no longer a clear favorite and the DWAGA even had the superior performance on some test functions.

As the behavior of the algorithms on each of these four test functions is so diverse, each of the four test functions will be examined in detail one at a time.

4.2.1 F1 Comparison Results

For F1 the combined Pareto front consists half from DWA and half from NSGA-II. As can be seen from the confidence intervals for coverage in table 4 the NSGA-II slightly outperforms DWA, but since the difference is this small it is important to also evaluate Diversity and spacing in order to be sure which algorithm is better. It can be seen in Table 5 that NSGA-II is better in both spacing and diversity and as a result NSGA-II should be considered the better performer on F1 (but DWA is very close).

4.2.2 F6 Comparison Results

For F6 the combined Pareto front consists 1/3 from DWA and 2/3 from NSGA-II. As can be seen from the confidence intervals, the NSGA-II outperforms DWA in coverage, but it can be seen that DWA also contributes good solutions since 1/3 is a decent proportion, and so we evaluate diversity and spacing. In the Diversity and spacing the NSGA-II outperforms DWA. When these 3 measurements are considered together it is clearly seen that NSGA-II performs better.

Table 4. Comparing[2] the Coverage_1 and Coverage_2 measurements for the NSGA-II and DWAGA algorithms

	Coverage_1(DWA)				Coverage_1(NSGA)			
	avg	std	Conf Interval		avg	std	Conf Interval	
F1	46.9%	3.2%	3.2%	7.3%	53.1%	3.2%	50.6%	55.6%
F6	33.4%	3.5%	3.5%	6.3%	66.6%	3.5%	63.9%	69.4%
F7	69.1%	1.7%	67.8%	70.4%	30.9%	1.7%	29.6%	32.2%
F8	59.0%	21.3%	42.5%	75.4%	41.0%	21.3%	24.6%	57.5%

	Coverage_2(DWA)				Coverage_2(NSGA)			
	avg	std	Conf Interval		avg	std	Conf Interval	
F1	75.1%	7.3%	69.4%	80.7%	88.9%	3.6%	86.1%	91.7%
F6	44.2%	6.3%	39.3%	49.1%	92.4%	2.2%	90.7%	94.1%
F7	93.8%	2.2%	92.0%	95.5%	54.0%	4.0%	51.0%	57.1%
F8	71.1%	25.3%	51.7%	90.6%	66.2%	34.0%	40.0%	92.5%

Table 5. Comparing[3] the Spacing and diversity for the NSGA-II and DWAGA algorithms on four bi-objective functions

	Spacing						
	Rank (D)	Rank (N)	s	p-value	bonf corr. p-value	Better	Statistically Significant
f1	45.5	15.5	2.37	4.1E-19	1.9E-17	NSGA	Yes
f6	42.0	19.0	3.20	1.4E-09	6.8E-08	NSGA	Yes
f7	8.0	45.5	2.44	4.1E-22	2.0E-20	DWA	Yes
f8	42.9	17.5	2.64	1.3E-13	6.0E-12	NSGA	Yes
	Diversity						
	Rank (D)	Rank (N)	Pooled Std. Dev.	p-value	bonf corr p-value	Better	Statistically Significant
f1	45.5	15.5	2.27	4.1E-19	2.0E-17	NSGA	Yes
f6	45.5	15.5	3.20	3.2E-13	1.5E-11	NSGA	Yes
f7	25.5	27.5	2.44	4.1E-01	19.6	DWA	No
f8	15.8	45.2	2.64	5.3E-16	2.6E-14	DWA	Yes

4.2.3 F7 Comparison Results

For F7 the combined Pareto front consists 2/3 from DWA and 1/3 from NSGA-II. As can be seen from the confidence intervals for Coverage measure, this time the DWA outperforms NSGA-II. To be certain that DWA is in fact better than NSGA-II we first looked at Diversity, but since results of this test are inconclusive (the two algorithms can't be statistically differentiated based on this test), spacing becomes the determining factor. Here the results are in DWA favour. Based on these three measurements one can conclude that DWAGA is the better method for solving F7.

This is an important result for the research in DWA because F7 has a concave Pareto front. It has been assumed that DWA would have problems with solving this

---

[2] The confidence intervals are formed using the normal parametric approach as the results were found to be normally distributed when using a normality plot.

[3] The results were found to not be normally distributed, so the T test was done on the ranks (a non-parametric test).

type of a function but not only did it solve the problem well but it also outperformed NSGA-II.

4.2.4 F8 Comparison Results

For F8 the combined Pareto front consists 3/5 from DWA and 2/5 from NSGA-II. As can be seen from the confidence intervals for Coverage measure it is inconclusive which algorithm is better. The T-test in Table 6 confirms this. As a result we look at Diversity of the two methods where DWA outperforms NSGA-II. So, based on these measurements we conclude that DWAGA performs better than NSGA-II on F8.

Table 6. T Test[4] for looking in more detail if there is an advantage in coverage for DWAGA. It can be seen that it cannot be determined that DWA has better coverage than NSGA-II.

T Test on NSGA Coverage – DWA Coverage for f8			
α	0.01	Diff(f8)	0.179
No. of Ind. tests	48	pooled std	0.0551
α / 48(see footnote[5])	0.00021	conf. interval	-0.053
N	30		0.412
T	4.22	t-score	3.2578
		p-value	0.0019
		p-value * 48 (see footnote[3])	0.0902

4.3   Results When Run on 3 Objective Problems

The Coverage_1 measurements in Table 7 indicate that for functions F12 and F13, the combined Pareto front consists almost entirely of the solutions found by DWAGA algorithm while the NSGA-II had found a smaller part of the Pareto front. The Coverage_2 results indicate that both algorithms find same quality of solutions because almost all solutions found by each algorithm are used in the combined Pareto front. The Coverage_3 results indicate that the DWAGA has identified a large number of solutions that the NSGA-II was unable to find. The DWAGA managed to find almost all the solutions that NSGA-II identified plus many more. As a result the DWAGA provided a better and more detailed representation of the Pareto front and outperformed the NSGA-II.

As can be seen in Table 8, the DWAGA is executing much faster than NSGA-II, which is a big benefit with the huge search spaces that are associated with multi-objective problems.

This shows a possible deficiency in the Pareto front style approach. When a search space gets large, the NSGA seems to have trouble finding many solutions and is negatively impacted in its performance time. For example by switching from 2 objectives to 3, the Pareto front has changed from a line to a plane. As the number of objectives increases, the size of the Pareto front increase geometrically in the size.

---

[4] The regular T test was used as the results were found to be normally distributed when using a normality plot.

Table 7. The Coverage_1, Coverage_2, and Coverage_3 measurements for the NSGA-II and DWAGA algorithms on two tri-objective functions

	Coverage_1(DWA)			Coverage_1(NSGA)		
	avg	std	95% Confidence Interval	avg	std	95% Confidence Interval
F12	0.9729	0.0029	0.9719	0.9739	0.5203	0.00539	0.5182	0.5223
F13	0.9007	0.0821	0.8713	0.9300	0.7108	0.02205	0.7029	0.7187

	Coverage_2(DWA)			Coverage_2(NSGA)		
	avg	std	95% Confidence Interval	avg	std	95% Confidence Interval
F12	0.9997	0.0005	0.9995	0.9999	0.9791	0.0075	0.9765	0.9818
F13	0.9859	0.0110	0.9820	0.9898	0.9889	0.0044	0.9873	0.9905

	Coverage_3(DWA)			Coverage_3(NSGA)		
	avg	std	95% Confidence Interval	avg	std	95% Confidence Interval
F12	0.4797	0.0054	0.4777	0.4816	0.0271	0.0029	0.0261	0.0281
F13	0.2892	0.0220	0.2813	0.2971	0.0993	0.0821	0.0700	0.1287

Table 8. The algorithm run-time measurements for NSGA-II and DWAGA

	Time (DWA)			Time (NSGA)		
	avg	std	95% Confidence Interval	avg	std	95% Confidence Interval
F12	333009	151033	278962	387055	612279	5148	610437	614122
F13	442595	92429	409520	475670	4257111	88272	4225523	4288699

Consequently, to find this Pareto front, an algorithm must find a proportionately greater number of solutions. Since the NSGA-II stores the Pareto front solutions in its population it requires an geometrically larger population size because once a population member finds an optimal solution it will keep that solution to the end, especially with elitism. This causes more and more of the population members to be used for storing solutions instead of exploring. Eventually near the end of the run only few population members will remain free to explore. In order to have the NSGA-II be able to explore a large search space and be able to store solutions that represent it well, it will require the possession of a very large population. This will cause the algorithm to run slowly, due to the fact that it has to perform fitness calculations as well as the time taken sorting this huge population.

This problem does not apply to the DWAGA, which has an archive to store all the best solutions. It can have a smaller population, which can be used only for the searching of new solutions and not have to try to maintain all the best solutions. This allows the algorithm to identify a very large solution space with a relatively small population. This seems to allow the DWAGA to scale better than NSGA-II for problems with higher number of objectives.

5   Conclusion

In this paper, we compared two EMOO methods against each other: the Non-dominated Sorting Genetic Algorithm (NSGA-II) and the Dynamic Weighted Aggregation (DWA) system. To make the comparison fair and to remove an extra factor from the analysis, the DWA has been layered on top of a GA instead of and ES algorithm that it was originated for (since the DWA can be easily applied to any EC system). Using various traditional EMOO measures, such as Coverage, Spacing and Diversity, we determined that the DWA could handle concave problems as advertised. Furthermore, while most of the bi-objective functions we tried were better handled by the NSGA-II, when tri-objective problems were used, the DWAGA outperforms the NSGA-II and runs much faster. We believe that the cause of the DWAGA's success at higher number of objectives is due to its use of an archive, alleviating the need of the storage of the Pareto front (which can grow exponentially with the number of objectives) within the population itself.

Acknowledgements

The Authors would like to acknowledge the Natural Science and Engineering Research Council of Canada for support of this research.

References

[1] Coello Coello, C.A.: A Short Tutorial on Evolutionary Multiobjective Optimization. In Zitzler, E., Deb, K., et. al. (eds.): *The First International Conference on Evolutionary Multi-Criterion Optimization* (EMO 2001). Springer, Berlin (2001) 21-40

[2] Jin, Y., Okabe, T., Sendhoff., B.: Adapting Weighted Aggregation for Multiobjective Evolution Strategies. In Zitzler, E., Deb, K., et. al. (eds.): *The First International Conference on Evolutionary Multi-Criterion Optimization* (EMO 2001). Springer, Berlin (2001) 96-110

[3] Jin, Y., Okabe, T., Sendhoff., B.: Dynamic Weighted Aggregation for Evolutionary Multi-Ojbective Optimization: Why Does It Work and How? In Spector L., et. al. (eds.): *GECCO 2001 - Proceedings of the Genetic and Evolutionary Computation Conference.* Morgan Kaufmann, San Francisco (2001) 1042-1049

[4] Srinivas, N. and Deb, K.: Multi-Objective function optimization using non-dominated sorting genetic algorithms. *Evolutionary Computation*, 2(3):221–248 (1995)

[5] Deb, K., Goel, T.: Controlled Elitist Non-dominated Sorting Genetic Algorithms. In Zitzler, E., Deb, K., et. al. (eds.): *The First International Conference on Evolutionary Multi-Criterion Optimization* (EMO 2001). Springer, Berlin (2001) 67-81

[6] E. Zitzler, K. Deb, and L. Thiele. Comparison of multiobjective evolution algorithms: empirical results. *Evolutionary Computation*, 8(2):173-195 (2000).

[7] J.D. Knowles and D.W. Corne. Approximating the nondominated front using the Pareto archived evolution strategies. *Evolutionary Computation*, 8(2):149-172 (2000).

[8] Ang, K.H., Chong, G., Li, Y.: Preliminary Statement on the Current Progress of Multi-Objective Evolutionary Algorithm Performance Measurement. In Eberhart R., Fogel, D.B. (eds.): *Proceedings of the 2002 Congress on Evolutionary Computation* (CEC'02). IEEE Press (2002) 1139-1144

[9] Yaochu Jin, Tatsuya Okabe and Bernhard Sendhoff.   Solving Three-objective Optimization Problems Using Evolutionary Dynamic Weighted Aggregation: Results and Analysis. In: *Proceedings of Genetic and Evolutionary Computation Conference.* pp.636, Chicago, 2003

[10] Czyzak, P., and Jaszkiewicz, A.: Pareto simulated annealing - a metaheuristic technique for multiple-objective combinatorial optimization. *Journal of Multi-Criteria Decision Analysis* vol. 7, pp. 34-47 (1998).

[11] Hansen, P.H.: Tabu Search for Multiobjective Optimization: MOTS. *Proceedings of the 13th International Conference on Multiple Criteria Decision Making*, 1997.

[12] Serafini P.: Simulated annealing for multi objective optimization problems. In Multiple Criteria Decision Making: Expand and Enrich the Domains of Thinking and Application, G.H. Tzeng, ed., Springer (1993).

[13] Ulungu, E., Teghem, J., Fortemps, P., and Tuyytens, D.: MOSA Method: A Tool for Solving Multiobjective Combinatorial Optimization Problems. *Journal of Multi-Criteria Decision Analysis*, vol. 8/4, pp. 221-236 (1999).

[14] Joshua Knowles and David Corne: Memetic Algorithms for Multiobjective Optimization: Issues, Methods and Prospects. 2003

A Backbone-Based Co-evolutionary Heuristic for Partial MAX-SAT

Mohamed El Bachir Menaï[1] and Mohamed Batouche[2]

[1] Laboratoire d'Intelligence Artificielle, Université de Paris8,
2 rue de la liberté, 93526 Saint-Denis, France
menai@ai.univ-paris8.fr
[2] Laboratoire LIRE, Département d'Informatique,
Université Mentouri, 25000 Constantine, Algérie
batouche@wissal.dz

Abstract. The concept of backbone variables in the satisfiability problem has been recently introduced as a problem structure property and shown to influence its complexity. This suggests that the performance of stochastic local search algorithms for satisfiability problems can be improved by using backbone information. The Partial MAX-SAT Problem (PMSAT) is a variant of MAX-SAT which consists of two CNF formulas defined over the same variable set. Its solution must satisfy all clauses of the first formula and as many clauses in the second formula as possible. This study is concerned with the PMSAT solution in setting a co-evolutionary stochastic local search algorithm guided by an estimated backbone variables of the problem. The effectiveness of our algorithm is examined by computational experiments. Reported results for a number of PMSAT instances suggest that this approach can outperform state-of-the-art PMSAT techniques.

1 Introduction

Many problems in artificial intelligence (AI) and operations research (OR) are optimization problems, where the objective is to find a best assignment to a set of variables such that a set of constraints are satisfied. Real world problems found in application areas including scheduling [4] and pattern recognition [12] contain hard and soft constraints. Hard constraints must be satisfied by any solution, while soft constraints specify a function to be optimized. Various approaches have been proposed to represent over-constrained problems. Freuder and Wallace [12] presented the concept of partial constraint satisfaction, where the objective is to maximize the total number of satisfied constraints. Borning *et al.* [7] introduced the notion of constraint hierarchies, where the distinction between hard and soft constraints is extended to a multiple level constraint hierarchy.

Boolean satisfiability (SAT) is among the most interesting AI formalisms for reasoning, planning and learning [23]. The SAT problem asks to decide whether a given propositional formula, in conjunctive normal form (CNF), has a model. The maximum satisfiability (MAX-SAT) problem is the optimization version of SAT

E. Talbi et al. (Eds.): EA 2005, LNCS 3871, pp. 155–166, 2006.
© Springer-Verlag Berlin Heidelberg 2006

which consists to find an assignment maximizing the number of satisfied clauses. The weighted MAX-SAT is a more general case, where each clause is associated with a positive weight. The goal is to minimize the sum of weights of violated clauses. Problems involving hard and soft constraints can be naturally encoded as weighted MAX-SAT. Each hard constraint can be represented by a weighted cost which exceeds the sum of the weighted cost of all soft constraints. However, a solution for a MAX-SAT instance may violate some clauses whose satisfiability is a necessary condition for the feasibility of the real solution. For example, a MAX-SAT solution for a university time tabling may contain collisions of different courses in the same room at the same time, the same lecturer can be scheduled in different rooms at the same time, and so on. Cha *et al.* [8] introduced the Partial MAX-SAT (PMSAT) to formulate independently hard and soft constraints. Hard constraints are called *mandatory clauses*; their satisfiability is required for any PMSAT solution. Other related problems to PMSAT are the DISTANCE-SAT defined by Bailleux and Marquis [2], and the sub-SAT introduced by Xu *et al.* [24]. The DISTANCE-SAT problem asks to check if there is a model of a CNF formula, that conflicts with an expected configuration on at most a given number of variables. The sub-SAT is a formulation for relaxed Boolean satisfiability, that allows violation of a given number of clauses in a CNF formula.

The current research on algorithms used to solve PMSAT is limited. Cha *et al.* [8] used a weighting-type stochastic local search to solve PMSAT by repeating each mandatory clause n times. In this way, the search always prefers a solution which satisfies all mandatory clauses, regardless of the level of remaining clause violations. However, this can lead to an important increasing of the total number of clauses when their number is initially large. They applied various strategies to escape from local minima such as LWM, RESTART and RESET [8]. LWM strategy consists to add weights to all unsatisfied clauses, and to continue the search when a local minimum is reached. RESTART strategy allows the algorithm to restart from a random initial assignment, while RESET consists to reset the weights given by the algorithm and to continue the search from the current assignment. In the reported experimental study [8], RESET outperforms LWM and RESTART on random instances. In [14], a new approach for solving PMSAT is described. It is based mainly on recycling a model of the mandatory clauses to satisfy as many clauses in the second formula as possible. The reported results show the overall superiority of this method in comparison to a weighting-type local search algorithm. A problem of practical significance in the design of SAT and MAX-SAT solvers, is how to identify and exploit the problem structure properties to improve their performance. Some interesting properties which influence the hardness of a SAT have been identified such as the *easy-hard-easy* phase transition [9, 15], and the *backbone variables* [16], a set of literals which are true in every model. The backbone of a MAX-SAT instance is the set of assignments of values to variables which are the same in every possible optimal solution [20].

The aim of this paper is to integrate a backbone guide moves to a co-evolutionary stochastic local search algorithm for solving the PMSAT problem.

In a first phase, both formulas of a PMSAT instance are solved as a single MAX-SAT instance using a backbone guided co-evolutionary search. In a second phase, the best assignment found is recycled to satisfy all mandatory clauses using the estimated backbone. The effectiveness of this algorithm is demonstrated empirically on some PMSAT instances derived from standard SAT instances. In the reminder of this paper, we explain in more details the proposed method for PMSAT and report on results of computational tests in which our algorithm is compared to related approaches. In the next section, we describe a co-evolutionary method used for MAX-SAT (Bose-Einstein Extremal Optimization). In section 3, we present a brief review of backbone variables and related notions. In section 4, we formalize a new method for PMSAT. In section 5, we report on experimental results. We finally conclude and plan for future work in section 6.

2   Bose-Einstein Extremal Optimization Method for MAX-SAT

Bose-Einstein Extremal Optimization (BE-EO) [13] is an approximative algorithm for solving the MAX-SAT problem. It is based on an adaptation of Extremal Optimization (EO) [5] heuristic to MAX-SAT. The search space is explored according to EO, while starting solutions are sampled using the Bose-Einstein probability distribution.

Extremal Optimization method is introduced by Boettcher and Percus [5] for solving hard optimization problems such as the Graph Partitioning. It was motivated by the Bak-Sneppen [3] model of biological evolution which describes the co-evolutionary process of species. In this model, optimal adaptation emerges naturally from the dynamics of species by elimination of badly adapted ones. Species are sites of a lattice and each one has an associated fitness value ranging from 0 to 1. A fitness represents a time scale at which the species will mutate to a different species or become extinct. A selection process against the worst adapted species is applied. At each update, the smallest fitness value is replaced by a new random one which impacts the fitness values of its neighbors. After a certain number of steps, a state of optimal adaptation (*Self-Organized Criticality*) is reached in which all species are intimately connected. When the system is driven back to a SOC state, any perturbation of this equilibrium involves large fluctuations in the configuration of fitness values (*critical avalanches*). The duration t of these avalanches follows a power-law distribution $P(t) \propto t^{-\tau}$ (τ close to 1). Extremal Optimization method is a conversion of the extremal dynamics of the Bak-Sneppen model into an approximative algorithm for optimization problems. The search process is characterized by hill-climbing large fluctuations (i.e. avalanches in the Bak-Sneppen model) allowing search diversification. It evolves to a SOC state where sub-optimal solution can be found (almost all species have optimal fitnesses).

The Bose-Einstein distribution is a quantum distribution function. It describes the probability distribution of an amount of energy between identical but indistinguishable particles with integer spin, called *bosons* (e.g. photons).

Szedmak [21] proved that this distribution function can improve the performance of stochastic local search algorithms for satisfiability problems. He demonstrated that the mean Hamming distance between a sample of initial solutions and the optimal solution is reduced when initial solutions are generated using the Bose-Einstein distribution rather than the uniform one.

Given a MAX-SAT instance of n Boolean variables $x_1, \ldots, x_n$, and m weighted clauses $(c_i, w_i)_{i=1,m}$. Each clause c_i is a disjunction of literals (a variable x_i or its negation $\neg x_i$), and $w_i \in \mathbb{N}$ is its weight. A MAX-SAT instance is a conjunction of clauses (CNF formula). The fitness λ_i of a variable x_i is defined as the negation of the fraction of the sum of weights of unsatisfied clauses in which x_i appears, by the total weights of clauses connected to this variable :

$$\lambda_i = \frac{-\sum_{j=1}^{m} w_j |x_i \in c_j, \Im(c_j) = 0}{\sum_{k=1}^{m} w_k |x_i \in c_k} \tag{1}$$

$\Im(c_j) = 0$ means that the clause c_j is unsatisfied. The cost contribution of a variable x_i is defined by $-\lambda_i$. The best solution S found to a MAX-SAT instance is associated to the minimum of the cost function $C(S) = -\sum_{i=1}^{n} \lambda_i$.

The algorithm BE-EO for MAX-SAT is outlined as follows [13].

Algorithm BE-EO/MAX-SAT

1. Randomly generate a solution S according to the Bose-Einstein distribution. Set $S_{max} \leftarrow S$.
2. If S satisfies all the clauses of the MAX-SAT instance, return $(S : \text{model})$.
3. Evaluate λ_i for each variable x_i.
4. Rank $x_i, (i = 1, n)$ from the worst to the best according to λ_i. Select a rank j such that $P(j) \propto j^{-\tau}$.
5. Flip the truth value of x_j in S.
6. If $C(S) < C(S_{max})$ then set $S_{max} \leftarrow S$.
7. If the number of steps does not exceed the given bound, return to step 2.
8. If the number of generated Bose-Einstein initial solutions does not exceed the given sample size, then randomly generate a solution S according to the Bose-Einstein distribution. Return to step 2.
9. Return (S_{max}).

Good performance is reported for BE-EO/MAX-SAT on some specific classes of weighted and unweighed MAX-SAT instances [13] outperforming WalkSAT [18] and a tabu search method.

3   Backbone Variables

The *backbone* of a problem instance is a set of variables having fixed values in all optimal solutions. These variables are critically constrained as the elimination of any one of them will exclude any optimal solution. Related notions to backbone in satisfiability are *backdoors* [23] and *spine* [6]. A backdoor is a variable subset such that if some particular truth values are assigned to these variables, the simplified instance is satisfiable and can be solved in polynomial time. Williams *et al.* [23] demonstrated that a concrete computational advantage can be obtained by exploiting backdoors. The spine of a set of clauses is a set of literals which are false in all models of a subset of satisfiable clauses [20].

Several researches dealing with competitive SAT and MAX-SAT solvers have made use of backbone variables. Monasson *et al.* [16] investigated the backbones of 3-SAT and $(2 + p)$-SAT, and conjectured that a backbone is an order parameter for the decision problems. Other studies [17, 19, 1] have demonstrated that the size of the backbone is correlated with the hardness of SAT problems. Slaney and Walsh [20] have studied backbones in optimization and approximation problems including graph coloring, traveling salesperson problem, number partitioning and blocks word planning. They showed that backbones are often an important indicator of hardness in optimization and approximation. Subsequently, heuristic methods which identify backbone variables, may reduce problem difficulty and improve performance. Dubois and Dequen [11] proposed a systematic search method which incorporates estimated backbone variables. Telelis and Stamatopoulos [22] designed a method for generating initial assignments to an iterated algorithm by sampling heuristically the backbone variables, and reported good results on some random MAX-SAT instances. Climer and Zhang [10] developed a technique for identifying backbones and *fat variables* (variables which are absent from every optimal solution). They exploited it for discovering backbone and fat arcs for instances of the asymmetric traveling salesperson problem (ATSP) and achieved performance improvements. Zhang *et al.* [25] improved the performance of the well known WalkSAT procedure [18] on some instances of SAT and MAX-SAT from SATLIB [27] using structure information of reached local minima.

4   Backbone-Based Co-evolutionary Heuristic for PMSAT

Given two CNF formulas f_A and f_B over a set of variables $X = \{x_1, \ldots, x_n\}$. The PMSAT problem $P = f_A \wedge f_B$ asks to satisfy all the clauses of f_A and as many clauses in f_B as possible. The number of satisfied clauses in f_B determines the quality of a solution to P.

We propose a two-phase algorithm for solving P. In a first phase, P is considered as a MAX-SAT instance and approximated using a variant of the algorithm BE-EO/MAX-SAT. A backbone variables sampling is integrated to BE-EO/MAX-SAT to guide the search towards potentially good solutions. If the best solution found S_{AB} does not satisfy f_A, then a second phase is performed to recycle S_{AB} to a model of f_A using the backbone information captured in the first phase. The backbone sampling may help to improve the performance of the second phase process, as it encapsulates information about the likelihood of each variable. However, exact backbone cannot be computed unless all optimal solutions are known. Hence, only an estimated pseudo-backbone is performed using information extracted from reached local minima.

The pseudo-backbone sampling used in this work is inspired by the sampling scheme presented in [22]. Let Ω be a set of solutions on X. $S(x_i)$ denotes the truth value of x_i in the solution S. A variable frequency of positive occurrences of x_i in all solutions of Ω, is defined by :

$$p_i = \frac{\sum_{S \in \Omega} S(x_i)}{|\Omega|} \tag{2}$$

assuming that all local minima are of equal quality. Else a weight cost may be assigned to each local minimum. $Q(S)$ denotes the contribution of a solution S, defined as the total number of satisfied clauses in f_A and f_B. A multiplier coefficient, equals to $|f_A|$, is added to $Q(S)$ to underline the priority of satisfying clauses of f_A. Let $\#sat_{f_A}(S)$ and $\#sat_{f_B}(S)$ be the number of satisfied clauses by S in f_A and f_B, respectively. $Q(S)$ is defined by :

$$Q(S) = |f_A| \cdot \#sat_{f_A}(S) + \#sat_{f_B}(S) \tag{3}$$

A more reliable definition of $p_i (i = 1, n)$ is given by :

$$p_i = \frac{\sum_{S \in \Omega} Q(S) \cdot S(x_i)}{\sum_{S \in \Omega} Q(S)} \tag{4}$$

Let X_α denotes the set of variables which appear in the set of clauses α. The main steps of the algorithm, called BBC-PMSAT, are described as follows.

Algorithm BBC-PMSAT

Phase 1: Solving $P = f_A \wedge f_B$ as a MAX-SAT instance

(a) Run BE-EO/MAX-SAT on P over X. Initialize Ω with reached local minima.

(b) Solve P using a variant of BE-EO/MAX-SAT (initial solutions are generated from Ω using variable frequencies p_i (Eqn. 4)).
    At a new local minimum S, if Ω holds a solution S^* such that $Q(S^*) < Q(S)$ (Eqn. 3), then replace S^* by S in Ω.

(c) Let S_{AB} be the best solution found after a preset number of steps. If S_{AB} satisfies f_A then return S_{AB} as a solution to P.

Phase 2: Recycling S_{AB} to satisfy f_A

(a) Let $f_A = f_{A_1} \wedge f_{A_2}$, where f_{A_1} is satisfied by S_{AB} and $X_{A_1} \cap X_{A_2} = \emptyset$ (simplification).

(b) Solve f_{A_2} over X_{A_2} as a SAT instance using a variant of the BE-EO/SAT (Phase 1, step (b)). Partial assignments to the variables of X_{A_2} are generated using variable frequencies p_i.

(c) After a preset number of steps, if a model S_{A_2} is found, then update S_{AB} and return it as a solution to P. Else return that no solution to P can be found.

5   Performance Evaluation

Since no public PMSAT instances are available, we generated them using SAT instances from DIMACS [26] and SATLIB [27] benchmark archives. We considered four sets of random and structured SAT instances of n variables and m clauses:

- uuf125-538* (100 random "phase-transition" hard 3-SAT instances of $n = 125$ and $m = 538$);
- f* (3 large random "phase-transition" hard 3-SAT instances: f600 ($n = 600$, $m = 2550$), f1000 ($n = 1000, m = 4250$), f2000($n = 2000, m = 8500$);
- par8-* (5 instances of SAT-encoded parity learning problem of $n = 350$ and $1149 < m < 1171$);
- flat* (10 instances of SAT-encoded graph coloring problem of $n = 300$ and $m = 1117$).

All SAT instances are chosen satisfiable in order to guarantee the generation of solvable PMSAT instances (f_A must be satisfiable). Random SAT instances are "phase-transition" hard ($\frac{m}{n} \simeq 4.3$ for random 3-SAT instances). Structured instances par8-* are also among the hardest DIMACS SAT ones. Random instances are generally used to control average problem difficulty by varying the ratio $\left(\frac{m}{n}\right)$ of clauses to variables, while structured instances are used to measure the effect of hidden structure on algorithm performance. PMSAT instances are generated using a partition of each SAT instance into two subsets F_A and F_B (representing f_A and f_B formulas, resp.) such that $|F_A| = \lceil \alpha m \rceil + 1$ and $|F_B| = m - |F_A|$, with $0 < \alpha < 1$. The program code is written in C and run on a computer (Pentium IV 2.9 GHz with 1 GBs of RAM) running Linux. BE-EO/MAX-SAT is run setting $\tau = 1.4$. All the results are averaged over 10 runs on each instance with a maximum of 300000 flips allowed per run. The total number of tries for each run of the algorithm BBC-PMSAT is shared between both phases of the algorithm. Let r be the first phase run length ratio of the total run length, #sat the number of solutions to PMSAT instances over 10 runs (it equals the average number of satisfied f_A instances) and v the relative error of a solution S given by:

$$v(\%) = \left(1 - \frac{\#sat_{f_B}(S)}{|F_B|}\right) \times 100 \qquad (5)$$

A key question regarding the algorithm BBC-PMSAT is how to evaluate its performance. The first objective is to determine the effect of the first phase run length ratio of the total run length. The second objective is to determine the impact of the pseudo-backbone variables size on the performance. The third objective is to determine whether or not BBC-PMSAT is competitive with its variant, called C-PMSAT, which does not integrate pseudo-backbone sampling. Additionally, BBC-PMSAT is compared to a weighting-type local search algorithm, called WLS, used by Cha et al. [8] with RESET strategy to solve PMSAT

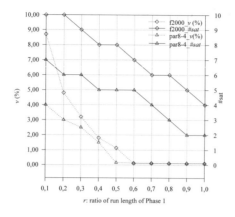

Fig. 1. Average error v (y-axis) and number of solutions $\#sat$ (additional y-axis) over 10 runs for instances f2000 and par8-4 ($\alpha = 0.3$) are plotted against the run length ratio of phase 1

instances. WLS/RESET solves PMSAT as MAX-SAT instance by repeating each clause in F_A, $|F_A|$ times.

Figure 1 presents average $\#sat$ and v over 10 runs obtained by BBC-PMSAT on the instances f2000 and par8-4, varying r from 10% to 100%. We observe clearly that the greater the value of r, the more the number of solutions $\#sat$ and the error v are reduced. An error v less than 1% is achieved after at least 50% of the total runtime length. However, allowing much more time to the first phase of the algorithm, means reducing the amount of time allowed to the second phase. Hence, the number of solutions $\#sat$ to PMSAT may decrease. For all

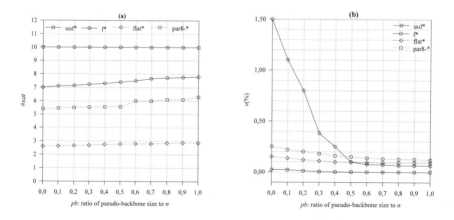

Fig. 2. ($\alpha = 0.3, r = 0.6$) **(a)** Average number of solutions $\#sat$ (y-axis) is plotted against the ratio of the pseudo-backbone size to the number n of variables for each instance class. **(b)** Average error v (y-axis) is plotted against the ratio of the pseudo-backbone sample size to the number n of variables for each instance class.

the instances, the average best performance in terms of average #sat and v is obtained with r ranging from 50% to 70%.

Figure 2 shows the number of solutions #sat and the average error v achieved by BBC-PMSAT on all the instances when varying pb, the ratio of backbone size to the number of variables n, from 0 to n ($\alpha = 0.3, r = 0.6$). As illustrated in figure 2.a, the number of solutions #sat is generally increasing with pb. It is constant in the case of uuf* which may have few backbone variables. For example, setting pb to $0.5n$, the gains achieved by BBC-PMSAT in terms of #sat on the instances f*, flat* and par8-* are 5.71%, 6.92% and 2.96%, respectively. For $pb = n$, the gains achieved on the same instances are 11.42%, 10.38% and 16.66%, respectively. Figure 2.b shows a fall in the average error v for all the instances: v decreases rapidly for the instances f*, until $pb = 0.5n$; v decreases relatively slowly for the remaining instances. For all the instances, the average best performance, in terms of quality of a solution (v), is obtained when $pb \geq 0.6n$. BBC-PMSAT performs

Table 1. Results of the algorithm BBC-PMSAT ($\alpha = 0.3, r = 0.6, pb = 0.7n$)

Instance	#sat	$v(\%)$	CPU time		Flips	
			Avg	Std	Avg	Std
uuf*	10	0	0.374	0.050	3881.9	556.0
f600	9	0.0113	6.092	1.815	28958.2	4623.1
f1000	7	0.0951	19.812	3.152	79575.1	8768.4
f2000	7	0.1135	44.655	3.547	138120.9	9868.5
flat*	2.8	0.0923	3.238	0.185	26926.8	1601.4
par8-1	7	0.0713	5.117	1.350	34877.1	9215.0
par8-2	7	0.1501	5.723	0.702	39681.0	4886.0
par8-3	6	0.1185	4.931	1.050	31657.5	7325.1
par8-4	5	0.1290	4.581	0.900	31917.0	6513.0
par8-5	5	0.1870	7.960	0.841	61275.4	4935.0
Average	6.58	0.0968	10.248	1.359	47687.0	5829.1

Table 2. Results of the algorithm C-PMSAT ($\alpha = 0.3, r = 0.6, pb = 0.7n$)

Instance	#sat	$v(\%)$	CPU time		Flips	
			Avg	Std	Avg	Std
uuf*	10	0.0231	0.514	0.200	6162.0	971.3
f600	9	0.0130	10.973	2.532	57457.3	7929.7
f1000	6	1.9150	22.650	1.650	115059.2	7021.6
f2000	6	2.5710	54.458	7.940	189310.3	35641.8
flat*	2.6	0.1504	4.154	1.201	32795.1	5831.0
par8-1	6	0.3511	7.516	0.324	60513.3	1472.1
par8-2	6	0.2510	7.380	0.508	58764.0	2410.5
par8-3	6	0.2315	5.763	1.810	44210.4	8355.0
par8-4	4	0.1712	4.716	0.152	36638.1	790.0
par8-5	5	0.2415	10.550	0.380	71652.1	1582.0
Average	6.06	0.5918	12.867	1.669	67256.1	7200.5

Table 3. Results of the algorithm WLS/RESET

Instance	#sat	$v(\%)$	CPU time		Flips	
			Avg	Std	Avg	Std
uuf*	9.8	0.0123	0.845	0.205	8316.4	1485.0
f600	9	0.0130	10.620	2.010	47150.6	18420.0
f1000	6	0.1235	29.271	2.055	136171.0	19040.0
f2000	5	1.5764	60.028	5.601	265124.5	41190.0
flat*	2.1	0.2351	6.068	1.295	43166.3	9147.6
par8-1	4	0.3133	10.408	0.520	79525.1	3290.0
par8-2	4	0.4840	9.120	0.180	70720.0	1123.1
par8-3	3	0.3586	7.151	0.642	53540.2	4058.0
par8-4	1	0.6315	5.154	0.210	39628.0	1620.5
par8-5	4	0.4099	10.250	1.261	73040.2	7850.0
Average	4.79	0.4158	14.891	1.397	81638.2	10722.4

particularly well on the instances par8-* which may have a large backbone size, making all the variables critically constrained.

Computational results performed by BBC-PMSAT, C-PMSAT and WLS/RESET are presented in tables 1, 2 and 3, respectively. The first column lists the benchmarks. Columns 2, 3 show the average number of solutions #sat and the average error v over 10 runs. Columns 4, 5 show the average CPU time and its standard deviation. Columns 6, 7 show the average number of flips and its standard deviation. BBC-PMSAT is tested using $\alpha = 0.3$, $r = 0.6$ and $pb = 0.7n$. Overall, BBC-PMSAT outperforms C-PMSAT and WLS/RESET on all the instances. The average gains in number of solutions are 8.58% and 37.36% w.r.t. C-PMSAT and WLS/RESET, respectively. In term of runtime cost, the average falls are 20.35% and 31.17% w.r.t. C-PMSAT and WLS/RESET, respectively. In conclusion, our results demonstrate that BBC-PMSAT can find high quality solution and performs faster than C-PMSAT and WLS/RESET.

6 Conclusion and Future Work

In this work, we introduced a backbone-based co-evolutionary algorithm for PM-SAT (BBC-PMSAT). This algorithm is based on a co-evolutionary stochastic local search method (BE-EO) which has been used successfully for solving a range of MAX-SAT instances. BBC-PMSAT approximates solutions to PMSAT in two main phases. In a first phase, PMSAT is solved as a MAX-SAT instance incorporating sampled pseudo-backbone variables to guide the search. In a second phase, the previously found solution is recycled to satisfy all the constrained clauses using estimated pseudo-backbone. BBC-PMSAT was compared to its variant without pseudo-backbone sampling (C-PMSAT) and to a weighting-type stochastic local search algorithm with RESET strategy (WLS/RESET) [8] for PMSAT. These algorithms were tested on four classes of PMSAT instances generated from standard SAT instances. The results indicate the effectiveness of using estimated pseudo-backbone variables. Indeed, BBC-PMSAT outperforms

C-PMSAT and WLS/RESET on all the instances in terms of average number of solutions and average runtime cost. The most significant gains are achieved on instances which may have large backbone size. The encouraging results obtained at this early stage, prove the high potential of this method. In future work, we plan to further investigate how the performance of BBC-PMSAT depends on the problem features and to continue computational tests on larger PMSAT instances.

References

[1] Achlioptas, D., Gomes, C., Kautz, H., Selman, B.: Generating satisfiable problem instances. In Proceedings of the 17th National Conference on Artificial Intelligence (AAAI-00), (2000) 256–261

[2] Bailleux, O., Marquis, P.: DISTANCE-SAT: complexity and algorithms. In Proceedings of the 16th National Conference on Artificial Intelligence (AAAI-99), (1999) 642–647

[3] Bak, P., Sneppen, K.: Punctuated equilibrium and criticality in a simple model of evolution. Physical Review Letters, 59, (1993) 381–384

[4] Beck, J.C., Fox, M.S.: A generic framework for constraint-directed search and scheduling. AI Magazine, 19(4), (1998) 101–130

[5] Boettcher, S., Percus, A.G.: Nature's way of optimizing. Artificial Intelligence, 119, (2000) 275–286

[6] Bollobas, B., Borgs, C., Chayes, J., Kim, J.H., Wilson, D.B.: The scaling window of the 2-SAT transition. Random Structures and Algorithms, (2001) 201–256

[7] Borning, A., Freeman-Benson, B., Wilson, M.: Constraint hierarchies. Lisp and Symbolic Computation, 5(3), (1992) 223–270

[8] Cha, B., Iwama, K., Kambayashi, Y., Miyasaki, S.: Local search for Partial MAX-SAT. In Proceedings of the 14th National Conference on Artificial Intelligence (AAAI-97), (1997) 263–265

[9] Cheesman, P., Kanefsky, B., Taylor, W.M.: Where the really hard problems are. In Proceedings of the 12th International Joint Conference on Artificial Intelligence (IJCAI-91), 331–337

[10] Climer, S., Zhang, W.: Searching for backbones and fat: a limit-crossing approach with applications. In Proceedings of the 18th National Conference on Artificial Intelligence (AAAI-02), (2002) 707–712

[11] Dubois, O., Dequen, G.: A backbone-search heuristic for efficient solving of hard 3-SAT formulæ. In Proceedings of the 17th International Joint Conference on Artificial Intelligence (IJCAI-01), (2001) 248–253

[12] Freuder, E., Wallace, R.: Partial constraint satisfaction. Artificial Intelligence, 58(1), (1992) 21–70

[13] Menaï, M.B., Batouche, M.: Efficient initial solution to extremal optimization algorithm for weighted MAXSAT problem. In Proceedings of the 16th International Conference on Industrial and Engineering Applications of Artificial Intelligence and Expert Systems (IEA/AIE-2003), LNAI 2718, Springer, (2003) 592–603

[14] Menaï, M.B.: Solution reuse in Partial MAX-SAT Problem. In Proceedings of IEEE International Conference on Information Reuse and Integration (IRI-2004), (2004) 481–486

[15] Mitchell, D., Selman, B., Levesque, H.: Hard and easy distributions of SAT problems. In Proceedings of the 10th National Conference on Artificial Intelligence (AAAI-92), (1992) 459–465

[16] Monasson, R., Zecchina, R., Kirkpatrick, S., Selman, B., Troyansky, L.: Determining computational complexity from characteristic 'Phase Transition'. Nature, 400 (1999) 133–137

[17] Parkes, A.J.: Clustering at the phase transition. In Proceedings of the 14th National Conference on Artificial Intelligence (AAAI-97), (1997) 240–245

[18] Selman, B., Kautz, H.A., Cohen, B.: Noise strategies for improving local search. In Proceedings of the 12th National Conference on Artificial Intelligence (AAAI-94), (1994) 337–343

[19] Singer, J., Gent, I.P., Smaill, A.: Backbone fragility and the local search cost peak. Journal of Artificial Intelligence Research, 12, (2000), 235–270

[20] Slaney, J., Walsh, T.: Backbones in optimization and approximation. In Proceedings of the 17th International Joint Conference on Artificial Intelligence (IJCAI-01), (2001) 254–259

[21] Szedmak, S.: How to find more efficient initial solutions for searching. RUTCOR Research Report 49-2001, Rutgers Center for Operations Research, Rutgers University, Piscataway, NJ, USA, (2001)

[22] Telelis, O., Stamatopoulos, P.: Heuristic backbone sampling for maximum satisfiability. In Proceedings of the 2nd Hellenic Conference on Artificial Intelligence, (2002) 129–139

[23] Williams, R., Gomes, C., Selman, B.: Backdoors to typical case complexity. In Proceedings of the 18th International Joint Conference on Artificial Intelligence (IJCAI-03), (2003) 1173–1178

[24] Xu, H., Rutenbar, R.A., Sakallah, K.: sub-SAT: A formulation for relaxed boolean satisfiability with applications in routing. In Proceedings of the International Symposium on Physical Design (ISPD-02), (2002) 182–187

[25] Zhang, W., Rangan, A., Looks, M.: Backbone guided local search for maximum satisfiability. In Proceedings of the 18th International Joint Conference on Artificial Intelligence (IJCAI-03), (2003) 1179–1186

[26] http://dimacs.rutgers.edu/Challenges/

[27] http://www.informatik.tudarmstadt.de/AI/SATLIB

Analysing Co-evolution Among Artificial 3D Creatures

Thomas Miconi and Alastair Channon

University of Birmingham,
Edgbaston B152TT,
Birmingham, UK
t.miconi@cs.bham.ac.uk

Abstract. This paper is concerned with the analysis of coevolutionary dynamics among 3D artificial creatures, similar to those introduced by Sims [1]. Coevolution is subject to complex dynamics which are notoriously difficult to analyse. We introduce an improved analysis method based on Master Tournament matrices [2], which we argue is both less costly to compute and more informative than the original method. Based on visible features of the resulting graphs, we can identify particular trends and incidents in the dynamics of coevolution and look for their causes. Finally, considering that coevolutionary progress is not necessarily identical to global overall progress, we extend this analysis by cross-validating individuals from different evolutionary runs, which we argue is more appropriate than single-record analysis method for evaluating the global performance of individuals.

1 Introduction

Coevolution has been introduced in artificial evolution as an alternative to traditional evolutionary methods based on fixed, explicitly defined fitness functions such as the genetic algorithm. The use of coevolutionary methods is based on the assumption that constant mutual adaptation between evolving individuals will lead to ever-increasing levels of fitness. This assumption of progress through mutual adaptation is the basis for *arms race* hypothesis [3]. Rosin & Belew [4] summarise the transposition of the "arms race" concept to artificial evolution:

> Since the parasites are also evolving with a fitness based on a competition's outcome, the success of a host implies failure for its parasites. When the parasites evolve to overcome this failure, they create new challenges for the hosts; the continuation of this may lead to an evolutionary "arms race" (...) New parasite types should serve as a drive toward further innovation, creating ever-greater levels of complexity and performance by forcing hosts to respond to a wider range of more challenging parasite test cases.

The assumption which underlies artificial coevolution, therefore, can be stated as follows: coevolution is expected to lead to an "arms race" (formally defined as

E. Talbi et al. (Eds.): EA 2005, LNCS 3871, pp. 167–178, 2006.

a sequence in which newer individuals consistently outperform their ancestors), which is expected to result in superior individuals. Unfortunately the fundamentally local nature of natural selection (based on differential gene propagation within a given, current environment which is local both in space and time) means that several problems may hinder this intuitive mechanism.

First, it is well-known that the "arms race" metaphor begs the question of *intransitivity* in the global fitness landscape: if an organism A can be said to be superior to B, and B is superior to C, it is not necessarily the case that A should always be superior to C. This may lead to the appearance of "cycles" [2] [5] or "circularities" [6] in the dynamics of evolution.

Moreover, the arms race concept refers to a historical progress, in which newer individuals outperform ancestral ones against their ancestral opponents: performance and progress are evaluated against the history of a particular evolutionary trajectory. However, such a progress is not necessarily related to global, overall progress towards superior individuals in the wider context of the whole search space. Nolfi & Floreano [2] have shown that these two notions of progress are not as correlated as it may seem. They performed two coevolutionary experiments based on a predator-prey scenario, with one important difference: in one run, coevolution occurred in a straightforward manner, by pitting individuals of a given generation against the champion of the previous generation (a method inspired by Sims [1], which we also use in the present article). In the other run, however, individuals of a given generation were evaluated not only against the current opposing champion, but also against the *previous* champions of the opponent population, following the "Hall of Fame" technique suggested by Rosin & Belew [4]. Unsurprisingly, the second type of experiment led to a more robust arms race, in that newer individuals were significantly better at outperforming their own ancestors. However, in some circumstances, when the authors compared the results of coevolution with a Hall of Fame against "naked" coevolution, they found that individuals evolved using the Hall of Fame were defeated by individuals evolved without it. While progress had been more straightforward and unambiguous, it had also been more limited in scope. This difference between historical progress with regard to a given evolutionary history, and overall superiority, is an important topic in this article.

2   Monitoring and Analysis of Coevolution

If progress can occur in competitive coevolution, it is important that it be properly detected. Several types of statistics have been proposed for analysing the results of coevolutionary processes, with a stress on the identification of progress.

First, Cliff & Miller's "Current Individual vs. Ancestral Opponents" method (CIAO) [7] and Nolfi & Floreano's "Master Tournament" method [2] both pit the champions of each generation against each other, and displaying the result as a grid of coloured dots, in such a way that dot (n, m) is coloured if the champion of generation n in one population defeats the opposing champion of generation m, and left blank otherwise. CIAO pits the champion of a population at generation

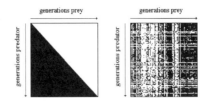

Fig. 1. Master Tournament matrices, taken from Nolfi & Floreano [2] in a predator-prey experiment. Black (resp. white) dots indicate a victory for the champion of the predator (resp. prey) population. The picture of the left represents an "ideal" situation of perfect progress, in which each champion is able to defeat all previous opponents. The picture on the right represents the results of a real experiment.

n against the champions of each *previous* generation in the opposing population, thus resulting in a triangle of dots. Master Tournament performs general confrontation between all champions of all generations, producing a square matrix of dots. The Master Tournament square can be seen as the collation of two CIAO triangles (one for each population) joined along their common hypotenuse, the diagonal of the square.

These methods have the advantage of providing reasonably complete information about an entire evolutionary run. However this completeness comes at a price. An obvious problem with these methods is their combinatorial complexity. Since N^2 evaluations are needed to obtain a complete table ($(N^2 - N)/2$ in the single-population case), as soon as N becomes even moderately large, calculating the figure is a time-consuming process. Of more concern to the analyst is the fact that the resulting figures are often somewhat obscure: although "ideal" conditions of progress lead to a very simple figure, these ideal conditions are rarely met in practice. Real experiments often produce disorderly arrangements of dots from which it may be difficult to extract any meaning at all.

A more recent technique for observing progress in coevolution has been proposed by Stanley & Miikkulainen under the name of Dominance Tournament [6]. Dominance Tournament was developed for single-population coevolution, but can be readily extended to multiple populations. In a dominance tournament analysis, one must keep track of every new individual that defeats all previously dominant individuals. Dominance is defined recursively: The first dominant strategy d_1 is the champion of the first generation; then, at every generation, the current champion becomes the new dominant strategy d_i if it can defeat all previous dominant strategies $d_{j<i}$. When two populations coevolve against each other, the method is adapted by specifying that a new dominant strategy must be able to defeat all dominant strategies from the other population.

Thus the dominance tournament method concentrates on a sequence of individuals which are seen as particularly important, due to their recursive superiority relationship. Dominance, in this context, is not synonymous with absolute superiority: some earlier individuals may be able to defeat the current dominant strategy. However such individual are seen as "idiosyncratic strategies", similar to parasites specialised against a (supposedly superior) host.

Dominance Tournament has the advantage of being much easier to compute than Master Tournament, since at any time the total number of dominant strategies against which candidates are to be tested is significantly lower than the total number of generation champions. It is also much easier to analyse, since it can be represented as a one-dimensional series of ticks along a time-coordinate axis, each tick corresponding to the appearance of a new dominant strategies. However, the massive simplification of the statistics eliminates a lot of information, and it is not clear exactly how precisely the Dominance Tournament captures the global trajectory of a given run.

Finally, both types of method must be applied to the history of a particular run: they essentially rely on "single record" analysis. They are useful in studying the trajectory of evolution and the presence (or absence) of coevolutionary progress. However it would be quite wrong to deduce anything from them about general progress in the sense of overall superiority over the whole search space.

3 Artificial Creatures

In the following sections we describe our own model for the evolution of artificial creatures in a physically realistic 3D environment. This model is broadly similar to the one introduced by Sims [1]. Besides minor technical modification, the most important difference between our system and Sims' is that our creatures are controlled by standard neural networks, based on classical McCulloch & Pitts neurons with sigmoid or radial activation functions, in contrast to Sims' creature which were controlled by functional networks, including arithmetic functions, tunable oscillators and logic operators (among others) as elementary building blocks. A complete description of (and justification for) the system can be found in a previous publication [8].

Morphology: As in Sims' model, the creatures are branching structures composed of rigid 3D blocks. The blocks (or "limbs") are connected to their parent limb by a hinge joint, except for the first ("root") limb. The genetic specification of a creature is given as a tree of nodes. Each of these nodes contain morphologic and neural information about one limb. Each node is responsible for storing the description of its limb's physical connection with its parent node's limb. The morphologic information in each genetic node specifies the dimensions of the limb (width, length and height), the orientation of this limb with regard to its parent (in the form of two parameters indicating polar angles with the xz and the xy planes, that is longitude and latitude, in the frame of reference of the parent limb), the direction of movement which may be either vertical or horizontal (that is aligned either with the y or with the z axis of the limb), and a boolean flag for reflection which governs symmetric replication along the xz plane of its parent. A limb also contains neural information, as described in the following paragraphs.

Creature control and neural organisation: Our creatures are controlled by neural networks. As in Sims' model, each limb contains a set of neurons. Genetic infor-

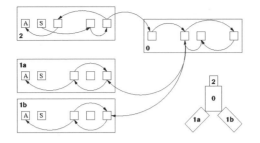

Fig. 2. Organisation of a fictional creature pictured in the bottom-right corner. Limb 0 has no sensor (S) or actuator (A). Limb 1 is reflected into two symmetric limbs 1a and 1b, which share the same morphologic and neural information.

mation about a given neuron specifies the *activation function* for this neuron, a threshold/bias parameter θ, and connection information. The activation function may be either a sigmoid ($\frac{1}{1+\exp^{-(\sigma+\theta)}}$) or the hyperbolic tangent $\tanh(\sigma + \theta)$ where σ is the weighted sum of inputs. The main difference between sigmoid and tanh is that the first has values in $[0, 1]$ while the latter has values in $[-1, 1]$. Connection information specifies, for each connection, the source of this connection (that is the neuron whose output is received through this connection) and a weight value. Neurons can only be connected with other neurons from the same limb, from adjacent limbs, or from the root limb. Each neuron may receive a variable number of connections, up to a maximum value (3 in the present experiments).

Sensor neurons and actuator neurons are handled specially. The first type of sensor neuron is a proprioceptive neuron, which measures the current angle formed by the hinge joint to which this neuron's limb is attached, scaled within the $[-1, 1]$ range. Two other types of sensors exist, each of them measuring respectively the x and y coordinates of the centre of a specific object (an inert cube) in the frame of reference of the limb, squashed through the tanh function. Every limb has exactly one proprioceptor, and may have any number of other sensors (within the maximum number of neurons for each limb). Actuator neurons command the movement of each limb, that is, its rotation around its joint. The output of an actuator indicates the desired angular velocity at this joint. Their inputs are defined similarly as other neurons, but their activation function is always a scaled hyperbolic tangent of the form $\tanh(\sigma + threshold)$. Each limb has exactly one actuator.

Expression of the genome: The creatures are constructed according to the information contained in the genetic nodes. A very simple developmental system translates the genotype into a corresponding phenotype, and may introduce additional complexity if the genetic information dictates it. Our system uses only one developmental feature, adapted from Sims: bilateral symmetry. In our model, each genetic node (corresponding to a limb) may possess a "reflection" flag, which means that when this node is read and the corresponding limb attached to its parent, a symmetric copy of this limb will also be created. Any further sub-limbs will similarly be duplicated in a symmetric fashion, which leads to the appearance of bilaterally symmetric branches. Our present design allows for only one type of symmetry, namely symmetry along the parent's xz plane. When a given limb is randomly generated, its reflection flag is set with probability P_{ref} (for this paper, $P_{ref} = 0.1$).

Genetic operators: We use three genetic operators, broadly similar to those used by Sims. *Crossover* is performed by simply aligning the genetic nodes of both parents in two rows, then building a new list of genetic nodes by concatenating the left part of one parent with the right part of the other. *Grafting* corresponds to the removal of a branch (that is a limb and all its sub-limbs), and its replacement by a branch taken from another individual. Connectivity information is adapted and maintained: the neurons of the trunk establish the same connections with the new branch as they had with the old one, and similarly the new branch has the same connection with its new trunk as it had with its previous trunk. *Mutation* occurs by sequentially modifying each parameter within a genome (from limb size to connection weight) with a given probability P_{mut}, and also removing a limb and adding a new, randomly generated limb, also with probability P_{mut} (in this paper, $P_{mut} = 0.04$).

4   Experiments and Results

4.1   The Evolutionary Algorithm

We use the same task as Sims [1]: two creatures compete for control of a single cube. The cube is placed in the center of the world, and the creatures start on each side of the cube. After a fixed amount of time has elapsed, distances d_1 and d_2 between the centre of the root limb of each competitor and the centre of the cube are computed. The score of each contestant is the difference between these distances, $d_1 - d_2$ for competitor 1 and $d_2 - d_1$ for competitor 2. Lower score correspond to superior creatures.

The evolutionary algorithm is also similar to Sims'. For every run, creatures are divided into two populations. At every generation, creatures of each population are evaluated against the current champion of the opposing population. The creature which obtains the best score becomes the new champion of this population. Survival rate is 50%, which means that half the population is replaced at every generation. Selection of parents occurs by direct tournament selection based on score. New individuals are created with equal probability by one of three operations: grafting between the two individuals, crossover between the two parents, or three successive applications of the mutation operator to one of the parents. Then the mutation operator is applied to the resulting creature and produces the final offspring. If the developed phenotype of an offspring creature contains two intersecting non-adjacent limbs, or too many limbs, the creature is deemed non viable, and the reproductive operation chosen is repeated as often as necessary until a viable creature is produced. Each run covers 500 generations.

The system produced a wide variety of behaviours, some of which are illustrated in Figure 3. In the top-left frame, one creature catches the cube in a

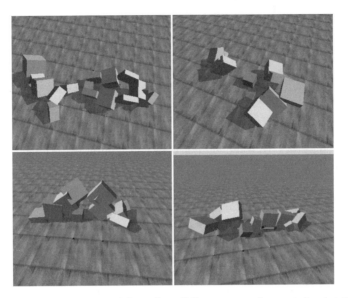

Fig. 3. Creatures evolved from four different runs. See text for details.

pinching motion and draws it towards its trunk limb before its opponent manages to reach it. In the top-right frame, two creatures use different methods to move toward the cube. In the bottom-left frame, one creature manages to push its opponent away from the cube, even though the other creature had reach the cube first. In the bottom-right frame, a two-armed creature is chasing the box that is being pushed aside by its opponent.

4.2   Coarse-Grained Master Tournament Matrices

To monitor the progress of evolution, we chose to introduce a modified version of the Master Tournament method. In our case the original method would be difficult to apply, since computing the whole Master Tournament matrix for 500 generations would be computationally prohibitive. Furthermore, as we mentioned in the introduction, Master Tournament grids are often difficult to read and analyse.

Our method consists in simplifying the Master Tournament by a "coarse-graining" operation. Instead of performing a full tournament between the champions of all N generations, we pick a fixed number k of champions and use this sample to produce a coarse-grain Master Tournament matrix. This means that we only perform tournaments between champions of generations which are integer multiples of the N/k fraction. In our example, we chose to condense our 500 generations into a 50x50 tournament grid, which means that we need to select the champions of every tenth generation (roughly) in each population. By sampling 10% of the generations, computing costs for head to head competitions are divided by 100.

Like any sampling process, coarse-graining incurs a loss of information. However, the information which is lost by coarse-graining is essentially short-term, small-scale information. When analysing the results of a coevolutionary experiment, we are usually interested in long-term trends, especially regarding evolutionary progress Coarse-graining adequately preserves this type of larger-scale information. In particular, the question of whether or not a given individual can consistently outperform older ones, which is the crucial aspect of the "arms race" concept, is not affected by coarse-graining. Moreover, coarse-graining can actually make a Master Tournament matrix more descriptive by suppressing spurious, irrelevant information: as we make clear in the following paragraphs, coarse-grained Master Tournament matrices may exhibit discernible, informative features which are often difficult to observe in full matrices.

4.3   Reading a Coarse-Grained Master Tournament Matrix

Figure 4-left shows a coarse-grained Master Tournament matrix for a particular run. Each (m, n) location is marked with a dark square point if the champion of population 1 generation $10 * m$ defeats the champion of population 2 at generation $10 * n$, or with a light cross mark otherwise. The $y = x$ line, drawn in a lighter shade, provides a time axis for the actual run. Points on this line indicate how the actual run went along, indicating the victorious population at each generation. On a coarse-grained Master Tournament matrix, vertical patterns are related to individuals from population 1, while horizontal patterns are related to individuals from population 2.

A first observation for this run is that the $y = x$ line goes through several regions of different colour. This means that the champions of the two populations successively outperform each other, an indicator of healthy competition. However, the particular patterns of this alternation provide a better insight about the course of evolution in this run.

Identifying similar phenotypes from their competitive profiles: The graph in Figure 4-left contain many similar lines and columns. In particular, it may be seen that many columns offer strikingly similar patterns of dark and light marks, although with appreciable variation. Each column, however, corresponds to the competitive profile of a champion in population 1: it accounts for its successes

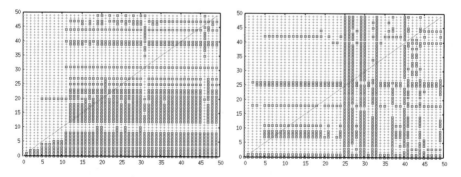

Fig. 4. Coarse-grained Master Tournament matrix for two different runs

and failures against every champion of population 2. Two identical columns denote two individuals that defeat the same opponents, and are defeated by the same opponents. It is not too far-fetched to assume that similarity in competitive profile is linked to similarity in phenotypes.

Similarity is not identity, and much variation can be seen. However there are at least two columns which offer a significantly different profile to the neighbouring columns, namely columns 31 and 46. These two columns can be said to represent different types from their neighbours, due to the difference in competitive profiles. In particular they are unique in being able to defeat the opponents in rows 41-43.

The significance of the high similarity in columns after 12 can be seen as an indication that, at least in population 1, evolution seems to have settled on a particular type of creature, which is marginally "fine-tuned" in the later course of coevolution. This capacity to indicate phenotypical convergence is an interesting property of coarse-grained Master Tournament matrices.

Evidence of breakdowns in arms race: The arms race concept implies that newer individuals are consistently able to outperform their ancestors. Breakdowns in the arms race are associated with a loss of adaptive function, since an ability (to defeat some individuals that could be defeated by ancestors) has been lost by the newer individual [9].

Breakdowns or interruptions in arms races are easy to locate on a Master Tournament matrix. Any dark mark occurring immediately above a light mark, or any light mark occurring immediately on the right of a dark mark, indicates such an interruption: it means that a given individual (from population 2 in the first case, from population 1 in the second case) was unable to defeat an opponent that could be defeated by its ancestor. Such breakdowns may be very short events, indicating a prompt recovery. Alternatively they may result in a long-term loss, or even permanent loss of the capacity to defeat some particular individuals.

Let us take the example of the first horizontal stripe of lighter marks, at rows 9-11. These rows correspond to particularly fit individuals which are able to defeat a large number of opponents (all of them for row 11). In particular, they have no difficulty defeating the champions of generations 12 to 18 in population 1, as can be seen from the fact that their rows are void of dark marks in the section between columns 12 and 18.

Yet the same graph shows that, from generation 12 onwards, the first bisectant encounters a series of dark marks, indicating superiority of the champions of population 1. This indicates that by generation 12, the current champion of population 2 had become unable to defeat individuals that earlier ancestors could defeat. How did this come to be ? If the champion of population 2 at generation 11 was good enough to defeat all opponents that population 1 would ever come up with, why was it displaced with one that would prove to be inferior?

Causes for breakdowns in arms races: This alternation between a lighter stripe and a sudden block of dark marks indicates a dramatic example of a breakdown in the arms race. In this particular case, the cause can be identified as *over-specialisation*. While it is true that the champion of generation 12 in population 2 was potentially able to defeat a large number of opponents, population 2 managed to come up with a new individual which was even better (that is obtained a lower differential distance to the cube) against the current champion of population 1. Unfortunately this change, while beneficial in the short term, proved fatal when population 1 managed to evolve a counter-strategy which defeated this specialised opponent. This allowed the newly evolved type of individual in population 1 to take the lead, even though previous champions from population 2 would have been able to defeat it.

Figure 4-right shows the results of a different run. This figure exemplifies several other informative patterns. In particular, let us look at the centre of the matrix, at row and column 25. At that point, we see that the first bisectant encounters a kind of wedge, composed of two stripes of dark marks - one vertical, one horizontal. Can we infer some meaning from this pattern ? The wedge shape indicates that a successful change in population 1 (indicated by the appearance of a different competitive profile, leading to a distinct, darker series of columns) has led to a dramatic breakdown in the arms race on the side of population 2. The appearance of this new champion in population 1 has *upset* the hierarchy in population 2: the previous champion was no longer the best possible candidate against this new opponent. Confronted with the new, successful champion of population 1, population 2 has settled on a new champion, which happened to perform better, or at least less badly, than others against this particular new opponent (though not well enough to actually defeat it). This new "champion of fortune", however, was not particularly well-rounded and performed badly against a large range of opponents. Innovation in population 1 has caused a *confusion* in population 2.

This idea of new individuals breaking down the arms race by upsetting the hierarchy and voiding previous adaptations in their opponents is not necessarily linked to wedge-like patterns, but simply to the appearance of a new type of opposing champions. For example, Figure 4-left contains several dark horizontal lines, apparently isolated. In particular, the individual in row 31 indicates that this champion suddenly lost much of its ancestors' aptitudes against opposing champions. What is the cause of this loss ? If we track the point at which this new, poorly performing champion occurs (by locating its intersection with the first bisectant) and observe the corresponding column, we notice that the individual from population 1 at column 31 has a subtly different pattern from its predecessors. The poor performance of population 2 at generation 31 is thus caused by the emergence of a new opponent which upsets the hierarchy in population 2 and propels an apparently poor individual to the rank of "champion".

These interruptions in the arms race (temporary or long-term) that can be observed on the coarse-grained Master Tournament matrix are an indication of the local nature of co-evolution. Because co-evolution is only concerned about the immediate present, it may directly induce a loss of ability against past or future opponents. This loss may occur spontaneously (as in over-specialisation) or may be provoked by a change in the opposing population (as in "confusion").

5   Cross-Validation of Coevolutionary Runs

Master Tournament matrices, however informative, can only describe performance within the context of a particular run. This is not necessarily sufficient to express the general level of performance of an individual in the larger context of the entire search space. In order to detect whether a given individual may really be called superior, it is not enough to confront it to the population against which it evolved. Such a test could be seen as a confusion between the training set and the test set. Given several evolutionary runs, if we want to obtain a more global view of each individual's performance, the most simple method we can use is simply to test each individual not only against its own opponents, but also against other populations of other runs. In other words, we expect that *cross-validating* individuals from different evolutionary runs would provide more reliable information about their global efficiency.

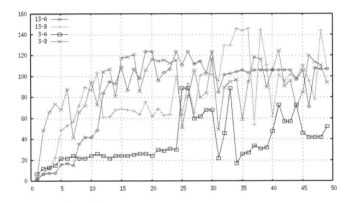

Fig. 5. Cross-validation of each individual in each of the four populations shown in Fig. 4. 13-A and 13-B are population 1 and 2 from the left-hand side matrix, while 3-A and 3-B are population 1 and 2 from the right-hand side matrix in Fig. 4.

Figure 5 shows, for each champion of all 50 generations in each population, the number of champions of all other opposing populations that it is able to defeat. 13-A and 13-B are population 1 and 2 from the left-hand side matrix in Fig. 4, while 3-A and 3-B are population 1 and 2 from the right-hand side matrix. This graph is interesting both for its similarities and its difference with the individual Master Tournament matrices in Figure 4.

Within this larger context, the best performing individuals are the champions of generations 34 and 36 from population 13-B, with a score coming close to the maximum 150, meaning that they can defeat almost all other champions. Looking at the corresponding rows in Fig. 4, we observe that they indeed obtain 'clean sheets' against all their opponents. However, this is also the case with rows 11 and 28-30, yet these ones obtain a much lower score on the cross-validation graph. This indicates a difference in performance that could not have been deduced from Master Tournament (coarse-grained or not) or Dominance Tournament analysis, nor indeed from any single-record analysis method alone.

Similarly, we see that population 3-A seems to perform rather poorly when compared to others. Specifically, after generation 10, all champions of population 3-A obtain much lower performance that champions in population 13-A. This is in contrast with the corresponding Master Tournament matrices, in which it can be seen that some champions of population 3-A are able to defeat all opposing champions from population 3-B (columns 25-26 and 33), while no champion in population 13-A shows such a perfect record. Again, a single-record analysis could not have detected this apparent superior performance of individuals from population 13-A.

References

1. Sims, K.: Evolving 3d morphology and behavior by competition. In: ALife IV : Proc. of the 4th Conference on Artificial Life, MIT Press (1994) 28–39
2. Nolfi, S., Floreano, D.: Coevolving predator and prey robots: Do "arms races" arise in artificial evolution? Artificial Life **4** (1998) 311–335
3. Dawkins, R., Krebs, J.R.: Arms races between and within species. Procs of the Royal Society of London, Series B **205** (1979) 489–511
4. Rosin, C.D., Belew, R.K.: New methods for competitive coevolution. Evolutionary Computation **5** (1997) 1–29
5. Watson, R.A., Pollack, J.B.: Coevolutionary dynamics in a minimal substrate. In Spector, L., Goodman, E.D., Wu, A., Langdon, W.B., eds.: Procs GECCO 2001, Morgan Kaufmann (2001)
6. Stanley, K.O., Miikkulainen, R.: The dominance tournament method of monitoring progress in coevolution. In: Procs GECCO 2002 Workshop, Morgan Kaufman (2002)
7. Cliff, D., Miller, G.F.: Tracking the red queen: Measurements of adaptive progress in co-evolutionary simulations. In: Proceeding of the European Conference on Artificial Life (ECAL-95). (1995) 200–218
8. Miconi, T., Channon, A.: A virtual creatures model for studies in artificial evolution. In: Procs of the 2005 IEEE Congress on Evolutionary Computation. (2005)
9. Ficici, S.G., Pollack, J.B.: A game-theoretic memory mechanism for coevolution. In et al, C.P., ed.: Proc. GECCO 2003, Springer (2003)

A Critical View of the Evolutionary Design of Self-assembling Systems

Natalio Krasnogor[1], Graciela Terrazas[1], David A. Pelta[2], and Gabriela Ochoa[3]

[1] Automated Scheduling, Optimisation and Planning Research Group,
School of Computer Science and Information Technology,
University of Nottingham
{nxk, gzt}@cs.nott.ac.uk
[2] Departamento de Ciencias de la Computacion,
ETSI Informatica, Universidad de Granada
dpelta@decsai.ugr.es
[3] Departamento de Ciencias de la Computacion,
Universidad Simon Bolivar
gabro@ldc.usb.ve

Abstract. The automated design of systems which self-assemble is a fundamental cornerstone of nanotechnology. In this paper we review some work in which we have applied Evolutionary Algorithms (EAs) for the *automated* design of systems self-assembly. We will focus in three important minimalist self-assembly problems and we discuss the difficulties encountered while applying EAs to these test cases. We also suggest some promising lines of work that could possibly help overcome current limitations in the evolutionary design of self-assembling systems.

1 Introduction

Self-assembly is a process that creates complex hierarchical structures through the statistical exploration of alternative configurations. These processes occur without external intervention. The specific system that is self-assembled (from a given set of components) is determined by the way the statistical exploration of conformations is performed. In turn, the exploration mechanisms are constrained by the individual components that undergo self-assembly and the conditions imposed upon them by their local environment. In general, components are autonomous, have no pre-programmed *master* assembly plan, and can only interact with their local environment and other components. Self-Assembly is a powerful autopoietic mechanism whose power, as a reusable engineering concept, lays in the fact that it is a distributed, not-necessarily synchronous, control mechanism for the bottom-up manufacture of complex systems. This control mechanism is distributed across a myriad of elemental components, none of which has either the storage or the computation capabilities to know and follow a master plan for the assembly of the intended system. Instead each component has a very limited behavioral repertoire which tells it what to do under a reduced set of well defined conditions. Self-Assembly processes are ubiquitous in nature. Understanding how nature produces self-assembled systems will represent an enormous

E. Talbi et al. (Eds.): EA 2005, LNCS 3871, pp. 179–188, 2006.

leap forward in our technological capabilities. Although major advances in the design of systems that exhibit self-assembly properties have been reported in the literature (e.g. [17, 16]), much less has been said about the *automated* design of self-assembly. In [8] the author tackles the problem of automated design of self-assembly for a very specific class of problems which are amenable to analytical solution. However, it is unrealistic to expect that each and every self-assembly system will have properties that make it agreeable to a hand-made design. Instead, as in many other industrial settings, we will need to resort to computer aided automated design of components, interaction matrices and assembly skeletons.

The complexity of self-assemble squares under a generalized model of tile assembly[13] was investigated in [1]. Several interesting results on the intractability of certain self-assembly processes were described. Although these papers point to promises and limitations of specific self-assembly processes it is important to remark that NP-hardness results have not, in the past, deterred the advance of other branches of science and engineering. On the contrary, NP-hard problems are regularly tackled (and solved to industrial standard satisfaction) with an arsenal of modern algorithmic techniques ranging from integer and linear programming, lagrangian relaxations to sophisticated metaheuristics like tabu search[6], simulated annealing[7] and memetic evolutionary algorithms[15].

A principled methodological approach for automated self-assembly design would be to systematically investigate automated design methods on (tunable) conceptual, highly idealized problems as it has been done in other domains like protein folding[4], traveling salesman problem[12], etc. To this end, in [9] we introduced a family of tunnable problems for self assembly. In this paper we complement that paper by reviewing some work in which we have applied Evolutionary Algorithms (EAs) for the automated design of systems self-assembly. We focus in three important minimalist self-assembly problems and we discuss the difficulties encountered while applying EAs into these problems. In this paper we also suggest some promising lines of work that could possibly help overcome current technological limitations.

2   Protein Structure Prediction and Wang Tiles as Paradigmatic Self-assembly Design Problems

In this section we introduce two problems which are paradigmatic self-assembly design problems, namely, the design of folding rules in protein structure prediction and the design of Wang tile families for the self-assembly of two-dimensional shapes.

2.1   Protein Structure Prediction

Proteins are hetero-polymers composed of amino acids. Under physiological conditions proteins fold into a three dimensional native state where they adopt their biological function. The protein structure prediction problem is concerned with the determination of the native state from the identity of the amino acids that

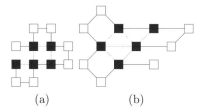

Fig. 1. HP protein embedded in the square lattice (a) and triangular lattice(b). Black boxes represent hydrophobic residues, while white boxes represent hydrophilic ones.

constitutes a given protein. That is, protein folding might be regarded as the self-assembly problem par excellence. The particular simplified model we are concerned with in this paper is the Hydrophobic-Polar model introduced by K. Dill[4]. The HP model (and its variants) abstracts the hydrophobic interaction process in protein folding by reducing the 20 naturally occurring amino acids into a binary alphabet, thus a protein becomes an hetero-polymer of non-polar or hydrophobic (H) and polar (P) or hydrophilic amino acids. An n amino acids protein is represented by sequence $s \in \{H, P\}^+$ with $|s| = n$. The sequence s is to be mapped to a lattice, where each residue in s occupies a different lattice cell and the mapping is required to be self-avoiding. Although simple to state, this problem remains NP-Hard[3].

The energy potential in the HP model reflects the fact that hydrophobic amino acids have a propensity to form a hydrophobic core. To capture this feature of protein structures, the HP model adds a value ϵ for every pair of hydrophobes that form a topological contact; a topological contact is formed by a pair of amino acids that are adjacent on the lattice but not consecutive in s. After normalization, the interaction energy between two non-polar amino acids is $\epsilon_{H,H} = -1$ while all other interactions (i.e. HP and PP) are 0. In this model optimally self-assembled native structures minimize an energy function that is a simple count of the number of HH contacts in the self-assembled conformation. Figure 1(a) and (b) shows sequences embedded in the square and the triangular lattices, with hydrophobic-hydrophobic contacts (HH contacts) highlighted with dotted lines. The conformation in Figures 1(a) and 1(b) show the embedding of the same protein instance into two different lattices, which result in energies of -4 and -6 respectively.

Automated Design of Protein Self-assembly. In this paper we will address the problem of automatically designing, by means of an evolutionary algorithm, the rules that are necesary to drive the dynamical process of folding towards the native state of specific proteins. We will employ two different computational abstractions to represent these folding rules. The first abstraction we use is that of a one dimensional uniform, contiguous neighborhood, cellular automata to simulate the folding process. In this case, the evolutionary algorithm is required to design the rules that define the cellular automaton, with the intention that by executing those rules the protein sequence embedded in the automaton will self-assemble into its native state. In the second computational abstraction we

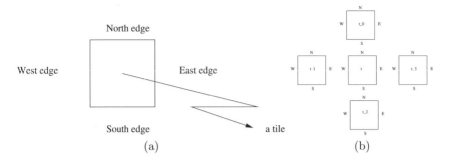

Fig. 2. (a) Schematic representation of a four edged tile. Each edge is distinguished by the labels *North, West, South, East*. (b) An example of a five tiles self-assembly.

represent the folding rules by an L-system grammar rather than by the rules of a cellular automaton. In this case the parallel interpretation of the L-system grammar drives the self-assembly of the protein structure into its target conformation.

2.2 Wang Tiles Self-assembly

Computation and self-assembly are connected by the theory of tiling, of which *Wang Tiles*[14] are a prime example. A Wang tile system is defined by a family of two dimensional square tiles embedded in the plane. Each side of a tile might have a specific glue type attached to it. When tiles move around in the plane, and two of them colide, they will either stay attached or they will separate and continue their brownian motion as independent entities. Whether they self-assemble or stay separated depends on the strength and compatibility of the glue types in their coliding sides. This process is initialized with a specific kinetic energy associated to the tile set (i.e. temperature). When tiles attach to each other they form complex shapes and the specific shapes which emerge are said to be self-assembled. This process can be mathematically described:

Let Σ be the set of symbols used to label the edges associated to each tile. This set of symbols encodes the "glue" types associated to each edge and includes the special case λ representing and edge with no glue. The set of tiles is $\mathcal{T} = \{t | t = (x_0, x_1, x_2, x_3)\}$ such that for any $k \leq 3 \, \exists a, a \in \Sigma, p >= 0 \; and \; x_k = a^p$. If $p = 0$ then a^0 is taken to be equivalent to λ, i.e., the no glue state for a given edge of the tile. A label a^p on an edge x_k encodes an "a" glue type with strength p.

We can associate x_0, x_1, x_2 and x_3 with the north, west, south and east edges respectively as shown in figure 2(a). Let also τ be the "temperature" parameter as in [1]. After coliding, two tiles t_i, t_j will self-assemble by their edges e_i, e_j if the glue types and strengths in those edges are equivalent and the glue strength larger than the temperature.

Given tiles t, t_0, t_1, t_2, t_3 they will self-assemble with t in the center (as shown in Fig 2(b)) if the glue strength of each attaching edges is bigger than 0 and the sum of all glue strengths bigger than τ. More precisely, t and t_i for $0 \leq i \leq 3$ will self-assemble if the following conditions hold:

$t = (x_0, x_1, x_2, x_3)$ and $t_i = (x_{i_0}, x_{i_1}, x_{i_2}, x_{i_3}, 0 <= i <= 3$ with $x_{0_2} = x_0, x_{1_3} = x_1, x_{2_0} = x_2, x_{3_1} = x_3$ and $|x_0| + |x_1| + |x_2| + |x_3| >= \tau$.

Please note that the conditions on x_k above can be succintedly written as $x_{k_g} = x_k$ with $g = (k + 2)\%3$ where $\%$ stands for the module operation. The reader must note that the labeling of edges as "north,west,south,east" is only a useful convention to simplify the exposition.

Automated Design of Wang Tile Families. The third and last automated design problem we will address is that of the automated design of $\mathcal{T}$ (i.e. a families of Wang tiles), which can self-assemble into a specific two dimensional shape, which in this paper is a square.

3   Evolutionary Algorithms for the Automated Design of Protein Self-assembly by Cellular Automata (CA)

Cellular automata have been used as models of physical and biological phenomena such as fluid flow, galaxy formation, earthquakes, biological pattern formation, etc. and as models of computation (see for example [18]). Briefly a CA consists of two components. The first one is a *lattice of N identical cells*, each of which have a state. Each cell is updated based on its current state and the state of its neighbors in the lattice. The neighborhood considered depends on the particular CA. The second component is the *transition rules* that give the updated state for each cell as a function of the neighborhood.

We used a CA to model the rules and dynamics which would drive a self-assembly process towards the native state of a given protein sequence. We had previously addressed this problem using a circular one-dimensional CA with only four states (1, 2, 3, 4), each one corresponding with the absolute moves **U**p, **D**own, **L**eft or **R**ight (relative to the position of the previous amino acid in the sequence) [10]. An example is shown in Figure 3 (a). Allowed rule radii were 1, 2 and 3. The evaluation of an individual involved: running the CA with the individual's value (set of rules), getting the final configuration of the automaton (the folded structure), applying this fold to the protein to obtain the energy value.

We also performed experiments with an extended set of rules which took into consideration the specific amino acids the rule was being applied to. This is shown in Fig 3 (b). To evolve the rule set that defined the CA we used a Genetic Algorithm. Implementation and parameter details are described in [10].

We have conducted extensive experiments but due to space limitations we only show the results for 3 instances in Table 1. We recorded the number of times (out of 10) that the optimum, optimum-1 and optimum-2 conformations were found.

The results may be analized from two points of view. One is oriented to answer the question: *is it possible to find a set of rules to reach the native state from this particular unfolded state?*. The answer is yes on two of the three cases, although it is clear that more experimentation should be done.

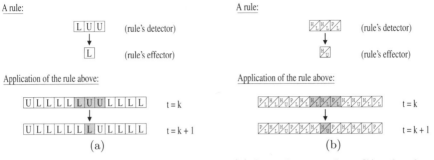

Fig. 3. (a) First approach to CA rule scheme (b) Second approach to CA rule scheme

Table 1. Number of runs in which the GA achieved the stated energy value. The 'Optimum' column displays the native energy value, while 'Opt', 'Opt + 1' and 'Opt +2' display the number of runs in which either the optimum energy was achieved or conformations with energies with a gap of one or two above that value was found.

Sequence	Length	Optimum	Opt	Opt + 1	Opt + 2
PHPPHHPPHPPHPPHHPPHP	20	-8	5		5
HHPPHPPHPPHPPHPPHPPHH	24	-9	1	2	5
PPHPPHHPPPPHHPPPPHHPPPPHH	25	-8			1

The other point of view focus on the quality of the search process, and here the results seems to deteriorate with the size of the instance. For the smaller one, 50% of the runs, lead to set of rules that allowed to achieve the optimal configuration. This percentage goes down to 10% in the second instance and in the third one, just one run allowed to obtain a configuration with energy 6.

One may conclude that: a) in principle, it is indeed possible to find set of rules for a cellular automaton which instigates the self-assembly of the native structure; and b) the search procedure should be enhanced.

4   Evolutionary Algorithms for the Automated Design of Protein Self-assembly by L-Systems

In [5] we introduced an L-systems' based evolutionary algorithm as the inference procedure for folded structures under the HP model in 2D lattices. The evolutionary algorithm attempts to find a set of rewriting rules (an L-system) that captures a target folded structure (which represents the native state for a given protein) on the selected lattice model.

The simplest class of L-systems, the D0L-systems, is deterministic and context free. We use D0L-systems to drive the self-assembly of the protein sequence.

Given a target structure (input), let say the one shown in Fig. 1(a), the evolutionary algorithm will evolve and L-system L (output) that, once evaluated, would produce a string (in internal coordinates) which matches the target structure (in the example, the end-product of the EA would be and L-system whose termination word is LRRLRRFLRR).

Table 2. Partial Results of the Automated Evolutionary Design of L-Systems for Protein Folding. The first column format $\frac{I}{S}$ denotes the protein sequence I with target self-assembled structure S, the second column shows the length of the protein sequence and the third column -following the same format as the first- shows the total number of runs of the EA and the number of successful runs.

Instance	Length	Success/Num. Runs
$\frac{HHHPPHPHPHPPPHPHPHPPH}{RRFRFRLFRRFLRLRFRR}$	18	$\frac{3}{40}$
$\frac{HHPPHPPHPPHPPHPPHPPHH}{RLLFLFFRRFLLFRRLRFFRRF}$	22	$\frac{1}{50}$
$\frac{PPHPPHHPPPPHHPPPPHHPPPPHH}{FFRRFFFLLFFFFRRFFFFLLFF}$	23	$\frac{1}{50}$

A Genetic algorithm was used to evolve the L-systems which would drive the self-assembly procees. Full details of the algorithm and experiments can be found in [5]. Table 2 shows some of the results we obtained evolving L-systems for the self-assembly of protein structures.

Similarly to the evolutionary design of CA rules for self-assembling, the automated design of L-systems met with partial success. On the one hand it is possible to show that the algorithm is capable of finding L-system which will induce the correct self-assembly behaviour. On the other hand however, the process is painfully slow and requires very many executions of the algorithm to obtain a successful L-system.

5 Evolutionary Algorithms for the Automated Design of Wang Tiles Self-assembly

We have applied a Genetic Algorithm to the automated design of the tile sets $\mathcal{T}$ which can self-assemble into a 2D square of 10x10 tiles. The GA used various parameters for crossover, mutation, population sizes, etc., which will be reported elsewhere. In order to evaluate an individual (i.e. assess its fitness) we placed it in a Wang tile self-assembly simulator. As the individual specifies various tile families, several instances of each family were placed in the simulator. Each tile was initialy placed on a randomly selected empty lattice position. Then, tiles move randomly for the duration of the simulation. Once the simulation finished the fitness function tried to identify (within the lattice) the shape with the most similarity to the target structure. This was done by a *Hamming distance* function defined as $H(L, S) = a_i$, where L was the simulation's final 2D lattice configuration and a_i is the maximum amount of tiles appearing within a square region S. The region was slided accross the lattice in order to find the better match ensuring that the fitness of an individual is equivalent to the minimal Hamming distance. Figure 4(a) shows a scanning example.

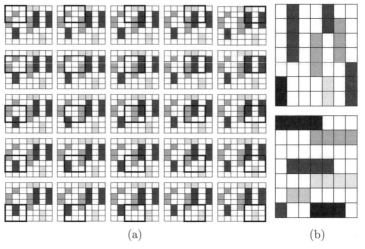

(a)   (b)

Fig. 4. (a) Scanning a lattice for a 3 × 3 square (b) Self-Assembled rows and columns

With the aim of determining which is the best set of parameters for both the GA and the Wang tile simulator we run an extensive set of experiments. After carefully selecting the best parameters the evolutionay algorithm was unable to evolve suitable tile sets that could self-assemble into the target structure. However, some intermediate structures were discovered by the algorithm. In this case, horizontal and vertical tiled strips (shown in Fig. 4(b)) were found.

6   Discussion and Conclusions

In previous sections we briefly sketched the application of evolutionary algorithms, more specifically genetic algorithms, to the automated design of components which could self-assemble into specific systems. Two of the showcases dealt with the design of rules, either for a cellular automaton or of an L-system, which could drive the process of protein folding (albeit in a very idealized model). In the third case we applied the GA to the design of tile sets and their glue types in order that they could self-assemble a target 2-dimensional shape. Although the application domain, the type of components and dynamic laws governing their use were different some common lessons could be drawn.

Firstly, in the three showcases large populations with short runs or small populations with long runs were required. That is, in the three cases studied the evolutionary design was computationally expensive. This requires a carefull consideration of the various parameters which define the GA behaviour as well as those parameters which are specific to the simulators. It may be possible that a co-evolutionary approach would be benefitial by simultaneously exploring the design space of system self-assembly and the parameter space of the GA.

Secondly, although in all three cases it was possible to achieve a moderate degree, yet not substantial, of success evolving the desired self-assembling system, the remarkable common fact is that intermediate self-assembled products -which are essential for the formation of the target system- were *always* discovered. That is, in L-systems and Cellular Automata we were able to find rules which allowed for the self-assembly of so called protein's "secondary structures". At the same time, the evolutionary design of Wang tiles was able to discover the equivalent of secondary structures in the form of self-assembled columns and rows. This common behaviour across three different domains and with differently customized evolutionary algorithms suggests an *evolutionary divide-and-conquer* methodology. That is, rather than trying to evolve from scratch the final design for a self-assembling system, we could instead evolve designs for *generalized secondary structures* and used those designs to bootstrap the final design. As an example consider the evolutionary design of Wang tiles to self-assembling a square. Instead of starting from completely random tile families we could seed the GA with those families known to form columns and rows as these features will certainly appear in any self-assembled square. Alternatively, in the case of L-systems we could evolve problem specific knowledge (e.g. specific rules for alpha-helices, beta sheets, etc) as to accelerate the design process of self-assembling rules for the whole protein structure.

A third lessons, which we will also be tested in future experiments, is what we named "intelligent freezing". During the evolutionary design of self-assembling systems it was possible to observe that certain *critical generalized secondary structures (CGSS)* were formed. Some of the runs that discovered CGSS managed to maintain them long enough as to profit from their discovery. On the other hand, some runs tampered with the CGSS destroying their essential features. Intelligent freezing would implement a mechanism to detect CGSS (eg. by tracking evolutionary activity waves[2]) an will protect these CGSS from being disrupted by genetic or other mechanisms (i.e. they will be frozen). Another interesting avenue of research would be to use what has been termed the "Parisian Genetic Programming" approach [11] as it has been very successful in a not unrelated inverse design problem.

In conclusion, although the automated design of self-assembling systems is at its infancy it is possible to achieve a modest degree of success with current evolutionary metaheuristics. On the other hand, as the size and complexity of the target self-assembling system increases, its likely that more robust and efficient EA will be needed. We have described three showcases of the application of genetic algorithms for systems self-assembly and we have suggested some promising avenues for further research.

Acknowledgements

N. Krasnogor would like to acknowledge many useful and enlightening conversations with G. Rozenberg, G. Paun, M. Gheorghe and P. Moriarty. He will also like to acknowledge the EPSRC (GR/T07534/01, EP/D021847/1) and BBSRC (BB/C511764/1) funding of his research on artificial intelligence as applied to bioinformatics and evolutionary chemistry. D. Pelta would like to acknowledge the funding for project TIC-2002-04242-C03-02.

References

[1] L. Adleman, Q. Cheng, A. Goel, M. Huang, D. Kempe, P. Moisset de Espanes, and P.W.K. Rothemund. Combinatorial optimization problems in self-assembly. In *Proceedings of the Annual ACM Symposium on Theory of Computing(STOC)*. ACM Press, 2002.

[2] M.A. Bedau and N.H.Packard. Measurement of evolutionary activity, teleology and life. In C.G. Langton, C. Taylor, D. Farmer, and S. Rasmussen, editors, *Artificial Life II*, volume 98-03-023, pages 431–461. Addison-Wesley, 1992.

[3] B. Berger and T. Leight. Protein folding in the hydrophobic-hydrophilic (HP) model is NP-complete. In *Proceedings of The Second Annual International Conference on Computational Molecular Biology, RECOMB 98*, pages 30–39. ACM Press, 1998.

[4] K.A. Dill. Theory for the folding and stability of globular proteins. *Biochemistry*, 24:1501, 1985.

[5] G. Escuela, G. Ochoa, and N. Krasnogor. Evolving l-systems to capture protein structure native conformations. In *Proceedings of the 8th European Conference on Genetic Programming (EuroGP 2005), Lecture Notes in Computer Sciences 3447, pp 73-84*. Springer-Verlag, Berlin, 2005.

[6] F. Glover, E. Taillard, and D. de Werra. A user's guide to tabu search. *Annals of Operations Research*, 41:3–28, 1993.

[7] S. Kirkpatrick, C.D. Gelatt, and M. P. Vecchi. Optimization by simulated annealing. *Science*, 220 no 4598:671–680, 1983.

[8] E. Klavins. Automatic synthesis of controllers for distributed assembly and formation forming. In *Proceedings of the IEEE Conference on Robotics and Automation*, 2002.

[9] N. Krasnogor and S. Gustafson. A family of conceptual problems in the automated design of self-assembly. In *Proceedings of the 2nd International Conference on the Fundations of Nanoscience: Self-Assembled Architecture and Devices, Utah, Snowbird resort, April 24-29*, 2005.

[10] N. Krasnogor, D.A. Pelta, D.H. Marcos, and W.A. Risi. Protein structure prediction as a complex adaptive system. In *Proceedings of Frontiers in Evolutionary Algorithms 1998*, 1998.

[11] P.Collet, E.Lutton, F.Raynal, and M.Schoenauer. Polar ifs + parisian genetic programming = efficient ifs inverse problem solving. *Genetic Programming and Evolvable Machines*, 1:339–361, 2000.

[12] G. Reinelt. Tsplib
(http://www.iwr.uni-heidelberg.de/iwr/comopt/soft/tsplib95/tsplib.html).
In *mirror site: gopher://softlib.rice.edu/11/softlib/tsplib*.

[13] P. Rothemund and E. Winfree. The program-size complexity of self-assembled squares. In *Proceedings of STOC*, 2000.

[14] H. Wang. Probing theorems by pattern recognition. *Bell Systems Technical Journal*, 40:1–42, 1961.

[15] N. Krasnogor W.E. Hart and J.E. Smith. *Recent Advances in Memetic Algorithms*. Studies in Fuzziness and Soft Computing Series - Springer, 2004.

[16] G.M. Whiteside and M. Boncheva. Beyond molecules: Self-assembly of mesoscopic and macroscopic components. *Proceedings of the National Academy of Science (PNAS)*, 99(8):4769–4774, 2002.

[17] G.M. Whiteside and B. Grzybowski. Self-assembly at all scales. *Science*, 295:2418–2421, 2002.

[18] S. Wolfram. *A New Kind of Science*. Wolfram Media Inc., 2002.

Algorithmic Self-assembly by Accretion and by Carving in MGS

Antoine Spicher, Olivier Michel, and Jean-Louis Giavitto

LaMI UMR 8042 CNRS – Université d'Evry, Genopole,
523 place des Terrasses de l'Agora, 91000 Evry, France
{aspicher, michel, giavitto}@lami.univ-evry.fr,
http://mgs.lami.univ-evry.fr

Abstract. We report the use of MGS, a declarative and rule-based language, for the modeling of various self-assembly processes. The approach is illustrated on the fabrication of a fractal pattern, a Sierpinsky triangle, using two approaches: by *accretive growth* and by *carving*. The notion of topological collections available in MGS enables the easy and concise modeling of self-assembly processes on various lattice geometries as well as more arbitrary constructions of multi-dimensional objects.

1 Introduction

Self-assembly is a process that creates incrementally complex hierarchical spatial structures. Nature presents a lots of examples, ranging from crystallization in physics to morphogenesis in developmental biology. There is no unified general theory of self-assembling, nor a unique definition. However, understanding the principles underlying self-assembly processing will open entire new opportunities for our technological capabilities. Self-assembled systems can be thought to be built of basic building elements (molecules, cells, etc.); together these basic elements exhibit a new, often highly, complex behaviour.

For a computer scientist, self-assembly processes are particularly inspiring because the dynamic organization of the involved entities emerge from many decentralized and local interactions that occur concurrently at several time and space scales. As a matter of fact, they have inspired several new computational models like *amorphous computing* [1] or *autonomic computing* [7].

The emergence of the global structure of self-assembled systems cannot be deduced from the individual composing elements. To obtain a deeper insight of these complex systems, simulation models are often the only available option. However, the modeling and the simulation of self-assembly can be very difficult to achieve, because of the representation of the underlying space and of the handling of complex spatial structures build in this space.

1.1 Self-assembly by Accretive Growth and by Carving

A central thema in the research in self-assembly processes is the organizational principles that can be used to structure a population of basic elements. The structure is incrementally built and often corresponds to a spatial structure. In this paper we will focus on the modeling of two kinds of self-assembly.

E. Talbi et al. (Eds.): EA 2005, LNCS 3871, pp. 189–200, 2006.
© Springer-Verlag Berlin Heidelberg 2006

Self-Assembly by Accretive Growth. One of the most fundamental kind of self-assembly is certainly processes where basic elements are united into a structure during a *growth process*. A growth process can be described as an iteration process. In such a process the output of an iteration step is used again as input for the next iteration step. In a growth process the form of a growing object in a certain growth stage is also determined by the form of the object in the preceding growth stage. In each growth stage, new basic entities (e.g., material) are added to this preceding growth stage.

We use the term *accretive growth* to qualify a growing process that takes place on the boundaries of the system. This kind of growth is to oppose to "intercalary growth" where the growing process is from the inside of the assembly.

Self-Assembly by Carving. Manca et al. have introduced a somewhat unusual type of computation strategy called *computation by carving* [9]. The idea is to generate a (large) set of candidate solutions of a problem, then remove the non-solutions such that what remains is the set of solutions. This idea to remove unwanted elements is also present in building shapes by space carving [8], an algorithm to compute a volume that is consistent with a set of photos of a 3D shape. Transposed in the domain of self-assembly, this leads to the idea to iteratively remove elements, starting from an initial shape.

1.2   DSL for the Simulation of Self-assembly

As noted above, the simulation of self-assembly can be very difficult to achieve. In this paper, we advocate the use of a domain specific language (DSL) for the modeling and the simulation, in an abstract and uniform setting, of accretive growth and carving.

DSLs are specially tailored programming languages designed for solving problems in a particular domain. To this end, a DSL provides abstractions and notations for the domain at hand. DSLs are usually small, and more declarative than imperative. Moreover, DSLs are more attractive for programming in the dedicated domain than general-purpose languages because of easier programming, systematic reuse, better productivity and flexibility. Our approach relies on two dedicated notions:

- dedicated data-structures, called *topological collections* are used to represent the space underlying a self-assembly process and/or the self-assembled system; and
- rewriting rules on topological collection, called *transformations*, are used to implement the local evolution rules usually used to specify the self-assembly process.

These two notions are studied in an experimental programming language called MGS. MGS is a vehicle used to investigate the notions of topological collections and transformations and to study their adequacy to the simulation of various biological and self-assembly processes [6, 4].

1.3 Organization of the Paper

The rest of this paper is organized as follows. The next section provides a quick introduction to MGS. Two kinds of topological collections are sketched: group-based data fields which are used to define various lattices used in the modeling of accretive growth, and abstract cellular complexes used to model arbitrary shape for carving. Section 3 presents three short and well-known examples of growth by aggregation processes in MGS. Section 4 shows the self-assembly of Sierpinsky triangles and section 5 build the same shape but using a carving process. The conclusion reviews some previous, related and future work.

2 A Short MGS Presentation

2.1 Transformations of Topological Collections

In this section, we present the notions needed to understand the MGS coding of the previous computation processes. MGS is a declarative programming language aimed at the representation and manipulation of local transformations of entities structured by *abstract topologies* [4]. A set of entities organized by an abstract topology is called a *topological collection*. Topological means here that each collection type defines a neighborhood relation specifying the notions of *locality, path* and *sub-collection*. A path is a finite sequence of elements e_i where e_{i+1} is a neighbor of e_i. A sub-collection B of a collection A is a subset of elements of A defined by some path and inheriting its organization from A. The *global transformation* of a topological collection C consists in the parallel application of a set of *local transformations*. A local transformation is specified by a rewriting rule r that specifies the change of a sub-collection. The application of a rewrite rule $\beta \Rightarrow f(\beta, ...)$ to a collection A:

1. selects a sub-collection B of A whose elements match the *pattern* β,
2. computes a new collection C as a function f of B and its neighbors,
3. and specifies the insertion of C in place of B into A.

The collection types can range in MGS from totally unstructured with sets and multisets to more structured with sequences, "group-based data fields" and "abstract cellular complexes". There are two kinds of patterns that can be used in a transformation.

Path Patterns. Path patterns match paths in a collection. A path pattern is a sequence of elements separated by a comma. The path pattern x,y defines a path of two elements, where y must be a neighbor of x. Arbitrary condition can be tested using guards inserted in a path pattern: (x / x>0), (y / y>x) matches two elements x and y such that the value of x is strictly positive and y is a neighbor of x and the value of y must be greater than the value of x.

Patch Patterns. Patch patterns allow the matching of arbitrary sub-collection. A patch pattern is specified using a set of clauses. We will present the patch pattern features we need on section 5.

2.2 Group-Based Data Field

Group-based data fields (GBF in short) are used to define topological collections with *uniform* neighborhood. A GBF is an extension of the notion of array, where the elements are indexed by the elements of a group, called the *shape* of the GBF [5]. The elements of the group are called the *positions* of the GBF. For example:

```
gbf Grid2 = < north, east >
```

defines a GBF collection type called `Grid2`, corresponding to the regular Von Neuman neighborhood in a classical array (a cell above, below, left or right – not diagonal). The two names `north` and `east` (together with their inverses `-north` and `-east`, always provided in a group structure) refer to the directions that can be followed to reach the neighbors of an element. These directions are the *generators* of the underlying group structure. The right hand side (r.h.s.) of the GBF definition gives a finite presentation of the group structure.

The list of the generators can be completed by giving equations that constraint the displacements in the shape:

```
gbf Hex2 = < east, north, northeast; east + north = northeast >
```

defines an hexagonal lattice that tiles the plane, see figure 1. Each cell has six neighbors (following the three generators and their inverses). The equation `east + north = northeast` specifies that a move following `northeast` is the same as a move following the `east` direction followed by a move following the `north` direction.

For convenience, we identify the type of a GBF with the presentation of the underlying group. A GBF g of type G can be formalized as a partial function g from the group specified by G to some set of values: g associates a value to some positions. In other word, the group elements act as indices of a generalized array. An empty GBF is the everywhere undefined function.

The topology of the collections of type G is easily visualized as the Cayley graph $\mathcal{G}$ of G: each vertex in the Cayley graph is an element of the group G

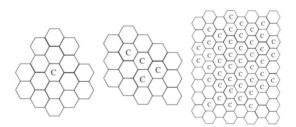

Fig. 1. Eden's model on an hexagonal mesh (initial state, and states after 3 and 7 time steps). This shape corresponds to the Cayley graph of `Hex2` with the following conventions: a vertex is represented as a face and two neighbors in the Cayley graphs share an edge in this representation. An empty cell has an undefined value. Only a part of the infinite domain is figured.

and vertex x and y are linked if there is a generator $\mathtt{u}$ in the presentation of G such that $x + \mathtt{u} = y$. A word (a sum of generators) is a path. Path composition corresponds to group addition. A closed path (a cycle) is a word equal to e (the identity of the group). An equation $v = w$ can be rewritten $v - w = e$ and then corresponds to a cycle in the graph. There are two kinds of cycles in the graph: the cycles that are present in all Cayley graphs and corresponding to group laws (intuitively: a backtracking path like $\mathtt{east} + \mathtt{north} - \mathtt{north} - \mathtt{east}$) and closed paths specific to the own group equations (e.g.: $\mathtt{east} - \mathtt{north} - \mathtt{east} + \mathtt{north}$). The graph connectivity (there is always a path going from P to Q) is equivalent to say that there is always a solution x to equation $P + x = Q$.

3   Growth Processes in MGS

Eden's Process. We start with a simple model of growth sometimes called the Eden model [3]. The model has been used since the 1960's as a model for such things as tumor growth and growth of cities. In this model, a 2D space is partitioned in empty or occupied cells (we use the value $\mathtt{true}$ for an occupied cell and left undefined the unoccupied cells). We start with only one occupied cell. At each step, occupied cells with an empty neighbor are selected, and the corresponding empty cell is made occupied.

The Eden's aggregation process is simply described as the following MGS global transformation: $\mathtt{trans\ Eden = \{\quad x,\ <undef>\ =>\ x,\ true\quad \}}$.

The Growth of a Snowflake. A crystal forms when a liquid is cooled below its freezing point. Crystals start from a seed and then grow by progressively adding more molecules to their surface. As an idealization, the molecules of a snowflake lie on an hexagonal grid and when a piece of ice is added to the snowflake, the heat released by this process inhibits the addition of ice nearby.

This phenomenon leads to the following cellular automata rule [16]: a black cell (value 1) represents a place of the crystal filled with ice and a white cell (value 0) is an empty place. A white cell becomes black if it has exactly one black neighbor, otherwise it remains white. The corresponding MGS transformation is:

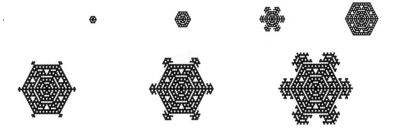

Fig. 2. Formation of a snowflake. The pictured states are the steps at time steps 1, 4, 8, 12, 16, 18, 20 and 23.

```
trans SnowFlake = {   0 as x / 1 == FoldNeighbor[+,0](x) => 1   }
```

The construct `FoldNeighbor` is not a function but an operator available only within a rule: it enables to fold a function on the defined neighbors of an element matched in the l.h.s. Here, this operator is used to compute the number of neighbors (the accumulating function is the sum and the initial value is 0). This transformation acts on a value of type `Hex2` and a possible run is illustrated in figure 2.

Diffusion Limited Aggregation. In a *diffusion limited aggregation* process, or DLA [15], a set of particles diffuse randomly on a given spatial domain. Initially one particle, the seed, is fixed. When a mobile particle collides a fixed one, they stick together and stay fixed. This process leads to a simple lattice gas automata that could be easily done in MGS using a topological collection and transformation:

```
trans dla = {
    'mobile, 'fixed   =>  'fixed, 'fixed
    'mobile, <undef>  =>  <undef>, 'mobile
}
```

We use two symbols `'mobile` and `'fixed` to represent respectively a mobile and a fixed particle (MGS's symbols are like Lisp's atoms). The two rules of the transformation deal with:

1. the aggregation: the first rule specifies that if a diffusing particle is the neighbor of a fixed one, then it becomes fixed (at the current position);
2. the diffusion: if a mobile particle is neighbor of an empty place (position), then it may leave its current position to occupy the empty neighbor (and its current position is made empty).

Note that the order of the rules is important because, following the rule application semantics of MGS, the first one has priority over the second. Figure 3 presents the final state of the application of the transformation `dla` on two kinds of topological collections: on the left, the neighborhood relationship is homogeneous and a GBF is used. On the right, the `dla` transformation is applied

Fig. 3. Example of DLA on two different topologies: an hexagonal mesh and a sphere. The plain hexagons and facets represent fixed particles. On the sphere, the empty positions are not drawn. The same transformation is used on the two collections.

on a meshed sphere. The elements are the facets, and two facets are neighbors if they share an edge. For more details, refer to [13].

4   Accretive Growth of Sierpinski Triangles

The Sierpinski triangles (ST from now on) is a fractal described by Sierpinski in 1915 and appearing in Italian art from the 13th century. It is also called the Sierpinski gasket or Sierpinski sieve [14]. The ST can be produced by taking the Pascal's triangle modulo 2 (see figure 4), or equivalently by iterating the bidimensional morphism defined on $\{0,1\}$ by $0 \longrightarrow \begin{smallmatrix} 0 & 0 \\ 0 & 0 \end{smallmatrix}$ and $1 \longrightarrow \begin{smallmatrix} 1 & 0 \\ 1 & 1 \end{smallmatrix}$. Starting from 1, we obtain:

$$
1 \longrightarrow \begin{matrix} 1 & 0 \\ 1 & 1 \end{matrix} \longrightarrow \begin{matrix} 1 & 0 & 0 & 0 \\ 1 & 1 & 0 & 0 \\ 1 & 0 & 1 & 0 \\ 1 & 1 & 1 & 1 \end{matrix} \longrightarrow \begin{matrix} 1&0&0&0&0&0&0&0 \\ 1&1&0&0&0&0&0&0 \\ 1&0&1&0&0&0&0&0 \\ 1&1&1&1&0&0&0&0 \\ 1&0&0&0&1&0&0&0 \\ 1&1&0&0&1&1&0&0 \\ 1&0&1&0&1&0&1&0 \\ 1&1&1&1&1&1&1&1 \end{matrix} \longrightarrow \cdots
$$

The formula for the binomial coefficient in Pascal's triangle is: $P(0,j) = 1$, $P(i,j) = 0$ for $i > j$ and $P(i,j) = P(i-1, j-1) + P(i-1, j)$ for the remaining cases. Considered modulo 2, this formula gives raise to the transformation below acting on a lattice `Grid2`:

```
trans ST1 = { <undef> |south> x |west> y => (x+y) mod 2, x, y }
```

In this rule, the comma is refined using a GBF generator: a |south> b means that b is a neighbor of a following the south direction. The transformation must be iterated on an initial lattice where the position $(0, j)$ are filled with 1 and positions $(i, 0)$ are filled with 0 for $i > 0$.

However, this transformation uses arithmetic operators (the + and mod). A more elementary computation is possible, turning the formula modulo 2 into a tiling process. Following [11] we consider 4 tiles corresponding to the two boolean

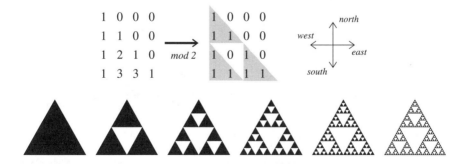

Fig. 4. *Upper line:* taking the binomial coefficients modulo 2 produces the shape of the ST. *Lower line:* ST can also be produced by iterating the carving of a triangle inside another triangle.

values a cell (i,j) receives from the cells $(i-1, j-1)$ and $(i-1, j)$. This tiling is easily coded and then simulated in MGS. We use the four 4 symbols 'T00, 'T10, 'T01 and 'T11 to represents the 4 types of tiles: tile 'Txy at position (i,j) means that x is the value of $P(i-1,j)$ and y is the value of $P(i-1, j-1)$. So the value 0 is represented by either 'T00 or 'T11 and the value 1 by 'T10 or 'T01. Finally, we use a transformation with 4 rules to specify the placement of the tiles:

```
trans ST2 = {
    <undef> |south> ('T00|'T11) as x |west> ('T01|'T10) as y
    => 'T01, x, y

    <undef> |south> ('T00|'T11) as x |west> ('T00|'T11) as y
    => 'T11, x, y

    ... two additional symmetric rules ...
}
```

The path pattern works as follow: the | operator in a pattern denotes an alternative: 'T00 | 'T11 matches the symbol 'T00 *or* the symbol 'T11; the **as** construct is used to bind the value of a pattern fragment to a variable: in ('T00 | 'T11) **as** x the pattern variable is bound to the actual value matched by the pattern.

5   Carving Sierpinski Triangles

Building a ST by carving is illustrated in figure 4. This process is also easily coded in MGS using *patch patterns* on *abstract cellular complexes*.

An abstract cellular complex is composed of elements of various dimensions (vertices, edges, surfaces, . . .) called *topological cells* of dimension n or n-cells [10]. These basic elements are organized following the *incidence relationship* that relies on the notion of boundary: let c_1 and c_2 be respectively a n_1-cell and an n_2-cell with $n_1 < n_2$, c_1 is incident to c_2 if c_1 belongs to the border of c_2. More especially, if $n_1 = n_2 - 1$, c_1 is called a *face* of c_2, and c_2 is a *coface* of c_1.

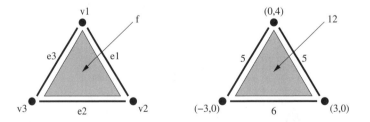

Fig. 5. On the left is an example of a cellular complex: it is composed of 3 0-cells (v_1, v_2, v_3), 3 1-cells (e_1, e_2, e_3), and a 2-cell f. The boundary of f is formed by its incident cells v_1, v_2, v_3, e_1, e_2 and e_3. Especially, the 3 edges are the faces of f, and therefore, f is the coface of e_1, e_2 and e_3. On the right, data are associated with the topological cells: positions are associated with vertices, lengths with edges and area with f.

This data structure generalizes the idea of graph, that is a complex composed of 0-cells and 1-cells. As the definition of a GBF collection uses the elements of a mathematical group as indexes, here n-cells are used as indexes to define a cellular complex based topological collection. Basically, a value is associated with each topological cell. This corresponds to the concept of *topological chain* in algebraic topology. This notion won't be detailed in the paper. An example of such a collection is given on figure 5.

Patch transformations have been created to handle any arbitrary cellular subcomplex. The main advantage of using these complexes is that we can handle cells of various dimensions to represent all the elements that compose the ST. In fact, in the previous representation, the ST were patterns appearing on a matrix of digits, that is, on a predefined space. Here the concrete geometric structure of the ST is specified and the building of the ST also builds "its own embeding space".

To represent the ST, we use an abstract cellular complex where the value of a vertex represents the coordinate of an embedding of the ST in the plane.

There are two transformations used to carve the ST. The first one, AV, adds a vertex in the middle of each edges (see figure 6):

```
patch AV = {
  ~v1 < e:[dim = 1] > ~v2
  => 'v:[dim = 0, cofaces = ('e1,'e2),
         val = { x=(v1.x+v2.x)/2, y=(v1.y+v2.y)/2, new=true }]
     'e1:[dim = 1, faces = (v1,'v)]
     'e2:[dim = 1, faces = (v2,'v)]
}
```

The keyword **patch** is used instead of the keyword **trans** to outline that the defined transformation uses patch patterns in its rules. In this patch transformation, v1 and v2 are not consumed (the ~ qualifier in front of an identifier) to allow the matching of all the edges incident to a same vertex. Indeed, if an element is matched by a pattern, it can't be matched in another one: two subcollections matched by the l.h.s. of some rules of a transformation cannot overlap. We say that the elements matched by a pattern are *consumed*. Here, if a vertex was matched and consumed together with one of its incident edges, no any other incident edges could be matched by the rule. A clause c1 < c2 means that cell c1 is incident to cell c2 and of lower dimension. The right hand side of the rule is a special form used to transform the matched edge e into two edges 'e1 and 'e2 incident to a new vertex 'v. A flag new distinguishes the newly created vertices.

The next step looks for all the hexagons and replaces them with three triangles (see figure 6):

```
patch RF = {
  f:[dim=2, faces = (e1,e2,e3,e4,e5,e6)]
  ~v1 < ~e1 > ~v2:[? v2.new] < ~e2 >
  ~v3 < ~e3 > ~v4:[? v4.new] < ~e4 >
  ~v5 < ~e5 > ~v6:[? v6.new] < ~e6 > ~v1
```

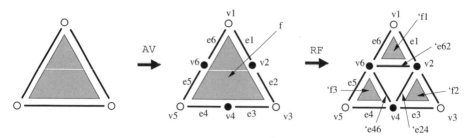

Fig. 6. Carving a triangle. The first transformation AV adds vertex in the middle of an every edge. The second transformation RV refines the central hexagonal face into three triangles.

```
=> 'e24:[dim=1, faces=(v2,v4)]
   'e46:[dim=1, faces=(v4,v6)]
   'e62:[dim=1, faces=(v6,v2)]
   'f1:[dim=2, faces=(e6,e1,'e62)]
   'f2:[dim=2, faces=(e2,e3,'e24)]
   'f3:[dim=2, faces=(e4,e5,'e46)]
}
```

In this patch, only the hexagon f is matched and consumed. We select its boundary without consuming it. Note the guards in the specification of the matched vertices: a flag is used to match only newly created hexagons.

6   Discussion and Conclusion

In this paper we have presented the use of a DSL language for the modeling and the simulation of two kinds of self-assembly processes: by accretive growth and by space carving. Despite their specificities, we are convinced that they are paradigmatic of a full class of self-assembly processes.

Most of the examples described in this paper relie on chemical processes. The sierpinsky gasket pattern has been really implemented using DNA molecules. Previously, the process has been designed and simulated using the *kinetic Tile Assembling Model* (kTAM) [11]. kTAM provides a complete framework for the description of such chemical reactions where a lot of physical parameters (like temperature, error rates, ...) are taken into account to allow accurate studies of crystallization processes. The DNA assembly of tridimensional fractal has been proposed and studied in [2], based on DNA trigonal tiles. Compared to this work, the MGS modelings presented in this work are much more abstract: the purpose is not to study the physical implementation using a DNA computing paradigm but to investigate the shape produced by some families of abstract self-assembly processes.

Obviously, the mechanisms provided by MGS allow the specifications of more complex and abstract operations, that could be very difficult to implement using polymerization and depolymerization reactions of kTAM for instance. These higher level features can be used in the domain of robotics self-assembly. For instance, [17] presents the elaboration of a self-reproducing machine. This

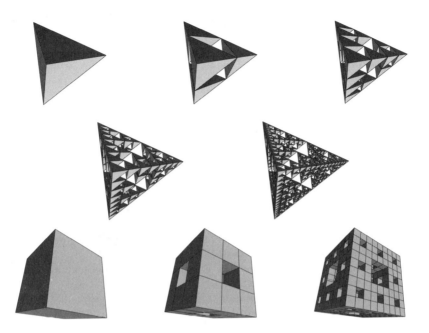

Fig. 7. On top, Sierpinski sponge building process: initial state and steps 1, 2, 3 and 4. At bottom, Menger sponge building process: initial state and steps 1 and 2.

machine is composed of elementary cubic modules. Each module is able to behave in different ways: pivoting, connecting or disconnecting with other modules, transfering data and power to its connected neighbors. The organization and the complex behaviors of the whole machine could be captured by a MGS modeling using topological collections and transformations. The modeling in MGS of such complex self-assembly processes, where we must specify the complex interaction of a few complex entities, is a part of our current work.

We insist on the expressivity brought by the notions of topological collections and their transformations. For example, the patch language used in section 5 is powerful enough to produce Sierpinski and Menger sponge (a generalization of carving a tetrahedron and a cube in 3D), see figure 7. MGS has also been succesfully used to model several biological growth processes, like the development of an epithelial sheet or a neurulation process [12], as well as the flock of birds or the subdivision of a triangulated surface.

References

[1] Abelson, Allen, Coore, Hanson, Homsy, Knight, Nagpal, Rauch, Sussman, and Weiss. Amorphous computing. *CACM: Communications of the ACM*, 43, 2000.
[2] A. Carbone, C. Mao, P. E. Constantinou, B. Ding, J. Kopatsch, W. B. Sherman, and N. C. Seeman. 3D fractal DNA assembly from coding, geometry and protection. *Natural Computing*, 3(3):235–252, 2004.

[3] M. Eden. In H. P. Yockey, editor, *Symposium on Information Theory in Biology*, page 359, New York, 1958. Pergamon Press.

[4] J.-L. Giavitto. Invited talk: Topological collections, transformations and their application to the modeling and the simulation of dynamical systems. In *Rewriting Technics and Applications (RTA'03)*, volume LNCS 2706 of *LNCS*, pages 208 – 233, Valencia, June 2003. Springer.

[5] J.-L. Giavitto and O. Michel. Declarative definition of group indexed data structures and approximation of their domains. In *Proceedings of the 3nd International ACM SIGPLAN Conference on Principles and Practice of Declarative Programming (PPDP-01)*. ACM Press, Sept. 2001.

[6] J.-L. Giavitto and O. Michel. Modeling the topological organization of cellular processes. *BioSystems*, 70(2):149–163, 2003.

[7] P. Horn. Autonomic computing: IBM's perspective on the state of information technology. Technical report, IBM Research, Oct. 2001. http://www.research.ibm.com/autonomic/manifesto/autonomic_computing.pdf.

[8] K. N. Kutulakos and S. M. Seitz. A theory of shape by space carving. *International Journal of Computer Vision*, 38(3):199–218, July 2000.

[9] V. Manca, C. Martin-Vide, and G. Paun. New computing paradigms suggested by dna computing: computing by carving. *Biosystems*, 52(1-3):47–54, Oct. 1999.

[10] J. Munkres. *Elements of Algebraic Topology*. Addison-Wesley, 1984.

[11] P. W. K. Rothemund, N. Papadakis, and E. Winfree. Algorithmic self-assembly of dna sierpinski triangles. *PLoS Biol*, 2(12):e424, 2004. www.plosbiology.org.

[12] A. Spicher and O. Michel. Declarative modeling of a neurulation-like process. In *Sixth International Workshop on Information Processing in Cells and Tissues (IPCAT'05)*, pages 304–317, York, August 2005.

[13] A. Spicher, O. Michel, and J.-L. Giavitto. A topological framework for the specification and the simulation of discrete dynamical systems. In *Sixth International conference on Cellular Automata for Research and Industry (ACRI'04)*, volume 3305 of *LNCS*, Amsterdam, October 2004. Springer.

[14] I. Stewart. Four encounters with sierpinski's gasket. *Mathematical Intelligencer*, 17:52–64, 1995.

[15] T. A. Witten and L. M. Sander. Diffusion-limited aggregation, a kinetic critical phenomenon. *Phys. Rev. Lett.*, 47:1400–1403, 1981.

[16] S. Wolfram. *A new kind of science*. Wolfram Media, 2002.

[17] V. Zykov, E. Mytilinaios, B. Adams, and H. Lipson. Self-reproducing machines. *Nature*, 435(7038):163–164, 2005.

Evolutionary Design of a DDPD Model of Ligation

Mark A. Bedau[1,2,3,*], Andrew Buchanan[1,2], Gianluca Gazzola[1,2],
Martin Hanczyc[1,2], Thomas Maeke[2,4], John McCaskill[2,4],
Irene Poli[2,5], and Norman H. Packard[1,2]

[1] Protolife S.r.l., Via della Libertà 12, Marghera, Venezia 30175, Italy
[2] European Center for Living Technology, S. Croce 1681, Venezia 30135, Italy
[3] Reed College, 3203 SE Woodstock Blvd., Portland OR 97202, USA
[4] Biomolecular Information Processing, Ruhr-Universitat Bochum,
c/o IZB Schloss Birlinghoven, D-53754 Sankt Augustin, Germany
[5] University of Venice Ca' Foscari, San Polo 2347, Venezia 30125, Italy
mark@protolife.net

Abstract. Ligation is a form of chemical self-assembly that involves
dynamic formation of strong covalent bonds in the presence of weak associative forces. We study an extremely simple form of ligation by means
of a dissipative particle dynamics (DPD) model extended to include the
dynamic making and breaking of strong bonds, which we term dynamically bonding dissipative particle dynamics (DDPD). Then we use a
chemical genetic algorithm (CGA) to optimize the model's parameters
to achieve a limited form of ligation of trimers—a proof of principle for
the evolutionary design of self-assembling chemical systems.

1 Evolutionary Design of Self-assembling Chemical Systems

Many familiar examples of supramolecular self-assembly—such as micelles and
vesicles—result solely from the dynamics of weak associative forces between
molecules. Such structures contain strong intramolecular covalent bonds that
are relatively fixed during the self-assembly process. Here we consider the self-
assembly of supramolecular structures formed through the dynamics of strong
bond formation in the presence of weak associative forces. Specifically, we focus
on the self-assembly that occurs during monomer to polymer ligation, as part
of the process of complementary templating. During the ligation process, weak
associative forces enable the template to act as a physical catalyst for the construction of the complementary polymer's strong bonds. We study templating
partly because it is critical in the growth, reproduction, and evolution of all
contemporary biological life, but mainly because it is one of the simplest forms
of molecular self-assembly that involves the dynamics of both strong bonds and

---

[*] Corresponding author.

E. Talbi et al. (Eds.): EA 2005, LNCS 3871, pp. 201–212, 2006.

weak associative forces. In addition this process results in the replication and transfer of chemical information.

Evolution in nature has created exquisite chemical systems for ligation. All fundamental processes in the cell such as DNA replication, transcription and translation are based upon template-directed ligation of monomers. Our goal here is to create an artificial evolutionary process that designs a chemical system that achieves a simple analog of ligation. Other kinds of artificial evolutionary processes have been used for chemical design; in particular, "directed" or "in vitro" evolution has been used to design molecules with specific desired functionality [1, 2, 3, 4, 5, 6, 7]. But our evolutionary design procedure is different from directed evolution in two crucial respects. First, rather than evolving a population of molecules (e.g., RNA) for a specific function, we evolve a population of experimental parameters that describe a complete chemical system or process. While directed evolution aims to optimize individual functional molecules, our procedure aims to optimize whole chemical systems or processes containing a number of chemical species engaged in myriad chemical reactions. Second, directed evolution involves chemical systems that contain molecules encoding the information that is evolving. By contrast, in our method the information that is evolving is encoded outside the chemical system (in an experimenter's lab notebook or inside a computer). Thus, our method can be applied to design virtually any kind of chemical system or process.

The work reported here concerns the evolutionary design of a chemical model, not a real chemical system. However, this is not a limitation of our method. The same method could be used to design real self-assembling chemical systems, ultimately including quite complex systems like artificial cells that involve the integration of different chemical systems for containment, metabolism, and genetics [8]. As it happens, the chemical systems we optimize here are analogous to the chemical system of non-enzymatic template-directed synthesis [9, 10, 11, 12, 13, 14, 15]. Like template-directed systems of ligation *in vivo* and *in vitro*, our system is supplied with a template molecule and an excess of monomers. It then evolves so as to optimize the assembly of monomers on the template to produce a ligated copy of the template. In Section 2 we describe our dynamic-bonding dissipative particle dynamics (DDPD) chemical model. A description of the chemical genetic algorithm (CGA) used to design chemical systems follows in Section 3. The results of applying the chemical genetic algorithm to DDPD models that achieve a simple form of ligation are presented and discussed in Section 4, followed in Section 5 by a discussion of the proper design of CGAs and their practical limitations. We conclude in Section 6 with a discussion of some different kinds of dynamics in evolutionary design or "programming" of chemical systems.

2   Dynamic-Bonding Dissipative Particle Dynamics (DDPD)

Our model of chemical reaction systems is based on the well-studied dissipative particle dynamics (DPD) framework [16, 17, 18, 19, 20, 21]. The DPD framework is a mesoscopic system simulator meant to bridge the gap between molecular dynamics (MD) models and continuous substance models. The extreme computational demands of MD models make them appropriate only for simulating small systems for brief intervals—orders of magnitude smaller than the time and length scales of interest here. Continuous substance models are inappropriate as models of molecular scale systems in which the discrete nature of particles impacts the dynamics of the system.

In DPD, the equations of motion are second order, with explicit conservation of momentum, in contrast to Langevin or Brownian dynamics. Solvent molecules may be represented explicitly, but random and dissipative forces are included in the dynamics to compensate for the dynamical effects of replacing the hard short-range potentials of MD by softer potentials in DPD simulations. This procedure allows a major accelaration of the simulation compared with MD.

Our work is based on a DPD implementation of a model of monomers and polymers in water. Some elements in the model represent bulk water (one model element representing many molecules). Other elements could represent hydrophilic or hydrophobic monomers. In some cases those elements are connected by explicit bonds, which are represented as springs that freely rotate about their ends. These complexes explicitly but very abstractly represent the three-dimensional structure of polymers. For example, amphiphilic molecules can be created by explicitly bonding a hydrophilic monomer "head" onto a hydrophobic "tail" (chain of hydrophobic monomers).

All the elements move in a two- or three-dimensional continuous space, according to the influences of four forces. A conservative force governs symmetric pairwise repulsion and attraction of elements. A dissipative force causes the kinetic energy of elements to move towards equilibrium with other elements in the region. A random force imparts kinetic energy to the elements in arbitrary directions. The strength of the random force is calibrated to balance the lessening of system energy due to the dissipative force, maintaining the temperature of the system around a more or less fixed point. All of these forces are considered to operate only within a certain local cutoff radius. The cutoff radius is a main mechanism for improving model feasibility. Elements which are strongly bonded to other elements are also influenced by the movement of those elements to which they are bonded, through the spring that connects them.

The DPD framework supports two distinct types of particle interaction. The first type of particle interaction is referred to as "strong bonds," which represent covalent chemical bonds. All strong bonds in DPD are specified initially, and subsequently cannot form or break. Strong bonds are modeled by a Hooke's law spring. One limitation imposed on the DPD simulations discussed here is that each element can have at most two strong bonds at a given time. The second type of particle interaction corresponds to weak forces such as van der Waals forces or hydrogen bonds. Weak interactions are modeled by the Lenard-Jones potential, with different parameter values possible for interactions between different particle types. In contrast to real systems, attractive forces are not limited to a pair of elements, but may simultaneously occur between a single element and many others. Orientation of individual elements also plays no role, as DPD elements are completely symmetrical. Thus, the pairing that occurs is a cooperative phenomenon.

DPD thermodynamic forces can create self-assembled structures held together with the weak associative forces. For example, a DPD system with amphiphiles in water can exhibit a wide variety of the known supramolecular amphiphilic phases, including monolayers, bilayers, micelles, rods, vesicles, and bicontinuous cubic structures [19, 22, 23, 24, 25].

We augment the DPD framework by making strong bonds dynamic. This dynamic-bonding DPD (or DDPD) is a DPD that is augmented with the following two rules:

- Bonds form (probability 1) if elements are within the bond-forming radius.
- Bonds break (probability 1) if bonded elements are outside the bond-breaking radius.

The strong bond strength parameter governs the strength of all strong bonds, whether or not they were present in the initial conditions. An obvious generalization is to allow lower probabilities in the two bonding rules.

Note that the temperature of the system changes when bonds form and break. However, the momentum in the system is constant, since the changes in the momentum of individual elements due to bonding events are always symmetrical with respect to the bonded particles.

Chemical amplification via templating is the basic mechanism of DNA replication, and also of simpler replicator systems such as von Kiedrowski's autocatalytic replicator system [26] and peptide replicators [27]. Monomers of a given type may participate in a weak interaction with monomers of a complementary type, and each may form strong bonds with a monomer of any type if the two are in the correct proximity and orientation. Given a template polymer made up of different types of units and a reservoir of free floating monomers, each unit of the template polymer can associate weakly with a complementary monomer. When and if the weak forces bring the units into the correct orientation and proximity with complementary units in the template polymer, strong bonds form between the monomer units producing a complementary polymer through the process of ligation.

If the paired complementary polymers are separated by a mechanism such as duplex melting due to temperature change or protein action, then each polymer may repeat the process, creating more templates and complements. By this means, the overall number of polymers in the system increases. Although this process results in the chemical amplification of polymers, the focus of the present work is simply ligation, and the optimization of parameters that result in the organization and ligation of monomers into polymers.

To keep the chemical system as simple as possible, we focus solely on the ligation of two types of monomers into trimers, and we prevent strong bonds from breaking. Pairs of opposite type units attract each other, while like type units are unlikely to become associated by weak forces, which is roughly analogous to complementary base pairing in the context of nucleotides.

We do not report here a complete analysis on catalytic efficiency of the templating process, with fitting of rate constants and comparison with background rates. Such analysis will be reported in future work.

3   Chemical Genetic Algorithm (CGA)

We now describe a "chemical genetic algorithm" (CGA) for designing chemical systems. We use the CGA to optimize the parameters of a DDPD model of ligation. The CGA could equally well be used to optimize parameters for other models, or for other chemical systems, or for other systems in general.[1]

The search space of our CGA is a subset of DDPD parameters. In particular, our genes are five chemical system parameters: (i) the strength of the attractive conservative force between complementary particle types, (ii and iii) the strength of the repulsive conservative force between the two types of like particles, (iv) the bond-forming radius, and (v) the bond strength. In the context of this paper, a genome is always a set of these five chemical systems parameters. Complete details regarding the specific values used can be found in the appendix.

The CGA search procedure starts by measuring the fitness of the genomes that form the first generation. Then the following loop is repeated until the experiment ends: The most recently produced instance of the most fit genome is used to create a subsequent generation of genomes, by mutations of the five genome parameters. These mutations are governed by a global mutation rate, which acts within a range and style of variation defined for each parameter. A candidate mutated genome is included in a subsequent generation only if it differs from each genome tested in any previous generation. Then the fitness of each genome in the new generation is measured.

A genome's fitness is measured by starting the DDPD with the genome's parameters and seeding the system with free monomers and template trimers. No complementary trimers are included initially. The fitness of a genome is the number of complementary trimers formed after a globally fixed number of model updates. Many generalizations and modifications of our search algorithm and fitness function could be explored.

4   Results of Evolutionary Design of a Ligation Model

We used the CGA to design chemical systems for complementary-bonding ligation dozens of times, all with roughly the same results. Figure 1 shows the time

---

[1] It is worth noting that our chemical genetic algorithm differs from another algorithm devised by H. Suzuki that has been given the same name [28]. Inspired by metabolic reactions of molecules responsible for the biological translation of genetic information, Suzuki's algorithm is an unusual genetic algorithm that includes analogues of a cell containing tRNAs, amino acids, and aminoacyl-tRNAs, as well as DNA. Our CGA, by contrast, is a ordinary genetic algorithm, but one that is applied to the problem of designing optimal chemical systems.

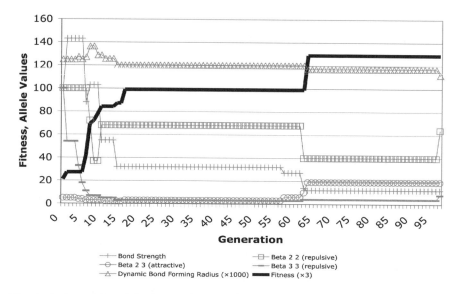

Fig. 1. A time series of the fitness (bold line) and allele values of the most fit chemical system in each generation of a typical CGA run. This shows the lineage of the most fit genomes through five-dimensional parameter space, indicating which allele (model parameter) changes correspond to each fitness increase. Fitness and many model parameters are scaled to improve visualization. Note that fitness increases overall.

series of the fitness and allele values of the most fit chemical system in a typical CGA run in this series. The fitness increases over time, in fits and starts (common with genetic algorithms), and the allele values in the genome of the most fit system change with each fitness increase.

Figure 2 shows the genealogical tree of each chemical system generated in the same CGA run. It vividly shows how the operation of the CGA allows only those measured as most fit to be parents. The top line of the genealogy corresponds to the lineage of the final optimal genome designed by the CGA.

Figure 3 again shows the maximum fitness in Figure 1, but now superimposed with twenty fitness measurements of some of the most fit systems. Each scatter plot was created by rerunning the DDPD parameters with twenty different random initial conditions. As might have been expected, the scatter reveals significant noise in our fitness measurements.

Figure 3 supports two conclusions. First, the overall increase in average fitness shows that the CGA is genuinely creating chemical systems with significantly better fitness. In other words, the CGA works as desired; this holds in general when we have used the CGA to program ligation systems. Second, increases in *measured* fitness do not always correspond to increases in *actual* fitness, because of noise in the fitness measurements, e.g., an initial configuration that creates an unusually large number of complementary trimers.

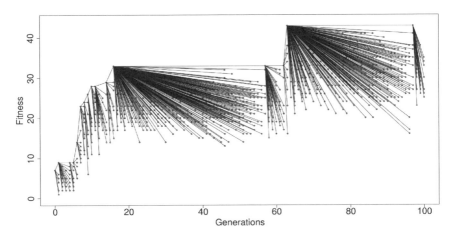

Fig. 2. A genealogical tree of each chemical system tested by the same CGA run shown in Figure 1. The generation of each system is indicated on the x axis, and its fitness on the y axis. Diagonal lines indicate parentage. Note that a parent is often not in the generation immediately preceding its children. Multiple systems in a given generation with the same fitness are shown as one point.

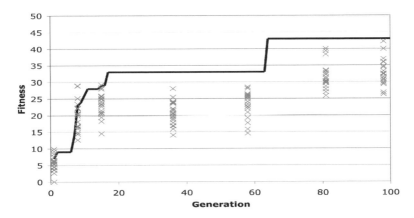

Fig. 3. A time series of the fitness of the most fit chemical system in each generation of the CGA run shown in Figures 1 and 2, overlaid with scatter plots of twenty fitness evaluations of some of those systems (with some random noise added, to distinguish identical fitness values). As expected, the fitness of the most fit is sometimes higher than any of the fitness values in the corresponding 20-value scatter plot, because the CGA generates many more than 20 trials.

5   Chemical Genetic Algorithm Design

Proper design of a CGA involves confronting trade-offs between the accuracy of fitness measurements and the evolutionary design time scale. More accurate measurements are always possible, but they take more time. The number of different systems that a CGA can evaluate is strongly limited by available time and technology. At the same time, the effectiveness of a CGA is limited by the accuracy of its fitness measurements.

5.1   Noise in Fitness Measurements

The true fitness of a genome that describes the parameters of a given chemical system is the system's propensity to produce templating reactions under a variety of initial conditions. The fitness function actually used here, however, is the number of complementary templates formed starting from a single random initial configuration. Specific attention is not given to how the complements were formed, what other chemical species were formed, or how many might form under different initial conditions. The scatter in Figure 3 shows that this is an imperfect measure of actual templating propensity. The noisiness of our fitness measurements strongly depends on the chemical system's experimental parameters, whether or not they are in the genome.

In general, the noisiness of each fitness function must be measured empirically. We could do this by applying the fitness function to a single genome in a variety of contexts, by varying such things as the number of model updates, U, used in one fitness measurement, the system size, S, and the initial density of monomers, M. The noisiness of each point in $U \times S \times M$ space could be measured with a scatter plot.

Any GA, including the CGA, can function properly only given sufficiently accurate fitness measurements. One could more accurately assess fitness by averaging repeated fitness measurements under different initial conditions, but this takes substantially more time. To ensure both proper CGA function and optimal CGA design speed, one should make the minimum number of fitness measurements necessary for the requisite level of significance in measured fitness. This raises a precise statistical question: How many fitness measurements are required to get an accurate enough measurement that the CGA can continually find better model parameters?

Fixing the values of all the simulation parameters and then repeating the simulation from different random seeds n times, one obtains n values of the fitness function, $X_1, \ldots, X_n$. Clearly, these variables are independent and identically distributed. The distribution of these variables depends on the details of the simulation, which depend on the parameters that govern the simulation, including U, S, and M, as well as the parameters encoded by the genome. Assume for the moment that this distribution is roughly approximated by a normal distribution. Then we could infer the mean value, μ, of the fitness function for n repetitions by means of standard confidence interval estimation techniques based on the t-distribution for the pivotal quantity $\frac{\sqrt{n}(\bar{x} - \mu)}{\hat{\sigma}}$, where $\bar{x}$ is the sample mean and $\hat{\sigma}$ is the estimate of the standard deviation. This would permit us to determine a sample size necessary to obtain an arbitrary desired accuracy, A. For a 95 percent confidence interval we can in fact derive from $A = \pm t_{n-1;0.025} \frac{\hat{\sigma}}{\sqrt{n}}$ the smallest sample size that leads to the desired accuracy.

This would permit us to determine a sample size necessary to obtain an arbitrary desired accuracy in estimated mean fitness, given the parameter values.

Of course, one would have to validate empirically whether the variation in fitness measurements is well approximated by a normal distribution. If not, through an analysis of the simulation process, one could derive a better approximation of this distribution and base the sample size calculation on this approximation.

5.2 Chemical Design Time Scales

The present discussion of time scales for simulation are based on use of a Macintosh Dual 2 GHz PowerPC with no parallelization (single-threaded code). The algorithm was not particularly optimized, but performs at speeds comparable to other DPD research codes on benchmark problems.

The running time of a CGA depends on the following numbers (with values for the present CGA results in parentheses):

- seconds per model update (10^{-1})
- model updates per fitness evaluation ($10^2 - 10^3$)
- systems evaluated ($10^3 - 10^5$)

Combining these numbers shows that each CGA run takes between $10^4 - 10^7$ seconds, that is, between hours and months. This spans the range of experiments worth and not worth attempting. Furthermore, we saw above that successful CGA operation might require averaging repeated fitness measurements before estimating a chemical system's actual fitness. Such repeated measurements would increase the running time of the CGA by an order of magnitude.

Thus, the time feasibility of our CGA designing DDPD parameters critically depends on the number of model updates required for each fitness evaluation and the accuracy of the evaluation. For example, the spontaneous self-assembly of vesicles in the DPD framework typically takes on the order of a week of user time, so the fitness function for a CGA designing vesicles would probably require about the same amount of time. Hence, the CGA would take years to evaluate the fitness of even hundreds of systems—which is clearly beyond the bounds of human patience.

We conclude by noting that the execution times discussed here may be significantly improved by hardware and clever coding, as well as DDPD enhancements that lead to more complex particle interaction primitives. On the other hand, full simulation of dissociation at thermal equilibrium could increase execution times.

6   Dynamics of Evolutionary Design of Chemical Systems

One can distinguish three kinds of dynamics involved in the evolutionary design of chemical systems. First, the DDPD model involves the dynamic of chemical species. This dynamic takes place in a continuous two- or three-dimensional space supporting spontaneous self-assembly processes. Bonds form and break; the concentrations of chemical species rise and fall; new species are created; old species go extinct. DDPD models achieve these dynamics by the addition of dynamic strong bonds.

The evolutionary design of DDPD parameters sufficient for ligation of trimer templates is a step toward a second kind of dynamic—evolution of informational polymers by natural or artificial selection. Modeling the evolution of informational polymers is a burgeoning field. The focus on the line of work presented here is a model in which the polymer evolution is produced by catalytic activity physically embodied in explicit spatial structures—an example of what could be called "embodied information processing."

The chemical GA itself is a proof of principle for a third kind of dynamic, specifically, the evolutionary design of a chemical system with prespecified functional properties. The scheme used here for the evolutionary design of chemical model parameters for ligation could be used to design model parameters for different self-assembled structures, such as micelles or lamellar sheets. The scheme could also be used to design the parameters of other kinds of models entirely. If those models are realistic, then the evolved model parameters could be used to design real chemical systems. The CGA can also be used to design real chemical systems directly, such as those that produce some desired kind of self-assembled structures. Designing a CGA to produce a specific kind of self-assembled system is a method for "programming" such systems.

Acknowledgments. Thanks to the European Center for Living Technology (ECLT) for hospitality during the work reported here. Thanks to Rüdi Fuchslin for conversations on DPD development. For comments on the manuscript, thanks to Peter Dittrich, Jim LaClair, Emily Parke, and Hideaki Suzuki. Thanks also to audiences at the ECLT in February and March of 2005, and at the Computer Science School in the University of Nottingham, where MAB presented this work. This work was funded in part by PACE (Programmable Artificial Cell Evolution), a European Integrated Project in the EU FP6 IST FET Complex Systems Initiative.

References

[1] Tuerk, C., Gold, L.: Systematic evolution of ligands by exponential enrichment: RNA ligands to bacteriophage T4 DNA polymerase. *Science* **249** (1990) 505–510.
[2] Ellington, A.D., Szostak, J.W.: In vitro selection of RNA molecules that bind specific ligands." *Nature* **346** (1990) 818–822.
[3] Irvine, D., Tuerk, C., Gold, L.: SELEXION. Systematic evolution of ligands by exponential enrichment with integrated optimization by non-linear analysis. *Journal of Molecular Biology* **222** (1991) 739–61.
[4] Chapman, K.B., Szostak, J.W.: In vitro selection of catalytic RNAs. *Current Opinions in Structural Biology* **4** (1994) 618–22.
[5] Rohatgi, R., Bartel, D.P., Szostak, J.W.: Nonenzymatic, template-directed ligation of oligoribonucleotides is highly regioselective for the formation of 3'-5' phosphodiester bonds. *Journal of the American Chemical Society* **118** (1996) 3340–3344.
[6] Wright, M., Joyce, G: Continuous in vitro evolution of catalytic function. *Science* **276** (1997) 614–617.
[7] Joyce, G.: Directed evolution of nucleic acid enzymes. *Annual Review of Biochemistry* **73** (2004) 791–836.

[8] Rasmussen, S., Chen, L., Deamer, D., Krakauer, D., Packard, N, Stadler, P., Bedau, M.: Transitions from nonliving to living matter. *Science* **303** (2004) 963–965.

[9] Joyce, G.F., Inoue, T., Orgel, L.E.: Non-enzymatic template-directed synthesis on RNA random copolymers. Poly(C, U) templates. *Journal of Molecular Biology* **176** (1984) 279–306.

[10] Acevedo, O.L., Orgel, L.E.: Non-enzymatic transcription of an oligodeoxynucleotide 14 residues long. *Journal of Molecular Biology* **197** (1987) 187–93.

[11] Zielinski, W.S., Orgel,L.E.: Oligoaminonucleoside phosphoramidates. Oligomerization of dimers of 3'-amino-3'-deoxy-nucleotides (GC and CG) in aqueous solution. *Nucleic Acids Research* **15** (1987) 1699-1715.

[12] Joyce, G.F., Orgel, L.E.: Non-enzymatic template-directed synthesis on RNA random copolymers. Poly(C,A) templates. *Journal of Molecular Biology* **202** (1988) 677–681.

[13] Hill, A.R. Jr., Orgel, L.E., Wu, T.: The limits of template-directed synthesis with nucleoside-5'-phosphoro(2-methyl)imidazolides. *Origins of Life and Evolution of the Biosphere* **23** (1993) 285–90.

[14] Liu, R., Orgel, L.E.: Enzymatic synthesis of polymers containing nicotinamide mononucleotide. *Nucleic Acids Research* **23** (1995) 3742-3749.

[15] Bohler, C., Nielsen, P.E., Orgel, L.E.: Template switching between PNA and RNA oligonucleotides. *Nature* **376** (1995) 578–81.

[16] Hoogerbrugge, P., Koelman, J.: Simulating microscopic hydrodynamic phenomena with dissipative particle dynamics. *Europhysics Letters* **19** (1992) 155–160.

[17] Groot, R., Warren, P.: Dissipative particle dynamics: bridging the gap between atomistic and mesoscopic simulations. *Journal of Chemical Physics* **107** (1997) 4423–4435.

[18] Marsh, C.: Theoretical aspects of dissipative particle dynamics. Ph.D. Thesis, University of Oxford, 1998.

[19] Shillcock, J., Lipowsky, R.: Equilibrium structure and lateral stress distribution from dissipative particle dynamics simulations. *Journal of Chemical Physics* **117** (2002) 5048–5061.

[20] Vattulainen, I., Karttunen, M., Besold, G., Polson, J.: Integration schemes for dissipative particle dynamics simulations: From softly interacting systems towards hybrid models. *Journal of Chemical Physics* **116** (2002) 3967–3979.

[21] Trofimov, S., Nies, E, Michels, M.: Thermodynamic consistency in dissipative particle dynamics simulations of strongly nonideal liquids and liquid mixtures. *Journal of Chemical Physics* **117** (2002) 9383–9394.

[22] Jury, S., Bladon, P., Cates, M., Krishna, S., Hagen, M., Ruddock, N., Warren, P.: Simulation of amphiphilic mesophases using dissipative particle dynamics. *Physical Chemistry and Chemical Physics* **1** (1999) 2051–2056.

[23] Yamamoto, S., Maruyama, Y., Hyodo, S.: Dissipative particle dynamics study of spontaneous vesicle formation of amphiphilic molecules. *Journal of Chemical Physics* **116** (2002) 5842–5849.

[24] Kranenburg, M., Venturoli, M., Smit, B.: Phase behavior and induced interdigitation in bilayers studied with dissipative particle dynamics. *Journal of Physical Chemistry* **107** (2003) 11491–11501.

[25] Yamamoto, S., Hyodo, S.: Budding and fission dynamics of two-component vesicles. *Journal of Chemical Physics* **118** (2003) 7937–7943.

[26] von Kiedrowski, G.: A self-replicating hexadeoxynucleotide. *Angewandte Chemie International Edition English* **25** (1986) 932–935.

[27] Tjivikua, T., Ballester, P., Rebek Jr., J.: A self-replicating system. *Journal of the American Chemical Society* **112** (1990) 1249–1250.

[28] Suzuki, H., Sawai, H.: Chemical genetic algorithms — Coevolution between codes and code translation. In Standish, R.K., Bedau, M.A., Abbass, H.A. (eds.), *Proceedings of the Eighth International Conference on Artificial Life (Artificial Life VIII)* (2002) 164–172.

Appendix

The DDPD parameters in the work reported here were as follows: The conservative force between particles i and j was given by an approximation of the Lennard-Jones potential $F_{ij}^C = \alpha(\frac{1}{r_{ij}^y} - \frac{\beta}{r_{ij}^z})$, where α is the maximum repulsive force, β is the factor for the attractive force, r_{ij} is the distance between the particles, and y and z are parameters governing the level of approximation. α was initially set to 100 and allowed to vary between 1 and 100 for like type interactions, while β was fixed at 1. For interactions between unlike types α was fixed at 1 and β allowed to vary between 1 and 100, being initially set to 5. In interactions with the "water" particle type, both α and β were fixed at 1. For all particle interactions, y was set to 0 and x to -1.

Every DDPD simulation ran for 500 iterations. The scaling factor for the dissipative and random forces, σ, was 3. The independent scaling factor for the random force was 1.73205 ($\sqrt{3}$). The integration interval, dt, was 0.01. The spring constant governing forces between bonded particles varied between 10 and 400, being initially set to 100. The minimal energy length for bonds was 0.01. The system was a 10 by 10 square initialized with 700 "water" particles, 90 free type-one particles, 180 free type-two particles and 10 chains of two type-one particles followed by a type-two, 1000 total particles, all randomly placed. The dynamic bond forming radius was chosen from 0.1 to 0.5 with a starting value of 0.125. Bonds were not dynamically broken, loops were not allowed to form, and the maximum length for dynamically formed polymers was fixed at 3. Dynamic bonds were allowed to form only between like particles of type-one or -two, and between type-one and type-two particles.

In the CGA, each generation had a population of 10 DDPD parameter files based on the parameter file with the highest fitness to that point, or the more recent file in case of ties. The five parameters that varied did so each with a mutation probability of 0.5. If mutated, the dynamic bond forming radius was chosen at random from ± 10% of the parent parameter. The range for all the other parameters was half to double the parent value. Crossover was not used.

Population Structure and Artificial Evolution

Arthur M. Farley

Computer and Information Science Department,
University of Oregon, Eugene Oregon 97403, USA
art@cs.uoregon.edu

Abstract. We investigate the effect that population structure has upon the course of artificial evolution. We represent an arbitrary population structure by embedding a population of individuals in a graph. Each individual resides at a vertex of the graph and can only choose a mating partner from among its neighbors in the graph. Each individual mates with the selected partner and is replaced by the resultant offspring in the next generation. We embed populations in a variety of trees and mesh-structured graphs and observe differences in rates of change of average fitness and percent polymorphism over successive generations. Results indicate that populations embedded in sparse random graphs having relatively low diameter yield results similar to those embedded in complete graphs.

Keywords: genetic algorithm, graphs, population structure, population dynamics.

1 Introduction

What effect does the structure of a population have upon the course of its evolution? In an unstructured population, every individual can mate with any other individual in the population. This is the implicit assumption of a *standard genetic algorithm* [5] [8]. In such an algorithm, a selection process associates mating probabilities with all pairs of individuals, which probabilities are based solely upon the relative fitnesses of the individuals in P. Mating pairs are then selected according to those probabilities. This selection process is the driving force behind improvements in average fitness over successive generations of a population.

Population structure is that aspect of the selection process that is independent of the fitnesses of individuals. *Population structure* consists of any a priori restrictions on the possibility that one individual of a population encounters another for purposes of mating. In human populations, such restrictions may derive from geographic location, religion, culture, or social class. Here, we consider a genetic algorithm where a priori restrictions are placed upon which individuals can possibly be selected as mating partners for any given individual. These restrictions correspond to the population's structure. To represent arbitrary population structures, we locate individuals at vertices of graphs and restrict individuals to selecting partners only from individuals located at neighboring vertices. We call the resultant genetic algorithm a *graph-embedded genetic algorithm*.

E. Talbi et al. (Eds.): EA 2005, LNCS 3871, pp. 213–225, 2006.
© Springer-Verlag Berlin Heidelberg 2006

There are several motivations for considering graph-embedded genetic algorithms. One is simply to observe the effects that different population structures have upon the course of evolution. Island models, wherein subpopulations are relatively isolated and individuals occasionally migrate, have been a population structure of interest in the study of both natural and artificial evolution [10]. Another motivation is the development and evaluation of methods for structuring the fine-grained parallelization of genetic algorithms [17]. In such a model, each processor in a computer network or multiprocessor manages one individual or several individuals as different processes; processes and processors exchange fitness values and genetic information as needed with their neighbors. How the interconnection structure among processors of a computer network or multiprocessor influences the course of the subsequently generated evolution is one question that is addressed by the research reported here.

1.1   The Model

We consider a generational genetic algorithm acting upon binary haploid individuals. We represent haploid individuals in terms of a single *genotype*, being a sequence of *genes*, each gene having a value chosen from a set of possible *alleles*. Each gene of a *binary genotype* has two possible alleles, which we represent as the set $\{0, 1\}$. A *binary haploid individual* holds a single binary genotype inherited from a single parent or created as a recombination of genotypes from two parents of the previous generation.

A population P consists of a set of $|P|$ indexed individuals $P[i]$, $0 \leq i \leq |P| - 1$. We represent the structure of a population P by a labeled graph $G = (V, E)$, where V is a set of vertices and E is a set of edges, each edge between an unordered pair of vertices (u, v). Each vertex is labeled with a unique integer $0 \leq i \leq |V| - 1$ and is referred to as $V[i]$. In a graph-embedded population P, individual $P[i]$ is located at vertex $V[i]$ of an associated graph G, where $|V| = |P|$. If (u, v) is an edge of G, then u and v are said to be *neighbors* in G. The *degree* of a vertex is equal to its number of neighbors. The *distance* between a pair of vertices in G is the minimum number of edges on a path between them. During graph-embedded evolution, individual $P[i]$ can only select a mating partner from among individuals located at neighbors of $V[i]$ in G. The probability that $P[i]$ selects an individual located at a non-neighboring vertex of $V[i]$ is 0.0.

We represent an environment by a *binary fitness function* mapping a binary genotype to a real value in the range 0.0 to 1.0. The value returned by a fitness function indicates an individual's degree of adaptation to the environment; the higher the fitness function value, the greater is the degree of adaptation. For our study, we use one general class of fitness function, *Step(genSize, stepSize)*, which maintains an arbitrary genotype having *genSize* genes, called the *target*. To determine an individual's fitness value, we compare its *genSize* genotype to the *target*; *stepSize* consecutive locations must be equal to the corresponding *target* locations to get credit for matching a step. Steps are contiguous and non-overlapping. As such, there are only *genSize/stepSize* possible matches. The fitness function returns the percentage of steps that are matched. The *Step* family of fitness functions corresponds to relatively simple environments for evolution, as they define unimodal fitness landscapes, i.e., having a single peak (the *target*). Increasing the *stepSize* adds a degree of difficulty for the evolutionary process, which then must accumulate *stepSize* consecutive values in an individual before the individual's fitness improves. This represents a degree of *epistatic interaction* between neighboring gene loci [7]. While unimodal, *Step* fitness functions with *stepSize* > 1 have local plateaus. These functions have been called *royal road* fitness functions, as they were thought to provide a favorable environment for the success of genetic algorithms [9].

We define a graph-embedded genetic algorithm, as follows:

```
Graph-Embedded Genetic Algorithm(pSize, gSize, f, gType)
    G = Generate-Labeled-Graph(gType, pSize);
    P = Generate-Initial-Population(pSize, gSize);
    Evaluate(P, f);
    until(done())
        {P = GenerateNewPopulation(P, G);
         Evaluate(P, f);
         }
}
```

We first create a labeled graph $G = (V, E)$ of a particular type *gType* with *pSize* vertices. We consider a range of graph types in our experiments, to be defined as we discuss particular experiments below. The fixed-size population P consists of *pSize* binary individuals. The process *GenerateInitialPopulation(pSize, genSize)* creates a random set of *pSize* binary haploid individuals, each individual consisting of a single genotype containing *gSize* alleles; each allele value is selected with equal probability from the set $\{0, 1\}$. With P embedded in G, individual $P[i]$ is located at vertex $V[i]$ of G. We refer to a population P embedded in a graph G of type *gType* to be a P_{gType} population.

The function *done()* returns true if a desired halting condition is met, e.g., either some predetermined number of generations have been considered or successive generations have not differed significantly. The process *GenerateNextPopulation(P, G)* creates a new population from an existing population. In our graph-embedded model of genetic algorithms, this process is as follows:

```
GenerateNewPopulation(P, N)
    {for (0 <= i < pSize)
        {partner = SelectEmbeddedPartner(i, G);}
         New[i] = Mutate(Recombine(P[i], partner));
         }
    for (0 <= i < pSize)
        {P[i] = New[i];}
    }
```

The process *SelectEmbeddedPartner(i,G)* is that element of our model where notions of natural selection and structured populations are captured. We consider a form of *fitness proportionate selection* [8] whereby individual $P[j]$ is selected to be a partner of $P[i]$ with probability f_j/F_p, where f_j is the fitness value of individual j and F_p is the sum of fitness values of all individuals located at vertices that are neighbors of $V[i]$ in G. Our graph-embedded model is a form of *replacement genetic algorithm*, as individual P[i] is replaced by its offspring in the next generation.

The process *Recombine(parent1, parent2)*, takes a pair of haploid individuals and creates a new haploid individual as a result of combining alleles of the two parent's genotypes. We apply *UniformRecombination(parent1, parent2)*, where one of the two alleles from the genotypes of the parents is selected with equal probability at each gene locus. Finally, the genotype of each new individual may be altered by the process *Mutate(p)*. Each gene location is considered in turn and is changed to the other allele value according to a given, constant, per locus *mutation probability*.

1.2   Measures and Parameters

To observe the effects of population structure on evolution, we consider two measures of a population at each generation. The first corresponds to a population's level of adaptation to an environment as measured by its *average fitness*, being the average of all individuals' fitness function values. We expect average fitness to be $1/2^k$ for a *stepSize* of k in a random, initial population. The other measure addresses the genetic diversity of a population. Our diversity measure is *percent polymorphic*, defined as the percentage of gene loci with less than 99 percent of their alleles being a single value in a population [11]. We expect percent polymorphic to be 1.0 in a random, initial population.

There are several parameters that will not change throughout the experiments reported here. We have set the population size to be 1000 for all experiments. This is a relatively small, finite population that makes the simulation experiments feasible, but that is hopefully large enough so that results are not dominated by small population effects, such as genetic drift [6]. Another factor not varied is genotype size, i.e., number of loci or genes in a genotype. We have set the genotype size to be 100, again a compromise between computational efficiency and genetic diversity. We set the mutation rate to be such that an average of 5 alleles over the whole population for each generation are modified by mutation, yielding a per locus mutation probability of $5 * 10^{-5}$, given the population and genotype sizes we have chosen. This rate is in the range of observed mutation rates in nature [11] and seems to provide good performance regarding average fitness with the *Step* fitness function. Our experiments are run for 2000 generations; each experiment is repeated 20 times to generate average values and standard deviations. In some cases, we only present results for the first 1000 generations to highlight the differences observed. We gather values for the two population measures every 50 generations; thus, we report average values at each of 41 generation points (i.e., including the initial population at generation 0).

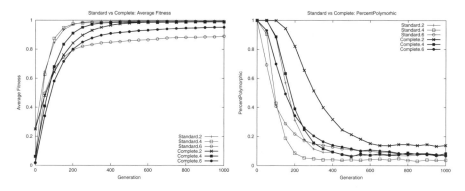

Fig. 1. Results under the standard and complete graph models

2 Experiments and Results

2.1 Complete Graphs

The *complete graph* χ_n is a graph on n vertices, such that every vertex in χ_n is connected to every other vertex by an edge of χ_n (i.e., every vertex is a neighbor of every other vertex). The complete graph is the most highly connected graph possible, having $n(n-1)/2$ edges. A graph-embedded genetic algorithm where the graph is complete corresponds to the standard replacement generational genetic algorithm in an unstructured population. As such, observations of graph-embedded evolution in population $P_{complete}$ will allow us to compare the population dynamics generated by the replacement genetic algorithm with that generated by the standard genetic algorithm. The results for $P_{complete}$ also provide benchmarks for comparison with the dynamics produced by populations that are not as well connected, i.e., more highly structured.

Figure 1 presents results for average fitness and percent polymorphic under the two models with *stepSize* of 2, 4 and 6. Qualitatively, both models yield similar population dynamics. Closer consideration indicates some interesting differences, however. For *stepSize*s 2 and 4, the standard algorithm reaches convergence at near optimal fitness prior to the $P_{complete}$ model. Fitness proportionate selection in the standard model can avoid selection of less fit individuals as parents, leading to faster convergence of fitness values and faster reductions in percent polymorphic. A *stepSize* of 4 results in faster convergence than a *stepSize* of 2 due to the initially greater range and variance in fitness values. With a *stepSize* of 6, we find evidence of premature convergence by fitness proportionate selection under the standard model. Average fitness converges at about 0.94, as genetic diversity is lost prematurely. Under the replacement model with $P_{complete}$, convergence is slowed; genetic diversity is maintained for a longer time, leading to higher average fitness at convergence.

2.2 Trees

We only consider connected population structures, which provide the potential for genetic information from any individual to impact the offspring of any other

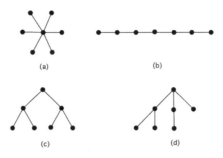

Fig. 2. Trees with 7 vertices: (a) star, (b) path, (c) CBTree and (d) MBTree

individual after some number of generations. A graph G is *connected* if and only if a path exists between any pair of vertices in the graph. *Trees* are *minimally connected* graphs, i.e., if any edge of a tree is removed, the resultant graph is no longer connected. As such, trees represent the opposite end of the spectrum of connected graphs from complete graphs. Trees contain no *cycles*, i.e., paths that start and end at the same vertex with no repeated edges. A tree of n vertices has $n-1$ edges. A *leaf* of a tree is a vertex having degree 1.

We can characterize graphs by a number of parameters, including number of leaves, maximum degree and diameter. The *diameter* of a graph is equal to the maximum distance between a pair of vertices in the graph. A *path* is the tree on $n > 1$ vertices having 2 leaves, a maximum degree of 2, and a diameter equal to $n-1$. A *star* is the tree on n vertices having $n-1$ leaves, a maximum degree of $n-1$, and a diameter of 2. Figures $2(a)$ and $2(b)$ present the star and path tree on seven vertices, respectively. A *path* minimizes the number of leaves and maximum degree and maximizes diameter, while the *star* minimizes diameter and maximizes maximum degree and number of leaves.

Figure 3 presents average fitness and percent polymorphic results for star and path compared to complete graph embedded populations, with *stepSize* of 4. In P_{star}, every leaf vertex has only the central vertex as possible partner. This forces in-breeding as evolution proceeds, resulting in loss of most genetic

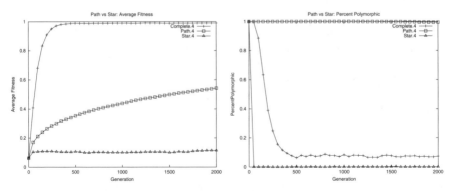

Fig. 3. Results with star and path populations

diversity from the population after relatively few generations and convergence at low average fitness values. In P_{path}, percent polymorphic remains relatively unchanged as genetic information moves slowly through the population. Average fitness steadily increases, almost linearly, after an initial, rapid increase, exceeding 0.50 after 2000 generations. These results suggest that individuals in different sections of the path have adapted to different aspects of the target genotype.

Now we consider several interesting classes of trees that fall between these extremes. A *complete binary tree CBTree* is a rooted tree, such that every vertex is either a leaf or is connected to two child vertices. Every vertex other than the root is connected to a parent vertex, as well. A *CBTree* has maximum degree of 3, $n/2$ leaves, and a diameter of (approximately) $2log_2n - 1$. To construct such a tree, we connect vertex labeled i to vertices labeled $2i + 1$ and $2i + 2$ as children. A *minimum broadcast tree MBTree* is a rooted tree that can be defined recursively as either a single, root vertex for $n = 1$, or two minimum broadcast subtrees of size $n/2$ whose root vertices are connected by an edge and is rooted at one of the two roots of the subtrees, for $n > 2$. MBTrees are so named because they allow messages to be broadcast in the minimum log_2n time units from the root under a synchronous, single-port model of communication [12]. A minimum broadcast tree has maximum degree of log_2n, approximately $n/2$ leaves, and a diameter of approximately $2log_2n - 1$. To construct such a tree for vertices labeled i up to j we connect vertex i to vertex $1 + (i + j/)2$ and then recursively construct trees for i up to $(i + j)/2$ and for $1 + (i + j)/2$ up to j. Figure 2(c) and 2(d) show an example of a *CBTree* and an *MBTree*, respectively.

We also consider two classes of random trees. A *random tree RanTree* will be a tree created by connecting vertex i to a vertex chosen uniformly at random from vertices labeled 0 up to $i-1$. A *power law tree PLTree* will be a tree created by connecting each vertex i in turn to a vertex j labeled 0 up to $i-1$ with connection probability to j equal to the current degree of j over the sum of current degrees. This is analogous to fitness proportionate selection in genetic algorithms, where a vertex's fitness corresponds to its current degree. Power law trees are so named as they have been shown to have power law degree distributions [1]. A random tree

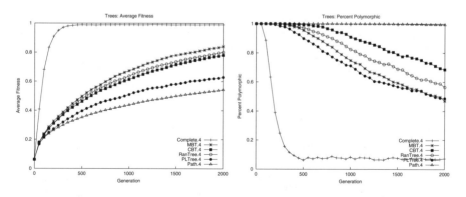

Fig. 4. Results with tree-structured populations

is expected to have lower maximum degree, fewer leaves, and greater diameter in comparison to a power law tree. Experimental results confirm these expectations. Generating 100 examples of both types of trees with 1000 vertices, we find that *RanTrees* have on average a maximum degree of 11, 500 leaves, and a diameter of 25, while *PLTrees* have a maximum degree of 64, 664 leaves, and a diameter of 18. A *RanTree* has average parameter values close to those of an *MBTree*.

Figure 4 presents average fitness and percent polymorphic results for these trees compared to populations embedded in complete graphs and paths for a *stepSize* of 4. Considering results for percent polymorphic first, we see that power law trees result in the highest rate of loss of diversity, closely followed by minimum broadcast trees. The complete binary tree, having low maximum degree, produces the lowest rate of decrease in genetic diversity of the new trees we consider. These results for trees show a direct relationship between rate of loss in percent polymorphism and maximum vertex degree. Average fitness results do not quite follow the opposite pattern. Minimum broadcast trees yield the greatest increase in average fitness, followed by random trees, which are followed closely by complete binary trees. Power law trees result in a significantly lower rate of increase in average fitness, but still well above the rate produced by paths and stars.

Putting the results for average fitness and percent polymorphic together and viewing them from an algorithm performance perspective, power law trees seem not to be favorable population structures for evolution, as they result in a relatively high loss of diversity yet a lower rate of gain in average fitness. Minimum broadcast trees most closely follow the pattern of complete graphs, yielding the highest rate of increase in average fitness with an accompanying higher loss of diversity. Random trees and complete binary trees perform well, providing relatively high rates of gain in average fitness accompanied by comparatively low rates of loss in diversity. All trees, having minimal connectivity and thus a reduced ability to propagate genetic information, yield increases in average fitness that fall significantly below those produced by evolution in $P_{complete}$.

2.3   Other Graphs

In this section, we consider a range of graphs having higher average vertex degree and greater connectivity. The first is the *circular lattice graph*, $CLG4$, wherein each vertex labeled i is connected to 4 other vertices $i + 1$, $i + 2$, $i - 1$ and $i - 1$ (label values taken *modn*, n being the number of vertices). Each vertex of a $CLG4$ has degree 4, and the graph has a diameter of about $n/4$. The $CLG4$ is of interest due to its large diameter and relatively high clustering coefficient. A graph's *clustering coefficient* is equal to the percentage of possible connections between neighbors of a vertex. In the circular lattice, as each vertex has degree 4, there are 6 possible edges between neighbors of which 3 are present, yielding a clustering coefficient of 0.50. In a random graph wherein all possible edges are equally likely, the probability of two neighbors having an edge between them is the same as the probability that any two nodes have an edge between them. For a random graph with 1000 vertices and 2000 edges, that probability is about 0.004.

A number of real-world networks, ranging from co-author and citation relationships in scientific publications to the power grid and the neural connection network of *C elegans*, have been found to have relatively high clustering coefficients [15]. These networks also have diameters on the order of *logn*. Networks with these two properties have been called *small world networks*. The lattice, while having a high clustering coefficient, has a diameter on the order of $n/4$ and is not a small world network. One means of constructing small world networks, introduced by Watts and Strogatz [16], is to randomly reconnect a small percentage of the edges of a circular lattice graph. We investigate populations embedded in two classes of *small world graphs*, $SWG10$ and $SWG20$, that are created by randomly reconnecting the other end of 10% and 20% of the edges in a $CLG4$, respectively. These small world graphs have average degree of 4, maintain a relatively high clustering coefficient, but have a sharply reduced expected diameter. While the circular lattice graphs have a diameter linearly related to the number of vertices, small world graphs constructed as above have been shown to reduce expected diameter to a logarithmic relationship. To take edge redirection to the extreme, we also consider connecting each vertex i to 2 other vertices chosen uniformly at random (i.e., redirecting 100% of the lattice edges), calling such graphs *random degree 4 graphs*, $Ran4Deg$. Each vertex of such a graph has degree at least 2, while the average degree is 4.

One graph structure that has received attention in the genetic algorithm community is the square grid graph [4] [13], with vertices located in rows and columns. We consider the 32 x 32 grid $GRID32$ (having 1024 vertices), each vertex having 4 neighbors (i.e., up, down, left and right) with boundary vertices having "wrap-around" edges to the other end of the same row or column. Note that these graphs have a clustering coefficient of 0.0 as neighbors are not neighbors. The diameter of such a square grid graph in general is approximately the square-root of n (i.e., the length of one side).

Figures 5 presents results for populations embedded in the above classes of graphs having average degree 4, comparing these to results for complete graph and $MBTree$ embedded populations. We see that the circular lattice, with its

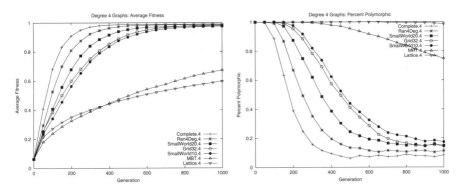

Fig. 5. Results with populations having degree four

maximal diameter for this class of graphs, results in its embedded population exhibiting the smallest changes in both average fitness and percent polymorphic. The $MBTree$, with half the edges but much smaller diameter, improves a population's ability to increase average fitness. What is most striking in our results is the change in population dynamics we observe when we turn the lattice into a small world graph by redirecting 10% of its edge. This reflects the significantly reduced diameter from the circular lattice that results from this small degree of redirection. The redirected edges act as "long-distance" edge, providing opportunity for interaction with other parts of the graph. The results for $GRID32$ slightly improve upon those of $SWG10$, in terms of average fitness. This again reflects the reduced diameter of $GRID32$ from that of $CLG4$; depending upon the labeling scheme in the grid, every row edge or every column edge is "non-local". As we redirect even more edges of the $CLG4$, we see that populations emdedded in $SWG20$ and $Ran4Deg$ yield results that move closer to those of $P_{complete}$, even though only about 0.004 of the edges found in a complete graph are used.

What happens as we continue to increase average vertex degree in the population structure? We next consider graphs having average vertex degree of log_2n, equal to 10 for our 1000 vertex graphs. *Minimal broadcast graphs* are graphs in which synchronous broadcast can be completed under the single-port communication model in the minimum log_2n time units in a graph with n vertices. Each vertex can communicate with at most one other vertex during a time unit, thereby at most doubling the number of vertices informed of a particular message. In [3], methods are introduced for creating sparse minimal broadcast graphs. Here, we consider a recursive method that yields a vertex degree and diameter of log_2n. An MBG graphs is constructed for labels ranging from i to j, as follows: if $j > 1$, then each vertex k in the range i to $(i+j)/2$ is connected to vertex $1+k-i+(i+j)/2$; we do the same recursively for ranges i to $(i+j)/2$ and $1+(i+j)/2$ to j (i.e., until $i = j$). We start with the full range of labels from 0 to $n-1$. We compare the MBG graphs to $Ran10Deg$ graphs, constructed as were $Ran4Deg$ graphs, by connecting each vertex to 5 other vertices chosen

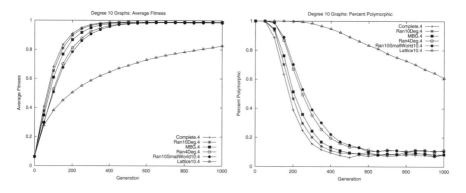

Fig. 6. Results with populations having degree ten

uniformly. We also consider two other classes of graphs constructed as per the degree 4 graphs discussed earlier. First is the circular lattice graph of degree 10, $CLG10$, wherein each vertex is connected to the next five vertices around a circle. Second is a small world graph of degree 10, $RanSmallWorld10$, in which 10% of the edges of $CLG10$ have been redirected randomly.

Figure 6 presents results for these classes of graphs having an average degree of 10, comparing their dynamics to results for $P_{RanDeg4}$ and $P_{complete}$. The increase in vertex degree does produce further increases in rates of increase of average fitness and of decrease of genetic diversity. First, we see the large change in dynamics that results from redirecting 10% of the edges when creating a small world graph from the circular lattice graph. Much of this change is due to a reduction in expected diameter. The population dynamics of evolution in $P_{RanDeg4}$ closely parallels that observed in $P_{RanSmalWorld10}$ despite having many fewer edges. As we saw for trees, average fitness values increase a little more rapidly in the randomly structured $P_{RanDeg10}$ than in P_{MBG}, event though the latter's structure is optimized for one-port broadcast. Quite striking is the result that the population dynamics created in $P_{RanDeg10}$ is nearly equivalent to that in $P_{complete}$, even though a $RanDeg10$ includes only about 0.01 of the edges found in the complete graph.

3   Conclusion

There has long been interest in the effects of population structure on natural evolution where populations are embedded in the structure of a physical ecosystem. Isolation of subpopulations has been a primary focus, producing ideas such as the earlier mentioned island model [18]. Independent subpopulation evolution is impacted by limited migration of individuals. Given a complex fitness function having many local maxima, i.e., the real world, subpopulation isolation can lead to species differentiation and variety. The basic island model could be approximately represented in our approach by cliques (i.e., complete subgraphs) with some individuals having a neighbor in another clique. The structure among cliques could be complete or a path, known as the stepping-stone model representing linear ecosystems such as a beach. The island model has been explored as one structure for parallelization of artificial evolutionary algorithms, as well [17].

There has been limited research on the impact of population structure on artificial evolution. DeJong and Sarma [2] [13] consider individuals embedded in a two-dimensional grid and vary the size of neighborhoods. They found, consistent with results reported here, that increasing neighborhood size (i.e., average vertex degree) and decreasing graph diameter leads to faster growth in average fitness. Such grids provide diameters that are on the order the square-root of n. Giacobini et. al. [4] found similar results when sites in the grid asynchronously select mating pairs from their neighborhoods.

Fine-grained, parallel, graph-embedded genetic algorithms are straightforward to implement. Each processor sends its resident individual along with its fitness value to all of its neighbors. After receiving individuals from neighbors, each processor selects a mating partner for its resident individual and performs recombination and mutation to create a new resident individual, and the cycle starts again. Messages can be stamped with generation numbers for synchronization purposes; time outs can alleviate unnecessary waiting due to failed connections or processors. Processors could be put in charge of more than one individual and could send only fitness values initially, sending an individual when it has been selected as partner by a neighbor, if individual genotype size is an issue.

We have investigated the effects that a range of population structures have upon artificial evolution. Population structures are represented as embeddings in graphs. We found significant differences among the structures, noting that random graphs of relatively low degree yield population dynamics approximating those observed in completely connected populations. Such random structures have relatively small diameters; their arbitrary connections seem to facilitate the random sharing of genetic material that typifies artificial evolution. The work reported here represents an initial foray into this research space. Investigating the interactions between fitness functions, selection mechanisms and various features of population structure appears to be an interesting area for future research.

References

1. Albert, R. and Barabasi, A.L., "Statistical mechanics of complex networks", *Review of Modern Physics*, 74, (2002), 47-97.
2. De Jong, K.A. and Sarma, J., "On decentralizing selection algorithms", in *Proceedings of the Sixth International Conference on Genetic Algorithms (ICGA95)*, Morgan Kaufmann, (1995) 17-23.
3. Farley, A.M., "Minimal broadcast graphs", *Networks*, 9, (1979), 313-332.
4. Giacobini, M., Alba, E. and Tomassini, M., "Selection intensity in asynchronous cellular evolutionary algorithms", in *Proc. of Genetic and Evolutionary Computing GECCO 2003*, LNCS 2723, Springer-Verlag (2003).
5. Goldberg, D.E., *Genetic Algorithms in Search, Optimization, and Machine Learning*, Addison Wesley, Cambridge,MA. (1989).
6. Hedrick P.W., *Genetics of Populations*, Jones and Bartlett, Sudbury,MA, (2000).
7. Kauffman, S.A., *The Origins of Order*, New York: Oxford University Press, (1993).
8. Mitchell, M., *An Introduction to Genetic Algorithms*, MIT Press, Cambridge, Ma., (1996).
9. Mitchell, M., Forrest, S., and Holland, J.H., "The royal road for genetic algorithms: Fitness landscapes and GA performance", in F.J. Varela and P. Bourgine, eds., *Toward a Practice of Autonomous Systems: Proceedings of the First European Conference on Artificial Life*, Cambridge, MA : MIT Press, (1992).
10. Maynard Smith, J., *The Theory of Evolution, 3rd Edition*, New York : Cambridge Univ. Press, (1993).
11. Maynard Smith, J., *Evolutionary Genetics, 2nd Edition*, New York : Oxford University Press, (1998).
12. Proskurowski, A., "Minimum broadcast trees", *IEEE Transactions on Computers*, 5(1981), 363-366.

13. Sarma J. and De Jong, K.A.,, "An analysis of local selection algorithms in a spatially structured evolutionary algorithm", in *Proceedings of the Seventh International Conference on Genetic Algorithms (ICGA97)*, Morgan Kaufmann (1997), 181-187.
14. Vose, M. , *The Simple Genetic Algorithm: Foundations and Theory*, Cambridge, MA : MIT Press, (1999).
15. Watts, D. *Small Worlds: The Dynamics of Networks Between Order and Randomness*, Princeton, NJ: Princeton Univ. Press, (1999).
16. Watts, D. and Strogatz, S.H., "Collective dynamics of small-world networks", *Nature*, 393 (1998), 440-442.
17. Whitley,D., Rana, S., and Heckendorn, R., "Exploiting separability in search: The island model genetic algorithm", *Journal of Computing and Information Technology*, 7, (1999) 33-47.
18. Wright, S. "Evolution in Mendelian populations", *Genetics*, 16, (1931), 97-159.

Outlines of Artificial Life: A Brief History of Evolutionary Individual Based Models

Stefan Bornhofen and Claude Lattaud

Laboratoire d'Intelligence Artificielle de Paris V,
LIAP5 – CRIP5, Université de Paris V,
45, rue des Saints Pères,
Paris 75006, France
{stefan.bornhofen, claude.lattaud}@math-info.univ-paris5.fr

Abstract. In the research field of Artificial Life, the concepts of emergence and adaptation form the basis of a class of models which describes reproducing individuals whose characteristics evolve over time. These models allow to investigate the laws of evolution, to observe emergent phenomena at individual and population level, and additionally yield new design techniques for computer animation and robotics industries. This paper presents an introductory non-exhaustive survey of the constitutive work of the last twenty years. When examining the history of development of these models, different periods can be distinguished. Each one incorporated new modeling concepts, however to this day all the models have failed to exhibit long-lasting, let alone open-ended evolution. A particular look at the richness of dynamics of the modeled environments reveals that only little attention has been paid to their design, which could account for the experienced evolutionary barrier.

1 Introduction

Artificial Life, or ALife, is the research field that tries to describe and study natural life by creating artificial systems that possess some of the properties of life. Its final aspiration is "understanding life by attempting to abstract the fundamental dynamical principles underlying biological phenomena, and recreating these dynamics in other physical media, such as computers, making them accessible to new kinds of experimental manipulation and testing." [1]. Besides the ambition of enriching the knowledge about Nature, Alife helps to find new design techniques. As computer simulations become more and more accurate, game, entertainment and robotics industries are constantly researching for new ideas to animate artificial characters.

The seminal novelty of ALife lies in its synthetic approach. Whereas traditional research is essentially analytic, breaking down complex systems into basic components, ALife attempts to construct complex systems from elemental units. The synthetic approach is based on two concepts, emergence and adaptation. A class of models which particularly applies these concepts describes reproducing individuals whose characteristics evolve over time. This paper refers to these models as "Evolutionary Individual Based Models" (EIBMs).

E. Talbi et al. (Eds.): EA 2005, LNCS 3871, pp. 226–237, 2006.

Considering the history of EIBMs in chronological order, four periods can be distinguished. They are thought of as overlapping stages of development in the art of individual based modeling by progressively incorporating new concepts. The beginnings of the first period reach back to the sixties. It comprises the discovery of new modeling techniques and the implementation of the first individual based models, but at that time extensive computer simulations were not yet feasible. At the end of the eighties, the progressing research culminated in the appearance of ALife as a distinct discipline. With the advent of computational power in the early nineties, allowing computers to run elaborate ALife systems at a tolerable speed, the second period incorporated evolution into individual based models, trying to capture population level phenomena by simple agents without particular morphologies in one ore two dimensional environments. The third period was marked by the adoption of environments with physical dynamics and directed the attention towards more elaborate phenotypic morphologies. Evolution was achieved by modifications of grammar-based genetic encoding schemes. The current fourth period of artificial embryology, since the end of the nineties, applies the discoveries of evolutionary developmental biology and models virtual creatures based on the concept of cell division by genetic regulatory networks. To this day, a great variety of extending or complementary research has been done with respect to each approach, but interestingly no further groundbreaking advances have been reported. It seems as if every model hits on limits of evolutionary complexity which prohibits the kickoff for long-lasting creative evolution in artificial worlds.

This paper serves a double purpose: to structure the history of EIBMs by classifying samples of the most influential works into periods, and to take advantage of this short survey to particularly review the dynamics, i.e. the rules and forces that produce motion or affect change within the environments. It will be suggested that the design of this component lags behind the advances in modeling the evolving individuals.

Section two presents the two important concepts of emergence and adaptation, both of which are present throughout the paper. Section three describes early EIBMs featuring simple evolving agents. Section four inspects models with physical dynamics and grammar-based genetic encoding. Embryological models are described in section five. Section six concludes with the synthesis of all the presented works.

2   ALife Concepts in Modeling

ALife researchers have been inspired by the creation as observed in Nature and developed the concepts of emergence and adaptation which are opposed to conventional human design techniques. This section describes the two concepts and their implementation in modeling.

2.1   Emergence

Models of complex systems, i. e. systems composed of a large number of interacting elements, are traditionally described by mathematical formulas like differential equations to manipulate some aggregate state variables of the system as a whole. This method allows a general and compendious way to analyze the behavior of a system.

However, as aggregate variables always oblige to deal with mean values, they have difficulties with heterogeneity in the system, and their high abstraction level often does not grasp the underlying reasons for the dynamics.

The Alife approach of modeling is "bottom-up engineering", thinking of complex systems as collections of distinct objects or individuals rather than continuous values. Individual based models can include refined representations of the individuals and their behavior. Emergence describes the phenomenon that simple local interactions between the entities of the system lead to a complex high level organization.

One of the earliest examples of emergence is Craig Reynolds' individual based model of boids [2]. Boids are autonomous agents simulating the flocking behavior of birds. Flocking arises as global behavior from the interaction of very few simple local rules. Placed into a virtual environment, the boids are programmed to follow three directives of "steering behavior" (figure 1):

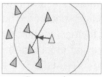
Separation

- *Separation*: to maintain minimum distance from other boids in the environment
- *Alignment*: to match velocities with other boids in the neighborhood.
- *Cohesion*: to move toward the perceived center of mass of boids in the neighborhood.

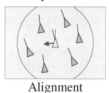
Alignment

These rules are entirely local, referring only to information accessible within a boid's own vicinity. Hence, the flock that forms is an emergent phenomenon.

Based on this algorithm, the boids can be enriched by more elaborate behaviors like obstacle avoidance or goal seeking. Obstacle avoidance allows the boids to fly through simulated environments while dodging static objects. Goal seeking behavior causes the flock to follow a scripted path. The boids render such an impressing realism at simulating flocking as well as other coordinated motion like fish schools or human crowds that they have been used in many cinematic animations such as the bat swarms of the motion picture "Batman Returns" [3].

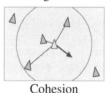
Cohesion

Fig. 1. Steering behavior

Within Nature, interaction of simple agents can be observed in insect communities like ants, bees or termites, and their cooperation strategies have inspired researchers to devise new optimization algorithms based on the concept of emergence [4].

2.2 Adaptation

The second principle of ALife modeling is adaptation, the capability of developing advantageous traits in response to a changing environment. Adaptation divides into lifetime learning and evolution which operate on different time scales.

Lifetime learning represents an individual's ability to interact and learn from its environment. In contrast, evolution is not defined for individuals, but in the context of entire populations. Evolution works with genetic information, called genotypes. Through a process of development, the "mapping function" translates genotypes into phenotypes which represent the individuals' manifestation within a simulated virtual environment.

Subsequently, the phenotype is evaluated by a fitness function which determines if the corresponding genotype is selected for further reproduction (figure 2).

Lifetime learning and evolution are profoundly interwoven. Their interplay leads to new and insufficiently understood phenomena like the "Baldwin effect" [5] which denotes that over time learnable traits of the phenotype are potentially assimilated into the genotype. Hinton and Nowlan clearly demonstrated this effect by a simple evolutionary simulation [6].

One of the most fundamental EIBMs are Richard Dawkins' biomorphs [7]. His purpose was to demonstrate an evolutionary model on the basis of selection and mutation as proclaimed by Darwin [8], and to point out the feasibility of discovering a desired genotype inside a huge genetic space. Biomorphs are two dimensional branching structures used as a graphic representation of a number of simple binary genes, controling features like depth of recursion, angles of branching and length of lines and allowing about 500 billion possible combinations. To produce offsprings, biomorphs use asexual reproduction by copying the parental genes with some probability of random mutation.

To evolve a biomorph, the user starts with a display of an initially given parent individual in the center of the screen and twelve children surrounding it. The user simply clicks on one of the children or the original parent to select it for survival and reproduction. Subsequently, the selected biomorph's genes are used to create a new generation, and the biomorph as well as its children are again displayed. These steps are repeated, and with every generation the biomorphs "adapt" to the given fitness function, that is to say, to the taste of the user. In spite of the simplicity of their genetic encoding, the resulting morphologies of biomorphs are surprisingly manifold. Some biomorphs may look like insects, microorganisms, trees or other familiar objects (figure 3).

The general concept of biomorphs has been extended into models which simulate the process of evolution not only based on user selection, but also on agent interactions within a more complex environment [9].

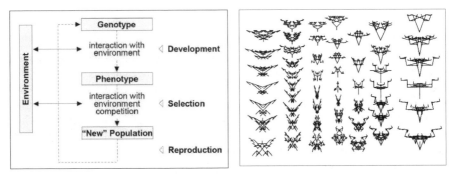

Fig. 2. The evolutionary cycle   Fig. 3. Evolved Biomorphs

3 Simple Agent Approach

Early implementations of EIBMs tried to capture the processes of evolution by modeling simple reproducing individuals acting on a small set of behavioral routines.

Their level of detail was reduced to a one or two dimensional environmental framework and individuals which lacked almost any morphology, but this is sufficient to observe ecological interactions between the organisms and the emergence of population level phenomena: the limitation of resources introduces a competition between the reproducers, and they become engaged in a struggle for existence. According to the principle of the "survival of the fittest", the organisms either develop successful strategies or die.

3.1 Tierra

In 1992, Tom Ray modeled evolution by the propagation of self-replicating programs running on a virtual machine, called Tierra [10]. These programs can be thought of as digital organisms whose genotype matches the phenotype, and whose physical environment consists of energy, i.e. CPU time, and limited space in memory.

An evolutionary run is started by introducing a hand-written ancestor program into the empty memory. To reproduce, the organism's code is executed. It writes a copy of itself into newly allocated memory space. Mutations introduce differences in the offsprings, and competition for memory causes an evolutionary process to begin.

During a run, organisms shrink by decreasing the length of their genotype, as shorter genes mean less genetic material to be copied more rapidly. Parasites occur, i.e. organisms that execute instructions of other programs, and even hyper-parasites develop which utilize instructions from parasites. At the same time, hosts evolve immunity to parasitism, forcing their parasites to evolve methods to get around the new defenses. These observations illustrate the "Red Queen principle" [11] which states that coevolving populations are due to continuing development in order to maintain their fitness relative to one another.

Tierra has been used to experimentally examine ecological and evolutionary processes such as host-parasite density dependent population regulation. Even if Tierra is modeled in an abstract virtual fashion, it finds many analogies to the real world and its diverse ecological communities.

3.2 Echo

John Holland's Echo system [12] is a simulator of virtual ecologies which is geared to more lifelike notions of space and time. It investigates mechanisms which regulate diversity and information-processing in systems comprised of many interacting adaptive agents, or "complex adaptive systems".

The surrounding environment is made up of a square toroidal lattice of sites which produces different types of regenerating resources, encoded by a letter. Agents are located at a site and possess a small set of simple interactions with their environment. They can relocate to another site, eat the resources and store them. At the same time, the environment charges a maintenance fee which can be considered as metabolic cost. An agent also features a small range of predefined inter-agent behaviors which are fighting, trading and mating. Fighting and trading allow for resource exchange. If an agent has collected sufficient resources to rewrite its genetic code, it reproduces asexually or, via mating, sexually.

This system exhibits emergent phenomena like the formation of agent communities and trading networks. Echo was used by environmental researchers to show that explicitly deriving differential equations was not necessarily the most accurate method for modeling food web complexity [13]. However, due to the high abstraction level of the Echo model, the degree of fidelity to real systems is uncertain.

3.3 Polyworld

An approach, more faithful to biological systems, was attempted by Larry Yaeger [14]. His virtual ecology, called Polyworld, brings together all the principle components of real living systems into one artificial system. Polyworld consists of a two dimensional plane with growing food bits. Just as in Echo, the agents interact, fight and mate, eat the food and relocate by expressing behavioral primitives. However, an agent's architecture exhibits more complexity. Its behavior is controlled by a neural network, determined from its genetic code. During lifetime, a Hebbian algorithm modulates the synaptic weights, so that the agents are able to learn. Moreover, organisms perceive their world through a sense of vision from their own point of view.

An evolutionary run is started with the introduction of a random population whose evolution is guided by a simple external fitness function rewarding the individuals' activities. If the population is on the verge of dying out, reproduction is regulated by the system. Evolved populations that exhibit behaviors which allow them to perpetuate their number by reproduction on their own are said to exhibit a "Successful Behavior Strategy".

A variety of species with recognizable behavioral strategies, like fleeing, grazing, foraging, following, and flocking, evolved from this model. These results met Yaeger's primary goal, that is to achieve the emergence of population level behavior from elementary naturalistic building blocks.

3.4 Discussion

In these early models, the multi-agent architecture of EIBMs is already visible. A number of evolving agents is placed in a non-evolving framework. The agents account for the "living" part of the environment, whereas the framework, representing their outside world, can be considered as the "non-living" part. It comprises the space where the agents' phenotypes are inserted and potentially holds accessory objects with simple dynamics, like obstacles or regrowing food bits. Interactions possibly occur among agents and between agents and the non-living component (figure 4).

Tierra's memory is one dimensional and features no explicit resources at all, so that space and CPU time are the only constraints for the individuals. Organisms do not migrate, they are bound to their initial location. Hence, the only significant interaction between organisms and non-living environment is the allocation of memory for offsprings. Echo and Polyworld incorporate the notion of food by modeling ingestion of nearby located resources and subjecting the agents to metabolism. Echo allows the agents to relocate to a discrete neighboring site. Polyworld features a two dimensional continuous flat world, in which agents express locomotion commands like "turn" or "move forward".

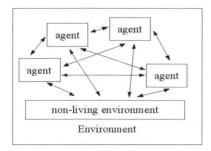

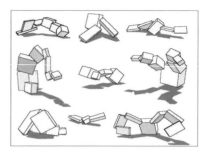

Fig. 4. Standard architecture of an EIBM   **Fig. 5.** Sims' evolved creatures for walking

Two major interactions between organisms and non-living environment, i.e. loco-motion and ingestion have been incorporated, but they are modeled as primitives and cannot be affected by evolution. The three presented models have proved that a high level of abstraction allows to grasp various population level phenomena. However, they yield little results on an individual scale. One of the main obstacles could be the lack of morphology that limits the agents' degrees of freedom. The models of the fol-lowing section particularly tackle this problem.

4 The Grammar-Based Approach

Then next generation of EIBMs has been augmented with environments of more physical accuracy. This improvement allowed a substantial gain of complexity of the individuals' phenotypes and their interactions with the outside environment. In this kind of models, a creature's morphology is made up of a number of pre-designed ele-mental units whose assembly is encoded in the genotype by a grammar-like record such as a nested graph or L-system [15]. The presented works still used their own implementations of physical dynamics. However, today a number of available physics engines relieve researchers of programming this component themselves [16].

4.1 Karl Sim's Block Creatures

In 1994, Karl Sims pioneered a new way of evolving both the morphology and behav-ior of virtual creatures [17]. Situated in a three dimensional world with realistic phys-ics, these creatures consist of collections of blocks, linked by flexible joints which are controlled by neural circuits. Joint angle sensors and touch sensors allow the creatures to obtain information from their environment. A creature's genotype is written as a nested directed graph which describes both its morphology and neural control archi-tecture. This representation provides modularity to the mapping from genotype to phenotype, and naturally leads to duplication and recursion of body parts.

Sims evolved several locomotion tasks like running and jumping on a flat surface, or swimming in a virtual marine environment. It turned out that different runs of evo-lution produced different solutions to the same problem. Some creatures were evoca-tive of real existing animals like a swimming snake or a walking crab. Others, equally effective at their tasks, used strange patterns of movement and form (figure 5). In an extending work [18], Sims studied competitive behavior. In a simple game, two opponents had to fight for possession of a cube that was placed halfway between them. The creatures not only evolved ways of reaching the cube quickly, but also of fending off their opponents.

4.2 Sexual Swimmers

Sexual Swimmers [19] is an artificial ecosystem which demonstrates the evolution of morphology and locomotion among a population of stick figures in a virtual two dimensional pond. A simple model of physics enables the agents to propel themselves through simulated water. Swimmers ingest regrowing food bits throughout the pond and reproduce by mating with other swimmers. As there is no explicit fitness function, selection is dictated by the swimmers' locomotion skills which allow them to quickly reach a desired goal, either food or mate. Moreover, the agents have basic perceptions, and the choice of a mate is influenced by preferences for morphological traits like color, length of limbs or degree of agitation.

When length is considered attractive, populations with elongated bodies and only few branching parts emerge. When selecting for color, swimmers of differing colors rarely mate with each other and population often breaks into distinct coloration groups. This work shows how the phenomenon of sexual attractiveness affects the course of evolution in respect of the creatures' body plan as well as locomotion style.

4.3 Framsticks

Framsticks [20] is a three dimensional virtual ecology, i.e. a project modeling creatures seeking food in their environment. Besides energy balls that can be ingested by the creatures, the outside world is enriched with a non-trivial topology and a water level. An agent is made up of connected sticks which can be specialized for various purposes like assimilation, strength, ingestion or sensors (figure 6). A neural brain computes excitations in neural nets, collects data from the sensors and sends signals to effectors that bend and rotate the connection points.

The Framsticks project proves that an increased level of complexity can yield the same results as those obtained in simpler population level simulations, while offering much more possibilities to investigate individual level behavior. Like in Sims' work, a number of locomotion techniques evolved from this model. Moreover, a comparison of different kinds of genotypes was published showing that evolution can be enhanced by the choice of well-designed genetic encodings [21].

4.4 Discussion

The environments of the presented models are characterized by the adoption of realistic physical dynamics. Ventrella's interest in population level phenomena dictated a relatively simple two dimensional pond. Three dimensional flat land and marine environments were modeled by Karl Sims, and the Framsticks project merged land and water worlds into various landscapes.

The design of possible interactions between the agents and their outside world still focuses on ingestion and locomotion. However, whereas ingestion remains to be modeled by behavioral primitives, the articulated morphology of a creature's phenotype allows for a new vision of locomotion. The behavioral building block of previous models is superseded by an emergent result of the agent's morphological activities.

This fact illustrates how, to achieve a given goal, evolution can exploit the dynamics of the environment. In the case of locomotion, evolution discovered that an organized behavior of the agent's morphological elemental units allows to relocate its phenotype. Nature offers a wide range of further demonstrations of this principle. For example, the forces among the molecules of the air lead to properties that allowed to evolve birds that flap their wings to fly, plants which disperse their seeds with the wind, or humans who stimulate their vocal cords to communicate.

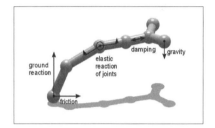

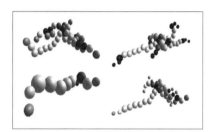

Fig. 6. Framsticks creature and its physics   **Fig. 7.** Evolved Blockpushers

5   The Embryogenic Approach

The grammar-based genotype encoding does not mirror the process of a real creature's biological embryogenesis, since the stage of development corresponding to the molecular chemistry is systematically skipped. The embryogenic approach, inspired by evolutionary developmental biology, attempts to evolve the morphology and neural architecture of virtual agents in a new, biologically more accurate fashion.

The mapping from genotype to phenotype takes place during a developmental phase. The genotype encodes "functional genes" which express the behavior of a cell like division, growth or death, and "regulatory genes" which generate substances that affect the activity of both gene types. The interplay between diffusing genetic information of adjoining cells forms a genetic regulatory network which directs the transformation of an agent from a single structural unit or cell into a multi-cellular organism.

5.1   Eggenberger's Evolved Morphologies

The constitutive work in this field was achieved by Eggenberger [22] who evolved static morphologies. The environmental framework consists of a discrete three dimensional lattice which constitutes both the diffusion space of various chemicals and the sites for the individual cells of the organisms' compound morphology. The lattice additionally contains substances whose concentration gradients provide a positional information to the cells.

Eggenberger demonstrated how artificial genetic regulatory networks can be modeled, and that it is possible to evolve artificial multicellular organisms in a way that they display high degrees of symmetry. Moreover, his work highlights that differential gene expression dissociates the complexity of information in the genotype from the complexity of the evolved phenotype.

5.2 Blockpushers

Taking up the idea of embryogenesis, Bongard and Pfeifer [23] developed a simulation system, called "Artificial Ontogeny", to evolve both the morphology and neural control of virtual creatures. Similar to Sims' work, these creatures exist on a flat plane within a three dimensional environment endowed with physical dynamics.

The ontogenetic process transforms a single structural unit in a continuous manner into an articulated agent composed of several units: After a unit splits from its parent unit, the two units are linked by a rigid connector. The new unit is attached to the rigid connector by a one degree of freedom rotational joint. In a similar manner, some or all units develop sensors, actuators and internal neural structure (figure 7). In order to evaluate its fitness, an agent is first grown and then tested against a given fitness function, that is to push a nearby block as far as possible.

The evolved blockpushers were found to solve the problem, showing that a minimal model of embryogenesis suffices to evolve agents that perform a non-trivial task in a virtual environment with physical dynamics. According to the authors, the obtained results "point to the high evolvability of the Artificial Ontogeny system" [23].

5.3 Discussion

The models of this section are characterized by a new approach with respect to the agent's genesis. For this purpose, the environments have been enriched with the capacity of diffusing substances in order to allow the propagation of gene products. In addition to the dynamics of realistic physics, the concept of diffusion is another example of environmental dynamics that allows emergent phenomena which are, in the case of embryogenesis, new ways of phenotypic shaping.

However, the quality of diffusion is only exploited during the process of the agents' developmental phase. Eggenberger abstains from complex physical dynamics as he is only interested in static phenotypes, whereas Bongard and Pfeifer adopt a physics-based three dimensional space in order to study morphological activity. Interactions between full-grown agents and the outside world do not seem to exceed those in the models of the last section.

The recency of the approach does not allow for final conclusions, but it is questionable whether further research will considerably surpass the results of grammar-based encoding schemes, as long as the environment is not endowed with new properties. Two approaches are suggested in the next section.

6 Synthesis and Conclusion

The study of EIBMs is becoming an increasingly important domain in Artificial Life research. EIBMs allow to investigate the laws of the evolution of autonomous agents at individual and population level. They are based on the two concepts of emergence, as the models are based on individuals, and adaptation, as evolution and possibly lifetime learning allow the individuals to enhance their fitness.

A short and non-exhaustive survey of influential EIBMs during the past twenty years has been presented in this paper. The works can be grouped into four periods which reflect a particular state-of-the-art. Successes have been made with respect to evolving both population level and individual level phenomena. Virtual ecologies achieve the formation of simple group behavior such as flocking or trading. As to evolution of individuals, simple locomotion behaviors can be readily bred.

From the view of creative and long-lasting evolution, it has to be recognized that in every model evolution ceases after initial progress. After all, current ALife approaches "do not seem to be as alive as we might hope" [24]. In search of a reason for this phenomenon, the history of EIBMs can teach a lesson: In early models, the main focus was placed on the emergent relationships between the evolving agents, whereas their outside world was somewhat considered as an uninteresting framework whose primary function was the supply of space and, at best, food bits. When the design of the environment switched to physical models, evolution was given the possibility to exploit dynamics not only among agents, but also between agents and environment, which resulted in the emergence of locomotion behaviors. However, after this incisive changeover, most of the attention returned to the agents. Even in more recent models the outside world remains not much more than an inert vacuum space whose sole purpose is to allow the agents to express their morphological activities. It stands to reason that if more care was accorded to the design of the environmental framework, evolution would not fail to discover ways to make use of its dynamics. This idea is indeed not a new one, since early pioneers in Artificial Life like John Holland already stated in 1962 that "the study of adaptation involves the study of both the adaptive systems and its environment" [25].

Starting from the current state-of-the-art, different ways of enriching the environment can be considered. As seen in the discussions of sections 4 and 5, the idea of creatures initiating dynamics in the environment might have been underestimated in current models. In extension to the embryogenic diffusion space, the environments could be enriched with several media whose properties can be exploited by evolution. If the media are able to propagate information, the approach could also provide new ways of communication among the creatures. Furthermore, since ingestion is still modeled as a behavioral primitive, a simple chemical model could complement the physical one and extend the creatures' metabolism to ingestion, digestion and excretion. Phenotypic evolution would occur not only at a functional, but also at a physiological level and affect the creatures' resource management. This approach is based on the idea that a fundamental criterion for Life is the presence of a metabolism. To be considered as "alive", any being, natural or artificial, should convert matter or energy of the environment into suitable forms for its organism [26].

These few ideas are only suggestions of how to reconsider the significance of all the components of a model in the research about life-at-it-is and life-as-it-could-be.

References

1. Langton, C.G.: Preface. Artificial Life II, Langton C.G. et al., eds., Volume X of SFI Studies in the Sciences of Complexity, Addison-Wesley (1992) 13-18
2. Reynolds, C.W.: Flocks, Herds, and Schools: A Distributed Behavioral Model. Computer Graphics, 21(4) (1987) 25-34

3. Sipper, M.: An introduction to artificial life, Explorations in Artificial Life (special issue of AI Expert), Miller Freeman, San Francisco, CA (1995) 4-8
4. Dorigo M., Stuetzle T.: Ant Colony Optimization, MIT Press, 2005
5. Baldwin, J.M.: A New Factor in Evolution. Am. Naturalist 30 (1896) 441-451, 536-553
6. Hinton, G.E., Nowlan, S.J.: How Learning Can Guide Evolution. Complex Systems, Vol. 1 (1987) 495–502
7. Dawkins, R.: The Blind Watchmaker. W.W. Norton, New York (1986)
8. Darwin, C.: On the origin of species. London, John Murray (1859)
9. Métivier M., Lattaud C., Heudin J.C.: A stress based speciation model in LifeDrop, Proceedings of the 8th Int. Conference on Artificial Life, Sydney, Australia (2002) 121-126
10. Ray, T.S.: An approach to the synthesis of life, Proceedings of Artificial Life II, Langton C.G. et al., eds., Addison-Wesley (1992) 371-408
11. Van Valen, L.: A New Evolutionary Law, Evolutionary Theory 1 (1973) 1-30
12. Holland, J.H.: The Echo Model, In: Proposal for a Research Program in Adaptive Computation, Santa Fe Institute (1992)
13. Schmitz, O.J., Booth, G.: Modeling Food Web Complexity: The Consequence of Individual-based Spatially Explicit Behavioral Ecology on Trophic Interactions, Yale Univ. (1996)
14. Yaeger, L.: Computational Genetics, Physiology, Metabolism, Neural Systems, Learning, Vision, and Behavior or PolyWorld: Life in a New Context. Artificial Life III. Ed Langton, (1994) 263-298
15. Lindenmayer, A.: Mathematical models for cellular interactions in development, Parts I and II, Journal of Theoretical Biology, Vol. 18 (1968) 280-315
16. Taylor, T., Massey, C.: Recent Developments in the Evolution of Morphologies and Controllers for Physically Simulated Creatures. Artificial Life vol. 7(1) (2001) 77-87
17. Sims, K.: Evolving Virtual Creatures. SIGGRAPH Proceedings, ACM Press (1994) 15-22
18. Sims, K.: Evolving 3D Morphology and Behavior by Competition. Brooks, R. & Maes, P. (eds.) Artificial Life IV: Proceedings of the Fourth International Workshop on the Synthesis and Simulation of Living Systems, MIT Press (1994) 28-39
19. Ventrella, J.: Sexual Swimmers (Emergent Morphology and Locomotion without a Fitness Function). From Animals to Animats. MIT Press (1996) 484-493
20. Komosinski, M., Ulatowski, Sz.: Framsticks: Towards a Simulation of a Nature-Like World, Creatures and Evolution. In: Proceedings of 5th European Conference on Artificial Life, Springer-Verlag (1999) 261-265
21. Komosinski, M., Rotaru-Varga, A.: Comparison of different genotype encodings for simulated 3D agents. In: Artificial Life Journal, 7 (4), MIT Press (2001) 395-418
22. Eggenberger, P.: Evolving morphologies of simulated 3d organisms based on differential gene expression, European Conference on Artificial Life (1997) 205-213
23. Bongard, J. C., Pfeifer, R.: Repeated Structure and Dissociation of Genotypic and Phenotypic Complexity in Artificial Ontogeny, in Spector, L. et al (eds.), Proceedings of GECCO 2001. San Francisco, CA: Morgan Kaufmann publishers (2001) 829-836
24. Brooks, R.A.: The Relationship Between Matter and Life, Nature, Vol. 409 (2001) 409–411
25. Holland, J.H.: Outline for a Logical Theory of Adaptive Systems, Journal of the Association for Computing Machinery (1962) 297-314
26. Farmer, J.D., Belin, A.: Artificial Life: The Coming Evolution, Proceedings of Artificial Life II, Langton C.G. et al., eds., Addison-Wesley (1992) 815-840

An Enhanced Genetic Algorithm for Protein Structure Prediction Using the 2D Hydrophobic-Polar Model

Heitor S. Lopes and Marcos P. Scapin

Bioinformatics Lab. (CPGEI), Centro Federal de Educação Tecnológica do Paraná,
Av. 7 de setembro, 3165 – 80230-901 Curitiba, Brazil
hslopes@cpgei.cefetpr.br, mpscapin@cpgei.cefetpr.br

Abstract. This paper presents an enhanced genetic algorithm for the protein structure prediction problem. A new fitness function, that uses the concept of radius of gyration, is proposed. Also, a novel operator called partial optimization, together with different strategies for performance improvement, are described. Tests were done with five different amino acid chains from 20 to 85 residues long and better results were obtained, when compared with those in the current literature. Results are promising and suggest the suitability of the proposed method for protein structure prediction using the 2D HP model. Further experiments shall be done with longer amino acid chains as well as with real-world proteins.

1 Introduction

A protein is a chain of amino acid residues that folds into a specific native 3-dimensional structure under natural conditions, just after being synthesized in the ribosome. The task of predicting this 3-D structure is called the protein structure prediction problem (PSP) and its resolution is of great importance for modern molecular biology.

Exhaustive search of the entire conformational space of a protein is not possible, even for the small ones. Simplified models, where amino acids are laid on a 2- or 3-dimensional lattice, have been proposed. Again, such models are feasible only for small proteins, due to its NP-completeness [1]. Consequently, heuristic optimization methods seem to be the most reasonable algorithmic choice to solve PSP, and, amongst them, many evolutionary computation approaches have been proposed [2], [3], [4], [5], and [6]. In this paper we present an improved genetic algorithm for PSP. Its most important feature is a new fitness function capable of directing the search towards good protein conformations. Using a benchmark, results show that our implementation achieves optimal or quasi-optimal solutions for small proteins.

2 2D HP Model

The 2D HP (2-dimensional Hydrophobic-Polar) model was introduced by [7] and it is the most widely studied discrete model for protein folding in the recent literature. It models the concept that the major contribution to the free energy of the native

E. Talbi et al. (Eds.): EA 2005, LNCS 3871, pp. 238–246, 2006.
© Springer-Verlag Berlin Heidelberg 2006

conformation of a protein is due to interactions among hydrophobic residues. They tend to form a core in the protein structure while surrounded by hydrophilic residues that interface to the environment.

In the HP model, the 20 standard amino acids are divided into two types, according to its affinity to water: hydrophobic (H for non-polar) or hydrophilic (P for polar). As it is a lattice model, the amino acid chain is embedded in a 2- or 3-dimensional square lattice and the movements of the chain are restricted to angles of 90°. In a legal conformation, the adjacent residues in the sequence must be adjacent in the lattice and each lattice point can be occupied by only one residue.

The free energy of a conformation is inversely proportional to the number of hydrophobic non-local bonds (or H–H bond). An H–H bond occurs if two hydrophobic residues occupy adjacent grid points in the lattice but are not consecutive in the sequence. Each such interaction contributes with –1 to the energy value.

3 Implementation

In this section, we describe in details the application of a genetic algorithm (GA) and the strategies proposed to improve its performance.

3.1 Chromosome Encoding

The dynamics and effectiveness of a GA is strongly influenced by the way solutions are represented. There are two ways of representing a chain in a lattice: either using absolute or relative coordinates. In the former, every amino acid uses Cartesian coordinates to define its position in the lattice. In the latter, the definition of an amino acid position takes into account the position of the previous one, with relative movements. Based on the results presented by [8], our implementation uses internal coordinates. Due to the 2-dimensional lattice used, there are only three possible moves, regarding the previous amino acid of a chain: (R)ight, (L)eft and (F)orward. These moves indicate that the next amino acid of the chain will be folded (together with the remaining forward chain) 90 degrees to the right, to the left or the chain will be stretched ahead.

Therefore, the GA will have a population of individuals with a single chromosome, each one representing a complete conformation. The chromosome is composed by a number of genes corresponding to the number of amino acids in the chain minus one (the starting amino acid of the chain), and every gene is defined over the alphabet $\{R, L, F\}$.

3.2 Initial Population

In this problem, a constraint to be handled is related to the self-avoidance of a conformation, i.e., whether illegal conformations are allowed during evolution or not. If not, it is necessary a procedure that guarantees the generation of only legal conformations in the initial population and in the application of the genetic operators. Another approach, called penalty method, allows the existence of unfeasible conformations during the evolution, but a penalty is added (to the fitness value of the individual) for every lattice point at which there is a collision of more than one amino acid. Our implementation uses the penalty method.

According to [3], the encoding in relative internal coordinates exhibits the problem that initial populations (randomly created) tend to have an increasing number of collisions as the length of the protein increases, making the GA waste efforts with illegal conformations before promising conformations can be found. Based on this statement, a different strategy was used to create the initial population aiming to minimize the collisions while generating a larger initial genetic diversity. This strategy divides the population into two parts that are generated differently. The proportion of each part is established by a user-defined parameter called *PopIniFull*. The first part of the population is randomly generated, as usual, and this is the part that the percentage of the *PopIniFull* parameter indicates. The second part is generated considering each individual as totally unfolded and then applying a number of random mutations between 3 and the total number of genes in the chromosome, uniformly distributed. Using this method, there will be a certain amount of individuals having few mutations that increases the diversity of the initial population and allows that the unfolded parts of the individuals help the evolution process.

3.3 Fitness Function

To evaluate an individual, it is necessary to translate its genotypical encoding, defined over the alphabet $\{R, L, F\}$, to obtain its Cartesian coordinates. This procedure allows knowing how the amino acids are disposed in the lattice, and then, the computation of an objective goodness measure of the conformation. In this work, we propose a new fitness function composed of three terms, as shown in Equation 1:

$$Fitness = NLB_H \times RG_H \times RG_P \qquad (1)$$

where NLB_H is the number of hydrophobic non-local bonds of the conformation and RG_H and RG_P are terms computed using the radius of gyration of the hydrophobic and hydrophilic residues, respectively, as explained below. The product of all terms in this equation indicates that all of them should be maximized.

3.3.1 Hydrophobic Non-local Bonds
It is believed that hydrophobic non-local bonds are the main force that drives the protein folding process. We are considering the problem as the maximization of the number of H–H bonds thus, for every hydrophobic non-local bond, NLB_H is added by 1. Since we are using a penalty method, NLB_H is decreased whenever a collision occurs. The penalty term, decremented from NLB_H, is composed by the number of grid points which are occupied by more than one residue, multiplied by the penalty weight which, in turn, is set according to the chain length: the longer the chain, the higher it is.

3.3.2 Radius of Gyration
The original HP model uses only the hydrophobic non-local bonds term to evaluate an individual but, according to [8], without a modified energy function, there will exist large plateaus in the energy landscape on which local search cannot find a descent direction, leading to a random search. This fact was also experienced in our preliminary implementation and, aiming to avoid this trap and enhance the fitness function, we propose the use of a new concept, called radius of gyration (*RG*).

RG of a solid body is the radial distance from a given axis at which the mass of a body could be concentrated without altering the rotational inertia of the body relative to that axis [9]. Hopefully, using *RG* in the fitness function the fitness landscape can be changed in such a way that the fitness function rewards more compact conformations with the same number of H–H bonds, bringing the evaluation closer to reality.

RG, in the scope of the PSP, indicates how compact a set of amino acids is: the more compact a conformation, the smaller is its radius of gyration. In this term of the fitness function, only hydrophobic residues were considered, rewarding the conformations that have smaller values of radius of gyration. This term is presented in Equation 2:

$$RG_H = MaxRG_H - \sqrt{\frac{\sum_{i=1}^{NH}\left[(x_i - \overline{X})^2 + (y_i - \overline{Y})^2\right]}{NH}} ,$$

(2)

where x_i and y_i are the Cartesian coordinates of the i-th hydrophobic residue, $\overline{X}$ and $\overline{Y}$ are the mean values of all hydrophobic x_i and y_i, respectively; NH is the number of hydrophobic residues in the chain; and $MaxRG_H$ is the radius of gyration of the amino acid chain totally unfolded. The second part of Equation 2 represents the radius of gyration of hydrophobic residues related to the point given by the mean coordinates, and it is subtracted from $MaxRG_H$ in order to maximize RG_H.

The term related to the hydrophilic radius of gyration in the fitness function has the opposite purpose as RG_H: it fosters the spreading of hydrophilic residues towards the edge of the conformation. This term is calculated in the same way as in Equation 2, except that, in this case, only hydrophilic residues are considered, and it is not subtracted from any other value (as in Equation 2). Using RG_H computed before, RG_P can be obtained using Equation 3:

$$DIFRG = \sqrt{\frac{\sum_{i=1}^{NP}\left[(x_i - \overline{X})^2 + (y_i - \overline{Y})^2\right]}{NP}} - RG_H ,$$

$$RG_P = \begin{cases} 1 & if\ DIFRG \geq 0 \\ \dfrac{1}{1 - DIFRG} & otherwise \end{cases}$$

(3)

In Equation 3, *DIFRG* computes the difference between the hydrophilic and the hydrophobic radii of gyration. A positive difference for *DIFRG* means that the hydrophobic residues are buried inside the conformation, while the hydrophilic ones are outside. Such situation is desired and in this case, the hydrophilic radius of gyration has no influence in the fitness function. However, if the opposite is true, meaning that the hydrophobic residues are more spread than the hydrophilic, which is not desired, this conformation will be penalized, decreasing its fitness value.

3.4 Genetic Operators and Local Improvement Strategies

In GA, genetic operators are used to create new individuals by means of modifying existing ones. Therefore, it is necessary a method for choosing individuals from the current population in order to apply the genetic operators. We used the tournament selection method that randomly selects a number of individuals from the population. These individuals compete in a tournament and the best one is chosen for the application of the operators. The first operator that is applied during the generation of a new population is the crossover operator. For this problem, this operator plays an important role since a piece of structure (conformation) that has been adequately folded can be of further use in the construction of a complete solution [10]. Two types of crossover were implemented: 1- and 2-point crossover and both are applied with the same probability during the evolution.

Another operator commonly used in GA is mutation. In this work, two different types of mutation were developed. The first is the simple mutation where each gene is tested, according to the mutation probability, to verify whether the actual value of the gene will be changed or not. The second type, called Improved Mutation, works as the simple mutation except by the fact that after each mutation is applied, the individual is reevaluated to check if its fitness has increased. In this case, the change is maintained, otherwise it is discarded. In order to guarantee some diversity during the evolution, 40% of the mutations are simple and 60% are improved mutations.

In our implementation, both crossover and mutation probabilities are not fixed during generations. They have an initial and a final value, respectively for the first and the last generation. The exact probability value in a given generation is a linear interpolation of the initial and final values.

A specially devised operator used in this work is named Partial Optimization. The basic idea of this operator is to randomly select two non-consecutive residues of the protein and fix their position in the lattice. Then, all the different possibilities of locating the intermediate residues maintaining the connectivity of the chain are calculated. The conformation that gives the maximum fitness among all of them is kept. This operator was inspired in a generalization of the 2-opt heuristics proposed by [11] for the traveling salesman problem. The number of intermediate residues to be permutated is a user-defined parameter named Partial Optimization size.

In preliminary tests, the GA frequently got trapped in local minima. Thus, it was necessary to implement a strategy, called Decimation, to make the GA overrides this situation. After each generation, the fitness of the best individual is checked in order to verify whether or not it has changed from the previous generation. If not, a counter is increased by 1. If so, the new best fitness is kept and the counter is reset to 0. When the non-improvement counter reaches 10, the decimation strategy is applied. The idea is to eliminate all individuals of the current population, except the best, and generate again a new population (in the same way explained in section 3.2), including the best individual previously found. Applying this strategy makes the population to have a large genetic diversity, hopefully allowing further evolution. A point that needs to be taken into account is the fact that all the newly generated population probably will have very low fitness values compared to the best individual previously found. Therefore, it is necessary to decrease the selective pressure giving more chance to all individuals to be selected. This is done by decreasing the tourney size in the selection method at the same time that the probability of applying the Improved Mutation is increased. This strategy decreases competitiveness between individuals and permits that all the population becomes, on average, a little better and contributes to the evolution. When this strategy is applied, the non-improvement counter returns to 0 and the verification of the best fitness change proceeds until the last generation.

4 Computational Experiments and Results

Several experiments were performed with the same instances used in [12], for five amino acid chains with 20, 36, 48, 64, and 85 residues. Such instances are not real-world proteins, but a benchmark for which the optimal folding with the 2D HP model is known. Despite of this, it would be interesting to evaluate our method comparing it with a similar one, over the same instances. According to [12], the maximum number of H–H bonds for those instances are: 9, 14, 23, 42, and 52, respectively.

For all the experiments, the parameter set used is shown in Table 1. It was not done a combinatorial experiment so as to find the most efficient set of parameters within the possible range. Instead, we conducted some preliminary tests with different combinations of parameters using a single instance. The set of parameters that performed best among the combinations tested was chosen as default. It is worth to note that possibly another set of parameters could perform better than those used here, but this investigation is subject of future research.

Table 1. Set of parameters for the genetic algorithm

Parameters	Values
Population size	500
Number of Generations	100
PopIniFull	30%
Tourney size	3%
Elitism	Yes
Crossover probability (initial / final)	50% / 70%
Mutation probability (initial / final)	5% / 10%
Partial optimization probability	4%
Partial optimization size	7 residues

Table 2. Comparison of results. Numbers in parenthesis indicates how many times the best score was found in 100 different runs and the bold values indicate the best result for a given instance.

Chain length	König and Dandekar [12]		Our implementation	
	Best solution	Mean value	Best solution	Mean value
20	9 (100×)	9.00	9 (100×)	9.00
36	14 (8×)	12.40	14 (6×)	**12.44**
48	23 (1×)	18.50	23 (2×)	**20.06**
64	37 (1×)	29.30	**40** (1×)	**33.58**
85	46 (1×)	40.80	**51** (2×)	**45.74**

As mentioned before, the penalty weight was (empirically) set according the length of the chain: 2, 2.5, 3, 3.5 and 4 for the 20-, 36-, 48-, 64- and 85-residue chains, respectively.

Tests were run 100 times and the individual with the highest number of hydrophobic non-local bonds from the last generation was considered the best of the run. The overall best individual for each instance is shown in Table 2, together with the number of times this solution was found within 100 runs. The mean number of H–H bonds of the 100 best individuals was calculated and also presented in that table, together with the results obtained by [12], for the purpose of comparison. Values in bold represents the best solutions.

For the 20-residue chain, as the global minimum was always reached, the performance measure considered was the mean number of energy evaluations needed to find the global minimum. König and Dandekar's implementation needed an average of 11824 energy evaluations while ours took 10830.

For the first three chains (namely, 20, 36 and 48 amino acids chains) our results were very similar to [12]. Both implementations were able to find the global optimum but, in average our implementation performed better for the 36- and 48-long amino acid chains. For the 64- and 85-long amino acids chains, our implementation obtained much better results than [12], either considering the best result or the mean value of energy function. For both instances, our best result was very close to the optimal solution known (42 and 52 H–H bonds, respectively).

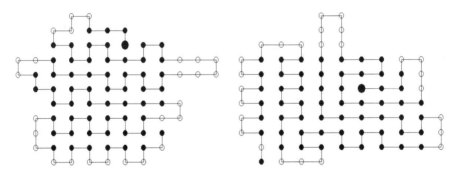

Fig. 1. Best conformations found for the 85-long amino acid chain

In general, our GA got similar results to [12] for the smaller chains and better results for the longer chains. It is important to consider that a difference of one bond from a conformation to another indicates a great improvement obtained by the algorithm and jumping from the closest local minimum to the global minimum can be considered a great achievement. From a solution to another with single bond more, it can mean a quite different folding.

The two best results found for the 85-residue chain are presented in Fig. 1, where the black dots are the hydrophobic residues and the white dots, the hydrophilic. The biggest dot is the beginning of the chain.

5   Conclusions

This paper presented novel strategies for using a genetic algorithm for the protein structure prediction problem using the 2D HP model. The use of the concept of radius of gyration in the fitness function took some smoothness to the fitness landscape, allowing better solutions to be found. Using this fitness function, two conformations with the same number of H–H bonds can be adequately discriminated. Also, the use of the partial optimization and improved mutation operators, together with the decimation strategy have enhanced the GA, allowing it to escape from local minima.

Besides the enhancements in the GA, it is important to emphasize the results obtained. While for short chains the results got no significant improvements (compared with [12]), for the long ones, significant local minima were found, suggesting that there is room for further improvement with longer chains. This subject shall be addressed in further experiments.

The two different solutions shown in Fig. 1 emphasize the difficulty of the PSP problem using a lattice model. The use of this model and the energy function based on the number of H–H bonds implicitly implicates a (strongly) multimodal fitness landscape with many equal-sized plateaus. This fact, by itself, requires efficient search strategies specially when using evolutionary computation techniques.

Exhaustive experiments aiming to find the best parameter set for the GA were not performed, even though the results achieved were very promising. Finding such set of parameters is computationally intensive and care must be taken on its generalization. Experience suggests that not only the size of the amino acid chain is important, but also, some implicit characteristic of the folded structure. These research directions will be explored in the near future.

Overall, results encourage the continuity of the work towards a more complex lattice model, and further tests with the use of real-world protein sequences.

Acknowledgments

Authors would like to thank CNPq for the financial support for H.S.Lopes (grants 305720/04-0 and 402018/03-6) and M.P.Scapin (grant 131355/04-0).

References

1. Berger, B., Leight, T.: Protein Folding in the Hydrophobic-hydrophilic (HP) Model is NP-Complete. J. Comp. Bio. 5 (1998) 27–40
2. Unger, R., Moult, J.: A Genetic Algorithm for Three Dimensional Protein Folding Simulations. In: Proceedings of the 5th Annual International Conference on Genetic Algorithms (1993) 581–588
3. Patton, A.L., Punch III, W.F., Goodman, E.D.: A Standard GA Approach to Native Protein Conformation Prediction. In: Proceedings of the 6th International Conference on Genetic Algorithms, Morgan Kauffman (1995) 574–581
4. Pedersen, J.T., Moult, J.: Protein Folding Simulations With Genetic Algorithms and a Detailed Molecular Description. J. Mol. Biol. 269 (1997) 240–259

246   H.S. Lopes and M.P. Scapin

5. Krasnogor, N., Pelta, D., Lopez, P.E.M., Canal, E.: Genetic Algorithm for the Protein Folding Problem: a Critical View. In: Proceedings of Engineering of Intelligent Systems (1998) 353–360
6. Day, R.O., Lamont, G.B., Pachter, R.: Protein Structure Prediction by Applying an Evolutionary Algorithm. In: International Parallel and Distributed Processing Symposium (2003) 155–162
7. Dill, K.A.: Theory for the Folding and Stability of Globular Proteins. Biochemistry 24 (1985) 1501–1509
8. Krasnogor, N., Hart, W.E., Smith, J., Pelta, D.A.: Protein Structure Prediction with Evolutionary Algorithms. In: Proceedings of the International Genetic and Evolutionary Computation Conference (1999) 1596–1601
9. Beer, F.P., Johnston, E.R.: Vector Mechanics for Engineers – Statics. New York: McGraw Hill, 1980
10. Unger, R., Moult, J.: On the Applicability of Genetic Algorithms to Protein Folding. In: 26th Hawaii International Conference on System Sciences, vol. I, IEEE Press (1993) 715–725
11. Croes, G.A.: A Method for Solving Traveling Salesman Problems. Oper. Res. 5 (1958) 791–812
12. König, R., Dandekar, T.: Improving Genetic Algorithms for Protein Folding Simulations by Systematic Crossover. BioSystems 50 (1999) 17–25

Incorporating Knowledge of Secondary Structures in a L-System-Based Encoding for Protein Folding

Gabriela Ochoa[1], Gabi Escuela[1], and Natalio Krasnogor[2]

[1] Department of Computer Science, Universidad Simon Bolivar,
Po. Box 89000, Caracas 1080-A, Venezuela
gabro@ldc.usb.ve, gabiescuela@netuno.net.ve
[2] School of Computer Science and I.T., University of Nottingham,
NG81BB, Nottingham, UK
Natalio.Krasnogor@nottingham.ac.uk

Abstract. An encoding scheme for protein folding on lattice models, inspired by parametric L-systems, was proposed. The encoding incorporates problem domain knowledge in the form of predesigned production rules that capture commonly known secondary structures: α-helices and β-sheets. The ability of this encoding to capture protein native conformations was tested using an evolutionary algorithm as the inference procedure for discovering L-systems. Results confirmed the suitability of the proposed representation. It appears that the occurrence of motifs and sub-structures is an important component in protein folding, and these sub-structures may be captured by a grammar-based encoding. This line of research suggests novel and compact encoding schemes for protein folding that may have practical implications in solving meaningful problems in biotechnology such as structure prediction and protein folding.

1 Introduction

Proteins are complex organic compounds made up of amino acids joined by peptide bonds[1]; they are essential to the structure and function of all living beings; and are amongst the most studied molecules in biochemistry. Proteins fold naturally into unique 3-dimensional structures, known as their native state or *tertiary structure*. The biological role of a protein will depend on this 3D conformation which in turn is determined by its amino acid sequence (also known as *primary structure*). Biochemists also distinguish *secondary structures* which are highly patterned sub-structures – mainly α-helices and β-sheets – that are locally defined, so there can be many secondary motifs present in a single protein (see Figure 1).

---

[1] An amino acid is any molecule that contains both amino and carboxylic acid functional groups. A peptide bond is a chemical bond formed between two molecules when the carboxyl group of one molecule reacts with the amino group of the other molecule, releasing a molecule of water.

E. Talbi et al. (Eds.): EA 2005, LNCS 3871, pp. 247–258, 2006.

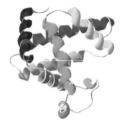

Fig. 1. A representation of the 3D structure of myoglobin, showing shaded α-helices

Genome projects are producing vast amounts of amino acid sequences, but understanding the biological role of these proteins will require knowledge of their structure. The problem of predicting the 3D conformation of a protein from its linear sequence, is known as the protein structure prediction problem (PSP). Although, biochemists use empirical techniques (e.g. magnetic resonance, and X-ray crystallography) on protein crystals in order to infer their conformations, these methods are costly and time consuming. Computational structure prediction methods will provide valuable information for the large amount of sequences whose structures will not be determined experimentally. Two classes of computational methods for the PSP are distinguished [1]. The first (e.g. threading and comparative modelling) rely on detectable similarities between the modelled sequence and known structures. The second class of methods, *de novo* or *ab initio* methods, predict the structure from sequence alone, without relying on similarity at the fold level between the modelled sequence and any of the known structures. Several heuristic search methods (e.g. monte-carlo methods, simulated annealing, and evolutionary algorithms) have been applied for de novo structure prediction [20, 4, 16, 11]. However, PSP is still an open problem and large instances are difficult to solve. A possible cause hindering the scaling of these techniques, are the current direct encodings used (see section 2.1). In the context of EAs applied to design, it has been argued that a *generative* or rule-based scheme, that specifies how to construct the phenotype, as opposed to a direct encoding of the phenotype; can achieve greater scalability through self-similar and hierarchical structures [7, 2, 8]. Moreover, a generative encoding would be a more compact representation of a solution. A first approach to a generative encoding for the PSP was presented in [6], where non parametric L-systems were evolved to capture protein tertiary structures. This approach although promising, met only partial success since the search process was slow and required many executions of the algorithm to obtain a successful L-system. Here we improved those results with two extensions: first, we consider parametric L-systems, and secondly, we incorporate knowledge about secondary structures in the form of predefined rules.

Previous work on evolving L-systems both for capturing blood vessels on the eye [9], and the growth process of trees [3], have had to rely on specific knowledge of the problem domain in order to enhance the algorithms' performance. This knowledge was, in both cases, incorporated in the form of predesigned fixed rules. On the other hand, previous studies on protein folding simulations with evolutionary algorithms, have used knowledge about secondary structure in order to improve the algorithms' behavior on large instances [13, 16, 10, 15]. Thus, the evidence gathered from previous research suggests that incorporating domain knowledge in the form of predesigned rules that capture secondary substructures, would enhance performance. In this paper we test this hypothesis and compare results with the non parametric L-systems (without knowledge about secondary structures) suggested in [6].

The next section will describe the model of folding used for the experiments in this paper, the HP model. This model although simple still captures the essential properties of protein folding. The currently used encodings for heuristic approaches to PSP in the HP model (which are all direct encodings), are also described. Thereafter, section 3 describes the formalism of L-systems. The proposed encoding, and evolutionary algorithm used are described in section 4. Section 5 presents and discusses our results, and finally section 6 offers our conclusions and hints for future work.

2   The HP Model

A major driving force in determining the tertiary structure of proteins is the hydrophobic effect. The idea behind this effect is that, energetically, protein folding is driven by two factors: *hydrophobic*(or "oily) groups "prefer" to "get away" from water, and *hydrophilic* (or polar) groups "prefer" to "dissolve" in water. Thus, the polypeptide chain folds such that the nonpolar amino acids are "hidden" within the structure and the polar residues are exposed on the outer surface. The HP model [5] captures this idea: only two types of monomers are distinguished: hydrophobic (H), and polar or hydrophilic (P). The set of valid protein structure conformations is the space of all self-avoiding paths (on a selected lattice, e.g., square 2D, triangular, cubic, diamond, etc.), with each amino acid located on a lattice bead. Hydrophobic units that are adjacent in the lattice but non-adjacent in the sequence (also called non-local H-H contacts) add a constant negative factor (generally =-1) and all other interactions are ignored. The native state is thought to be the global energy minimum.

Fig. 2. Native structure in the square 2D lattice for the primary sequence $HPHPPHHPHPPHPHHPPHPH$. White boxes corresponds to H, and black to P amino acids. The arrow indicates the starting point, and the dotted lines the non-local H-H contacts.

2.1   Problem Encoding

In the HP model, the structures can be represented by Cartesian coordinates, internal coordinates or distance geometry. We consider here internal coordinates, which can be absolute or relative. Under the absolute encoding, the structures are represented by a list of absolute moves. In a 2D square lattice, for example, a structure is encoded as a string in the alphabet $\{\mathbf{U}p, \mathbf{D}own, \mathbf{Le}ft, \mathbf{R}ight\}$. When using relative coordinates, each move is interpreted in terms of the previous one, like in LOGO turtle graphics; a structure is encoded as a string in the alphabet $\{\mathbf{F}orward, Turne\mathbf{Le}ft, Turn\mathbf{R}ight\}$. Figure 2, shows the optimal folding of an example protein, the structure is coded either as $RDDLU\,LDLDLUU\,RU\,LU\,RRD$ (absolute encoding) or $RF\,RRLLRLRRF\,RLLRRF\,R$ (relative encoding). The number of non-local $H - H$ contacts is nine. That is, the folding energy is -9.

3   L-Systems

L-systems are a mathematical formalism proposed by the biologist Aristid Lindenmayer in 1968 as an axiomatic theory of biological development. More recently, L-systems have found several applications in computer graphics [19, 18]. Two principal areas include generation of fractals and realistic modelling of plants. Central to L-systems, is the notion of rewriting, where the basic idea is to define complex objects by successively replacing parts of a simple object using a set of rewriting rules or productions. The rewriting can be carried out recursively.

The essential difference between traditional formal language grammars and L-systems lies in the method of applying productions. In formal languages productions are applied sequentially, whereas in L-systems they are applied in parallel, replacing simultaneously all letters in a given word. This difference reflects the biological motivation of L-systems. Productions are intended to capture cell divisions in multicellular organisms, where many division may occur at the same time.

A formal definition of L-systems is as follows [18]: Let V denote an alphabet, V^* the set of all words over V, and V^+ the set of all nonempty words over V. A $L-system$ is an ordered triplet $G = \langle V, \omega, P \rangle$, where V is the $alphabet$, $\omega \in V^+$ is a nonempty word called the $axiom$ and $P \subset V \times V^*$ is a finite set of $productions$. If a pair (a, χ) is a production, we write $a \rightarrow \chi$. The letter a and the word χ are called the $predecessor$ and $sucessor$ of this production respectively. It is assumed that for any letter $a \in V$, there is at least one word $\chi \in V^*$ such that $a \rightarrow \chi$. If no production is explicitly specified for a given predecessor $a \in V$, we assume that the $identity\ production$ $a \rightarrow a$ belongs to the set of productions P.

The derivation process of an L-system can be formally stated as follows: Let $\mu = a_1 \ldots a_m$ be an arbitrary word over V. We will say that the word $\nu = \chi_1 \ldots \chi_n \in V^*$ is $directly\ derived$ from (or $generated$ by) μ, and write $\mu \Rightarrow \nu$, if and only if $a_i \rightarrow \chi_i$ for all $i = 1, \ldots, m$. A word ν is generated by G in a derivation of $length$ n if there exists a $developmental\ sequence$ of words $\mu_0, \mu_1, \ldots, \mu_n$ such that $\mu_0 = \omega, \mu_n = \nu$ and $\mu_0 \Rightarrow \mu_1 \Rightarrow \ldots \mu_n$.

L-systems can be classified into context-free and context sensitive, according to whether production rules refer only to an individual symbol, or to a particular symbol only if it has certain neighborhood. L-systems can be also be deterministic or non-deterministic, according to whether there is exactly one production for each symbol, or there are several, and each is chosen with a certain probability during each iteration. Finally, L-systems can be parametric if there are numerical parameters associated with the symbols or productions.

4   Method

4.1   The Proposed Encoding: PFL-System

The encoding proposed is a simplified parametric, context-free L-system. The alphabet will depend on the lattice and coordinate system used. For the experiments reported here, we selected the square 2D lattice with relative coordinates. Thus, the terminal symbols are $\{F, L, R\}$. Two non-terminal symbols: A and H are included, they represent the predecessors of two predefined rules that capture secondary structures. Thus, the l-system's alphabet is $V = \{F, R, L, A, H\}$ (see Table 1). The axiom ω is a nonempty word in V^{+}, each symbol in the axiom has a parameter associated that determines the number of times it is repeated. The maximum for these repetition values are displayed in Table 1. These values were selected empirically for the set of (relatively short) instances studied in this paper, they are likely to depend on the instances length and complexity. The two prefixed rules are $A = RRLL$ and $H = LLRR$, and represent a single coil of a right-oriented and a left-oriented α-helix respectively (Figure 3). The secondary structure known as β-sheet is represented in the 2D Square HP model as a strings of Fs, so this substructure is also easily captured by the proposed encoding (symbol F with a parameter $n > 1$) (Figure 3). We termed our encoding PFL-system, where P stands for parametric, and F for fixed rules.

Table 1. L-system's symbols and their interpretation

Command	Description	Max. n	Symbol
forward(n)	move forward n times	4	F
right(n)	move right n times	2	R
left(n)	move left n times	2	L
right helix(n)	right helix n times	2	A
left helix(n)	left helix n times	2	H

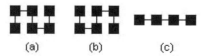

(a)   (b)   (c)

Fig. 3. Secondary structures in the 2D HP model: (a) right-oriented α-helix, $A = RRLL$; (b)left-oriented α-helix $H = LLRR$, (c) β-sheet , $F^{n}, n\,2$

4.2   Evolutionary Algorithm

In order to test whether the proposed encoding can capture a target folding in the 2D HP model, we used an EA as the inference procedure for exploring the L-system's space. Given a target structure in direct encoding (internal relative coordinates) the EA will evolve a generative encoding (L-system) that, once derived, would match closely the original target. The EA implemented was generational with linear ranking selection and elitism. As the variation operators, a recombination and three mutation operators were implemented. A mate selection strategy [17] (dissasortative mating) was also implemented as a mechanism for increasing the population genetic diversity. Dissasortative mating was implemented as follows: when selecting two individuals for a crossover, the first parent was selected as usual. To chose the second parent, a set of s (scan size) individuals were selected using the GA fitness-based selection method. Thereafter, the similarity between each of these s phenotypes and the first parent was computed, the phenotype with less similarity was chosen. For the experiments reported here, Hamming distance was used as the similarity measure, and the scan size s was set to 5. Two stopping criteria were considered: (i) if an individual arises with the maximum fitness, that is, its L-system grammar exactly represents the target folding; or (ii) a preset maximum number of generations is reached. The initial population, genetic operators, and fitness evaluation are described below.

Initialization. L-systems has two predefined production rules: A and H, and the axiom is a variable length word $\omega \in V^+$. A new individual is created by generating a random axiom of 5 to l symbols; where l is slightly larger (about 5%) than the string length of the target folding. In producing the axioms, the probability of generating a terminal symbol $\{F, L, R\}$ is 0.95 whilst that of generating a non-terminal symbol $\{A, H\}$ is 0.05. These values were empirically selected and more exhaustive studies should be performed, since the algorithm behavior was found to be sensitive to these probabilities. Moreover, the most effective settings are likely to depend on the particular instance under study.

Mutation. Three mutation operators were implemented: (i) addition, (ii) deletion, and (iii) modification of a single symbol in the axiom of an individual. The modification operator may alter either a symbol or its associated parameter. When a mutation is to be performed, 60% of times it will be a modification, 30% an addition, and 10% deletion.

Recombination. Recombination takes two individuals, $p1$ and $p2$ as parents and creates two offspring $o1$ and $o2$. Recall that individual's axioms are of variable length; a single cross point is randomly selected considering the length of the shorter axiom (lets consider it to be $p1$). $o1$ is of the same length as $p1$ and inherits from it the left sub-sequence (before the cross point); and the right sub-sequence from $p2$. $o2$ is of the same length as $p2$, and inherits from it the sub-string before the cross point, then it inherits from $p1$ all the symbols after the cross point, finally any remaining symbols to complete the length of $o2$ are taken from $p2$. Thus the proposed crossover has reminiscences with both 1-point

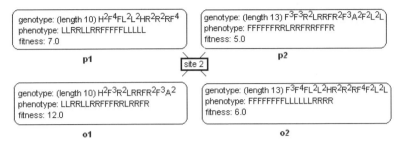

Fig. 4. Example of a crossover operators between two individuals of different lengths

and 2-point crossover. It will be a 1-point crossover if the two parents have the same length. An example of this operator is detailed in Figure 4.

Derivation and Fitness Function. For computing an individual's fitness, its L-system is derived. Phenotypes are directly derived from the axiom, that is, a single derivation step suffices for producing the phenotype from the genotype since the production rules are fixed, and contain only terminal symbols. The derived string will be truncated as soon as the length of the target folding is reached. This means that the rightmost part of the axiom may be discarded. The fitness value will be the number of matches between the produced phenotype and the target folding, that is a generalized Hamming distance. So, the minimum fitness is 0 and the maximum is the length of the desired folding.

5   Experiments and Results

Two sets of experiments were carried out. The first set compared the performance of the newly proposed encoding against the D0L-system implemented in [6]. The same group of proteins instances (see Table 3), and similar EA parameter settings (see Table 2) were employed for the sake of comparison. These four instances are available at http://www.cs.nott.ac.uk/~nxk/hppdb.html; and their foldings were obtained using MAFRA (Memetic Algorithm FRAmework) [14]. Notice that these foldings are not necessarily optimal, but are close to the optimal solution.

The number of successes (runs that produced the target folding exactly) out of 50 runs, is shown for each encoding (Table 4). Also a summary of the secondary

Table 2. Parameter values used for the experiments

Parameter	Value
Max. Number of Generations	2000
Population Size	50
Mutation Rate	0.05
Recombination Rate	1.0
Mating Strategy	Disassortative (5)

Table 3. Benchmark protein instances for the 2D HP model. L stands for the folding length, which is also the maximum attainable fitness of our EA approach.

Name	Protein Sequence	Target Folding	L
Ins18a	HPHPPHHPHPPPHPHHPPHPH	RFRRLLRLRRFRLLRRFR	18
Ins18b	HHHPPHPHPHPPPHPHPHPPH	RRFRFRLFRRFLRLRFRR	18
Ins22	HHPPHPPHPPHPPHPPHPPHPPHH	RLLFLFFRRFLLFRRLRFFRRF	22
Ins23	PPHPPHHPPPPHHPPPPHHPPPPHH	FFRRFFFLLFFFFRRFFFFLLFF	23

Table 4. Comparing No. of successful runs (runs that produced the target folding exactly) using both D0L-systems as proposed in [6], and the parametric L-system with fixed rules (PFL-system) proposed here

Instance	Secondary Structures	D0L-system	PFL-system
Ins18a	A, H	5/50	14/50
Ins18b	None	3/50	1/50
Ins22	None	1/50	1/50
Ins23	F^3, F^4, F^4	1/50	49/50

structures found by the algorithm is included. Notice that for the instances where secondary structures were present (Ins18a and Ins23), the new encoding (PFL-system) produced a significant higher rate of success. Whereas for the other two instances where there were not α-helices or β-sheets, the performance was comparable with that of the previously proposed encoding.

In order to have a dynamic view of the two encodings' performance, the best fitness (averaged over 50 runs) was plotted for each generation on a selected instance (Ins18a) (see Figure 5, Left). The best performance over the whole run is clearly produced by the parametric L-system with fixed rules (PFL-system).

The encoding proposed in [6], was unable to capture the folding of instances longer than twenty or so amino acids. For example, for an instance of length 34

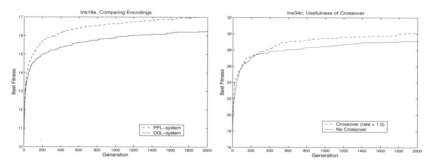

Fig. 5. Best-performance-trace curves. **Left**: Ins18a with two encodings; D0L-systems as proposed in [6], and the parametric L-system with fixed rules (PFL-system) proposed here. **Right**: Ins34c with and without recombination. The curves show the average of 50 (Ins18a) and 20 (Ins34c) runs.

Table 5. Benchmark protein instances of length 34. L stands for the folding length, which is also the maximum attainable fitness of our EA approach.

Name	Protein Sequence	Target Folding	L
Ins34a	PPPHHPPHHPPPPPHHHHHHHHPPHHPPPPHHPPHPP	FFLFRRLLRRFFFRRFLLFRLLFRFFLFLLRFRR	34
Ins34b	HPHHHHHPHHPPHPHHHHPHPPHHPPHHPPHHHHHHH	RRFRLLRRLLFLFRRFLFLLRRLLRRLLFLFRRF	34
Ins34c	HHHHHHHHPHPHHPHPPHHPPHHPPHPHHPHPPHHPH	LLFRLRRFRLLRFRRLLRRLLRRFRLLRFRRLLR	34

Table 6. Results for the benchmark instances of length 34 with and without recombination. The frequencies of obtained maximum fitness values are shown. The maximum possible fitness is 34 (exact match).

Name	Crossover	Frequencies of Obtained Fitness Values	Average	Secondary Structures
Ins34a	On	27:1, 28:3, 29:6, 30:5, 31:4, 32:0, 33:1	29.6	
Ins34a	Off	28:4, 29:5, 30:7, 31:4	29.55	A, F^3
Ins34b	On	29:4, 30:5,31:9,32:2	30.45	
Ins34b	Off	28:2, 29:4, 30:6, 31:4, 32:3, 33:1	30.25	H, A, A
Ins34c	On	28:2, 29:4, 30:6, 31:7, 32:0, 33:1,	30.1	
Ins34c	Off	27:1, 28:5, 29:9, 30:1 ,31:3, 32:1	29.15	H^2, A^2

(Ins34a), the best fitness statistics obtained after 20 runs were: average = 24.05, best run = 27.0, worst run = 19.0. In order to asses whether the newly proposed encoding had better scaling properties; a second group of experiments explored three instances of length 34 (see Table 5). Ins34a was obtained from the same source than the shorter instances described above, whereas Ins34b and c, and their foldings were taken from [21]. The parameter settings for these experiments are the same as before (see Table 2), but the number of replicas were 20 instead of 50 Furthermore, runs with and without crossover were carried out in order to asses the usefulness of this operator in this context.

Results suggest that the PFL-system encoding has better scaling properties than the previous D0L-system. Although the perfect match (34) was not found in any run, most runs arrive very close to the solution. Table 6, shows the frequencies of obtained maximum fitness for each instance with and without crossover. Crossover seems to be helpful to the evolutionary search although

Table 7. Best obtained individual, represented in PFL-system encoding, for each of the benchmark instances studied

Instance	Best Solution (PFL-system Encoding)	Encoding Length
Ins18a	$RFARLR^2FRHFR$	11
Ins18b	$R^2FRFRLFR^2FLRLRFR^2$	15
Ins22	$RL^2FLF^2R^2FL^2FR^2LRF^2R^2F$	15
Ins23	$F^2R^2F^3L^2F^4R^2F^4L^2F^2$	9
Ins34a	$F^2LFAR^2F^3R^2FL^2FRL^2FRF^2LFL^2RFR^2$	21
Ins34b	$R^2FRHL^2FLFR^2FLFL^2A^2FLFR^2F$	19
Ins34c	$L^2F^2R^2RFRL^2RFR^2H^2FRLLRFA^2$	18

the differences are not substantial. Some instances seems to benefit more from crossover than others.In order to have a dynamic view of the algorithm behavior with and without recombination, Figure 5 (Right) shows best-performance curves over the whole run for Inst34c. Clearly the recombinant GA outperform the GA with mutation only.

As a summary of results, the best solutions obtained with the PFL-system, for all the instances studied, are shown in Table 7. Notice that the inclusion of parameters helps in having a more compact representation of a folding, also the occurrence of secondary structures is captured and easily identified with this representation.

6   Discussion

An encoding scheme for protein folding in the HP model, inspired by parametric L-systems, was proposed. The encoding also incorporates problem domain knowledge in the form of predesigned production rules that capture the most commonly known secondary structures: α-helices and β-sheets. The ability of this encoding to capture protein native conformations was tested using an EA as the inference procedure for discovering L-systems. Given a target folding, the EA explores the space of possible L-systems (genotypes) until identifying one whose derivation (phenotype) closely matches the target folding.

This newly proposed encoding was found to improve our first attempt of using L-systems as a generative representation for protein folding [6], where problem domain knowledge was not incorporated. The suitability of the new encoding, however, seems to heavily depend on the particular instance under study. Instances with high frequencies of α-helices and β-sheets, would have a clear advantage. Longer proteins and 3D lattices should be addressed. Furthermore, two somehow opposite but complementary extensions could be suggested. First, incorporating other known secondary structures such as β-turns, β-hairpins, etc., as prefixed rules. Secondly, enabling the EA to discover their own production rules, that could be in principle stored and thereafter used in further runs with new instances. We have evidence on a related bioinformatic problem [12] that enabling the evolutionary algorithm to systematically and vigorously discover new "building blocks" (as the ones we described in this paper) can substantially improve the algorithm performance.

Finally, we believe that this proposed line of research opens up the possibilities for novel and compact encoding schemes of protein structures, that have potential implications in solving meaningful biotechnology problems such as structure prediction and protein folding.

Acknowledgements

Natalio Krasnogor acknowledges EPSRC (GR/T07534/01, EP/D021847/1) and BBSRC (BB/C511764/1) for funding his research on protein structure prediction, comparison and self-assembly.

References

[1] David Baker and Andrej Sali, *Protein structure prediction and structural ge-nomics*, Science **294** (2001), 93–96.
[2] Peter J. Bentley, *Exploring component-based representations - the secret of cre-ativity by evolution?*, Fourth International Conference on Adaptive Computing in Design and Manufacture (ACDM 2000) (I. C. Parmee, ed.), 2000, pp. 161–172.
[3] Luis DaCosta and Jacques-Andre Landry, *Generating grammatical plant models with genetic algorithms*, Proceedings of the 7th International Conference on Adap-tive and Natural ComputiNG Algorithms (ICANNGA, LNCS, Springer Verlag, 2005.
[4] T. Dandekar and P. Argos, *Folding the main chain of small proteins with the genetic algorithm*, J. Mol. Biol **236** (1994), 844–861.
[5] Ken A. Dill, *Theory for the folding and stability of globular proteins*, Biochemistry **24** (1985), 1501.
[6] Gabi Escuela, Gabriela Ochoa, and Natalio Krasnogor, *Evolving L-systems to capture protein structure native conformations*, Proceedings of the 8th European Conference on Genetic Programming, Lecture Notes in Computer Science, vol. 3447, Springer, 2005, pp. 74–84.
[7] L. J. Fogel, P. J. Angeline, and T. Bäck (eds.), *Shape representations and evolution schemes*, MIT Press, 1996.
[8] Gregory S. Hornby and Jordan B. Pollack, *The advantages of generative gram-matical encodings for physical design*, Proceedings of the 2001 Congress on Evo-lutionary Computation CEC2001, IEEE Press, 2001, pp. 600–607.
[9] G. Kókai, Z. Tóth, and R. Ványi, *Modelling blood vessels of the eye with paramet-ric L-systems using evolutionary algorithms*, Proceedings of the Joint European Conference on Artificial Intellingence in Medicine and Medical Decision Making (AIMDM-99) (Berlin), LNAI, vol. 1620, Springer, 1999, pp. 433–442.
[10] Natalio Kranogor, *Studies on the theory and design space of memetic algorithms*, Ph.D. thesis, University of the West of England, Bristol, UK, 2002.
[11] N. Krasnogor, B. P. Blackburne, E. K. Burke, and J. D. Hirst, *Multimeme algo-rithms for protein structure prediction*, Lecture Notes in Computer Science **2439** (2002), 769–779.
[12] Natalio Krasnogor and Stephen Gustafson, *The local searcher as a supplier of building blocks in self-generating*, Workshop Proceedings of the 2003 Genetic and Evolutionary Computation Conference, GECCO 2003, 2003.
[13] Natalio Krasnogor, D. Pelta, P.E. Martinez-Lopez, P. Mocciola, and E. de la Canal, *Enhanced evolutionary search of foldings using parsed proteins*, Proceed-ings of the Argentinian Operational Research Simposium (S.I.O. 97), 1997.
[14] Natalio Krasnogor and Jim Smith, *MAFRA: A java memetic algorithms frame-work*, Data Mining with Evolutionary Algorithms (Alex A. Freitas, William Hart, Natalio Krasnogor, and Jim Smith, eds.), 2000, pp. 125–131.
[15] Neal Lesh, Michael Mitzenmacher, and Sue Whitesides, *A complete and effec-tive move set for simplified protein folding*, Proceedings 7h Annual International Conference on Research in Computational Molecular Biology (RECMB), 2003.
[16] F. Liang and W. Wong, *Evolutionary monte carlo for protein folding simulations*, Journal of Chemical Physics **115** (2001), no. 7, 3374–3380.
[17] Gabriela Ochoa, C. Mädler-Kron, R. Rodriguez, and K. Jaffe, *Assortative mating in genetic algorithms for dynamic problems*, Applications of Evolutionary Com-puting, EvoWorkshops2005, LNCS, vol. 3449, Springer Verlag, 2005, pp. 605–610.

[18] P. Prusinkiewicz and A. Lindenmayer, *The algorithmic beauty of plants*, Springer, New York, 1990.

[19] Alvy R. Smith, *Plants, fractals, and formal languages*, Computer Graphics **18** (1984), no. 3, 1–10.

[20] I. Unger and J. Moult, *Genetic algorithms for protein folding simulations*, Journal of Molecular Biology **1** (1993), no. 231, 75–81.

[21] Berrin Yanikoglu and Burak Erman, *Minimum energy configurations of the 2-dimensional hp-model of proteins by self-organizing networks*, Journal of Computational Biology **9** (2002), no. 4, 613–620.

The Electromagnetism Meta-heuristic Applied to the Resource-Constrained Project Scheduling Problem

Dieter Debels[1] and Mario Vanhoucke[1,2]

[1] Ghent University, Faculty of Economics and Business Administration,
Hoveniersberg 24, Ghent 9000, Belgium
[2] Vlerick Leuven Gent Management School,
Operations & Technology Management Centre, Reep 1,
Ghent 9000, Belgium
{dieter.debels, mario.vanhoucke}@ugent.be

Abstract. Recently, an electromagnetism (EM) heuristic has been introduced by Birbil and Fang (2003) to solve unconstrained optimization problems. In this paper, we extend the EM methodology to combinatorial optimization problems and illustrate its effectiveness on the well-known resource-constrained project scheduling problem (RCPSP). We present computational experiments on a standard benchmark dataset, compare the results of the different modifications on the original EM framework with current state-of-the-art heuristics, and show that the procedure is capable of producing consistently good results for challenging instances of the problem under study. We also give directions for future research in order to further explore the potential of this new technique.

1 Introduction

The problem under study is the well-known resource-constrained project scheduling problem (RCPSP). The RCPSP can be stated as follows. A set of activities N, numbered from 1 to n ($|N| = n$), is to be scheduled without pre-emption on a set R of renewable resource types. Activity i has a deterministic duration $d_i \in$ IN and requires $r_{ik} \in$ IN units of resource type k, $k \in R$, which has a constant availability a_k throughout the project horizon. We assume that $r_{ik} \leq a_k$ for $i \in N$ and $k \in R$. The dummy start and end activities 1 and n have zero duration while the other activities have a non-zero duration. The dummies also have zero resource usage. A is the set of pairs of activities for which a finish-start precedence relationship with time lag 0 exists. We use S_i (P_i) to denote the set of immediate successors (predecessors) of activity i and S_i' (P_i') to refer to the set of all (immediate and transitive) successors (predecessors) of this activity. We assume graph G(N,A) to be acyclic. A schedule S is defined by an n-vector of start times $s = (s_1, ..., s_n)$ which implies an n-vector of finish times e ($e_i = s_i + d_i, \forall i \in N$). A schedule is said to be feasible if the precedence and resource constraints are satisfied. The objective of the RCPSP is to find a feasible schedule such that

E. Talbi et al. (Eds.): EA 2005, LNCS 3871, pp. 259–270, 2006.

the schedule makespan e_n is minimized. In this paper we report results for the application of a recent heuristic technique, electromagnetism (EM) [4]. EM is an 'evolutionary' algorithm that was originally developed for the optimization of unconstrained continuous functions. As we modify the technique to solve the RCPSP, we show that EM can also be used for combinatorial problems.

2   The Electromagnetism Meta-heuristic

Birbil and Fang [4] propose a so-called electromagnetism optimization heuristic for unconstrained optimization problems, i.e. the minimization of non-linear functions. These optimization problems with bounded variables can be modeled as depicted at the left side of Fig. 1. At the right side, we show how the RCPSP can be reformulated as an unconstrained optimization problem. To obtain a Euclidean solution space, we opt for a schedule representation in random-key (RK) format [17]. To transform an RK vector $x \in IR^n$ into a schedule $S = \sigma(x)$ with an associated makespan $e_n(\sigma(x))$, a schedule generation scheme (SGS) is necessary. We make use of the serial SGS, as it is sometimes impossible to reach an optimal solution with the parallel SGS [14]. In the remainder of the paper we assume that a higher RK value corresponds to a lower priority of the activity. By setting $l_i = 0$ and $u_i = n$, we assume that each priority element of an RK vector is a real value between 1 and n. In order to decrease the solution space of the search process, we adapt this vector with new lower and upper values to $l_i = |P_i'| + 1$ and $u_i = n - |S_i'|$. In doing so we increase the likelihood that the obtained solution corresponds to a precedence-feasible priority structure where each activity has a lower priority (i.e. a higher RK value) than its predecessors.

Consider the example project presented in Fig. 2. This project network contains 9 non-dummy activities for which the duration is given above the node and the resource requirement for the single resource below the node. The corresponding lower and upper values (l_i and u_i) for the RK value of each activity i are given between brackets. Assume that the resource availability equals 2, then Fig. 3 depicts two feasible schedules for the example project.

The EM heuristic assumes a multidimensional solution space where each point x represents a solution. A charge is associated to each point, related to

General	RCPSP
minimize $f(x)$	minimize $e_n(\sigma(x))$
s.t.	s.t.
$x \in [l, u]$	$x \in [l, u]$
where $[l, u] = \{x \in IR^D \mid l_i \leq x_i \leq u_i, i = 1, \dots, D\}$	where $[l, u] = \{x \in IR^n \mid l_i \leq x_i \leq u_i, i = 1, \dots, n\}$
with D the dimension of the solution space	with n the number of activities

Fig. 1. Formulation of unconstrained optimization problems

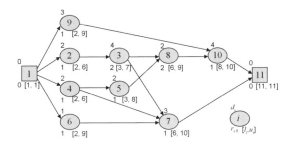

Fig. 2. Example project network with 9 non-dummy activities

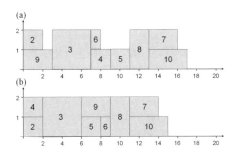

Fig. 3. Two feasible schedules for the example project

the objective function value $f(x)$ associated with the solution point x. As in evolutionary algorithms, a population is created in which each solution point will exert attraction or repulsion on other points, of which the magnitude is proportional to the product of the charges and inversely proportional to the distance between the points (Coulomb's Law). The principle behind the algorithm is that inferior solution points will prevent a move in their direction by repelling other solution points in the population, and that attractive points will facilitate moves in their direction. The generic pseudo-code for the EM algorithm is as follows:

> Algorithm EM
> $iter$:=1
> while *stop condition not satisfied* do
>   *compute_forces*
>   *apply_forces*
>   *local_search*
> $iter$++
> endwhile

The function EM contains three subroutines (*compute_forces*, *apply_forces* and *local_search*), that are iteratively applied as long as the stop condition is not satisfied. The total force exerted on each point by all other points is calculated in the function *compute_forces* and depends on the charge of the point under consideration as well as on the points exerting the force. The charge of the kth population point x^k is determined by its objective function value $f(x^k)$ in relation to the objective function value $f(x^{best})$ of the current best point in the population. For a minimization problem, the charge q^k of point x^k is determined according to eq. 1. In the first term we calculate the value of q^k as given by Birbil and Fang [4], and in the second term we translated the formulation to the RCPSP context. Note that m represents the population size.

$$q^k = e^{\left(\frac{-D\cdot\left(f\left(x^k\right)-f\left(x^{best}\right)\right)}{\sum_{l=1}^{m}\left(f(x^l)-f(x^{best})\right)}\right)} = e^{\left(\frac{-n\cdot\left(en\left(\sigma((x^k))\right)-en\left(\sigma((x^{best}))\right)\right)}{\sum_{l=1}^{m}\left(en(\sigma((x^l)))-en(\sigma((x^{best})))\right)}\right)} \quad (1)$$

The total force exerted on a point by all other points is calculated in a similar way with Coulomb's law and is inversely proportional to the distance between the points and directly proportional to the product of their charges. The set of force vectors $F^k(k=1,\ldots,m)$ exerted on the corresponding point x^k is determined as shown in eq. 2. The point with a relatively good objective function value attracts the other one, the point with the inferior objective function value repels the other. In $\|\text{dist}(x^l,x^k)\|$ we measure the normalized distance between two points (schedules) x^k and x^l. The distance equals the sum of the absolute deviations of the priority values x^k and x^l of each activity i, i.e. $\text{dist}(x^l,x^k)=\sum_{i=1}^{n}|x_i^l-x_i^k|$. In order to normalize the distance measure to the interval $[0,1]$, we set $\|\text{dist}(x^l,x^k)\|=\text{dist}(x^l,x^k)/\text{dist}_{max}$ with dist_{max} the maximum of all distances between each pair of points, i.e. $\text{dist}_{max}=\max_{l=1,\ldots,m;k=1,\ldots,m}\text{dist}(x^l,x^k)$. Thus, points with a better objective function value attract point x^k, while points with an inferior objective function value repel x^k.

$$F^k = \sum_{l=1,l\neq k}^{m} \left\{ \begin{array}{l} (x^l - x^k)\left(\frac{q^l\cdot q^k}{\|\text{dist}(x^l,x^k)\|}\right) \text{ if } f(x^l) < f(x^k) \\ (x^k - x^l)\left(\frac{q^l\cdot q^k}{\|\text{dist}(x^l,x^k)\|}\right) \text{ if } f(x^l) \geq f(x^k) \end{array} \right\} \quad (2)$$

The movement according to the resulting forces is performed in *apply_forces* and is shown in eq. 3. The move is based on the normalized force vector $\|F^k\| = F^k/\max_{i=1,\ldots,n}(F_i^k)$. Thus, the original force vector F^k only identifies the direction of the move. The magnitude of each move is determined by a randomly selected parameter λ, generated from a uniform distribution from the interval $[0,1]$ and also by the lower value l_i and upper value u_i for the priority value x_i^k belonging to the ith activity of population element k.

$$x_i^k = \left\{ \begin{array}{l} x_i^k + \lambda\|F_i^k\|(u_i - x_i^k) \text{ if } F_i^k > 0 \\ x_i^k + \lambda\|F_i^k\|(x_i^k - l_i) \text{ if } F_i^k \leq 0 \end{array} \right\} \quad (3)$$

After the application of the forces on the population elements, *local_search* aims to improve the newly obtained solution points. In the original version [4], a local search technique that explores the immediate (Euclidian) neighbourhood of individual points is proposed. However, for the RCPSP it is beneficial to use the iterative forward/backward scheduling technique [21] as a simple and effective local search technique. To obtain an improved schedule, the technique iteratively performs backward and forward passes. A backward pass transforms a left-justified schedule in a right-justified schedule by scheduling the activities backwards in decreasing order of their finish times. A forward pass transforms a right-justified schedule in a left-justified schedule by scheduling the activities forwards in increasing order of their start times. In doing so, the schedule makespan of each intermediate schedule is never higher than the makespan of the previous one.

To the best of our knowledge, the EM philosophy has only been used for scheduling projects by [7]. However, these authors present a *scatter search* algorithm for the RCPSP, and seed their algorithm with very basic principles taken from the electromagnetism philosophy. More precisely, they restrict the use of the EM philosophy to the description of the hybrid two-point/electromagnetism crossover operator. However, a closer look to this hybrid crossover reveals that

- Forces are only calculated based on *one* other population-element. This is not in line with the basic EM philosophy in which a point exerts a force on *all* other points.
- The forces are not related to the distance between solutions. This is in contradiction to the EM philosophy in which the magnitude of the force is inversely proportional to the distance between points, in order to follow the law of Coulomb.

In section 3 of the current paper, we present a step-wise adaptation of our EM algorithm to cope with the RCPSP, following the framework as proposed by [4].

3   Computational Experiments

We have coded the procedure in Visual C++ 6.0 and performed computational tests on an Acer Travelmate 634LC with a Pentium IV 1.8 GHz processor using the well-known PSPLIB dataset [19]. This set contains the subdatasets J30, J60, J90 and J120 with problem instances of 30, 60, 90 and 120 activities. In section 3.1, we describe a step-wise adaptation of the algorithm of section 2 in order to improve the performance. In doing so, we rely on specific characteristics of the RCPSP. Section 3.2 compares the performance of our EM heuristic with other state-of-the-art results.

3.1   Using Problem-Specific Characteristics of the RCPSP

Table 1 reports the results for our step-wise improvement scenarios as discussed in the following subsections, based on a run with 5,000 schedules. The column labelled "Basic" reports the results for the basic EM meta-heuristic of section 2. The following columns report the results for the different modifications on this basic EM algorithm. More details are given in the following subsections. The rows labelled "Avg.Dev.Lb" report the average deviation from the optimal solution (J30 instances) or from the critical path based lower-bound (J60, J90 and J120 instances). The rows labelled "Avg.CPU" indicate the average computation time to solve a problem instance (in seconds). For each adaptation, we have finetuned

Table 1. Results for the step-wise improvement scenarios for the basic EM meta-heuristic

Dataset	Basic	SRK	MUT	NBH	SUB
Avg.Dev.Lb					
J30	0.20%	0.22%	0.10%	0.11%	0.12%
J60	11.94%	11.50%	11.47%	11.43%	11.29%
J90	11.81%	11.06%	11.06%	11.05%	10.89%
J120	36.50%	34.32%	34.32%	34.21%	33.98%
Avg.CPU					
J30	0.09s	0.07s	0.08s	0.07s	0.06s
J60	0.16s	0.15s	0.15s	0.14s	0.13s
J90	0.26s	0.23s	0.23s	0.20s	0.19s
J120	0.36s	0.36s	0.36s	0.31s	0.34s
m					
J30	40	35	30	30	25
J60	30	40	25	30	25
J90	40	35	35	30	25
J120	40	35	35	30	15
p_{mut}					
J30	-	-	0.10	0.10	0.10
J60	-	-	0.01	0.01	0.01
J90	-	-	0.00	0.00	0.00
J120	-	-	0.00	0.00	0.00

the algorithm by setting the population size m to an optimal value. These values are given in the rows "m". As an example, the table reveals that the basic version of the algorithm reports the best results for a population size of 40 for the J30, J90 and J120 instances and a population size of 30 for the J60 instances. The rows labelled "p_{MUT}" are used to display the percentage of mutation, which will be discussed in section 3.1.2.

3.1.1 Topological-Order Representation or Standardized RK (SRK)

In the RK representation, each solution corresponds to a point in the Euclidian n-space, so that geometric operations can be performed on its components. Since this is one of the cornerstones of the EM method, we have adopted the RK representation in our EM-heuristic. However, the RK representation suffers from the fact that one schedule can have an infinite number of schedule representations. To deal with this problem, we propose to use a topological-order (TO) representation of schedules [30,31]. A TO representation in RK format of a schedule is any RK vector x for which $s_i < s_j$ implies $x_i < x_j$. To incorporate the TO condition, we change the RK representation to the so-called standardized RK (SRK) as suggested by [7]. More precisely, we first rank the activities according to their start times in the schedule, and then replace their priority values by the place in the ranking. In doing so, the SRK vector fits very well into the EM framework, since each vector element will have a value between l_i and u_i. As an example, the SRK vector for the schedule of Fig. 3(a) is $\{1, 2, 4, 5, 7, 5, 9, 8, 2, 9, 11\}$. Note that the SRK-value for each activity i always lies between $l_i = |P'_i| + 1$ and $u_i = n - |S'_i|$ and that we can only transform an RK vector x into SRK format $\pi(x)$ after the schedule generation. The results

Table 2. Input data for the example: the start RK vector and the two forces

Activity i	1	2	3	4	5	6	7	8	9	10	11
x_i^0	1	2	4	5	7	5	9	8	2	9	11
$\lambda^1 \lVert F_i^1 \rVert$	0	0.26	-0.21	-0.09	-0.15	0.03	0.57	-0.05	0.14	-0.17	0
$\lambda^2 \lVert F_i^2 \rVert$	0	-0.51	-0.37	-0.19	-0.44	0.02	-0.08	-0.18	0.09	0.57	0

of the incorporation of the TO representation by using the SRK can be seen in table 1 by comparing the columns labelled "Basic" and "SRK". The results show a beneficial effect for the J60, J90 and J120 sets and a negative effect for the J30 instances.

Although the SRK representation embeds the logic that early scheduled activities have a high priority, it also has a major drawback. The execution of a force on an SRK vector $x^0 = \pi(x^0)$ modifies the priority structure of the vector to an RK-vector x^1 which will be transformed by means of the serial SGS and the local search method to a schedule with a corresponding SRK notation $\pi(x^1)$. It is, however, possible that the resulting schedule (and consequently, the resulting SRK notation) is not different from the original one, i.e. $\pi(x^0) = \pi(x^1)$ while $x^0 \neq x^1$. This effect might prevent to exploit the advantages of the basic philosophy of the EM approach, which focuses on a gradual shift to other regions of the solution space. Due to the transformation from x^1 to $\pi(x^1)$, this gradual shift from x^0 to x^1 will be cancelled out, having an effect on the next moves of the meta-heuristic. More precisely, our tests revealed that 79% of the moves, performed on solutions in SRK format, result in a schedule for which $\pi(x^0) = \pi(x^1)$ for the 30-activity networks. This value decreases to 65% for the J60 instances, 56% for the J90 instances and only 14% for the J120 instances. Thus, the cancel-out problem is particularly relevant for small problem instances, as the solution space is too small to escape from a solution point.

Consider the example project of Fig. 2 and the two corresponding schedules of Fig. 3. The start vector x^0 corresponds to the schedule of Fig. 3(a). $\lambda^1 \lVert F^1 \rVert$ and $\lambda^2 \lVert F^2 \rVert$ are used to calculate the first and second move and are given in table 2. Table 3 displays the calculations of two moves in a sequence, based on the RK vectors while table 4 displays the calculations of these moves based on the SRK. In the first move of table 3, from x^0 to x^1, the RK-representation changes, but the corresponding schedule remains unchanged and is equal to the schedule of

Table 3. Illustration of the execution of two moves based on a RK vector

activity i	1	2	3	4	5	6	7	8	9	10	11
x_i^0	1	2	4	5	7	5	9	8	2	9	11
$\Delta x_i^0 = \lambda^1 \lVert F_i^1 \rVert (u_i - x_i^0)$		1.03				0.11	0.57		0.96		
$\Delta x_i^0 = \lambda^1 \lVert F_i^1 \rVert (x_i^0 - l_i)$	0		-0.21	-0.27	-0.62			-0.3		-0.17	0
$x_i^1 = x_i^0 + \Delta x_i^0$	1	3.03	3.79	4.73	6.38	5.11	9.57	7.7	2.96	8.83	11
$\Delta x_i^1 = \lambda^2 \lVert F_i^2 \rVert (u_i - x_i^1)$						0.09			0.54	0.67	
$\Delta x_i^1 = \lambda^2 \lVert F_i^2 \rVert (x_i^1 - l_i)$	0	-0.53	-0.29	-0.53	-1.48		-0.37	-0.3			0
$x_i^2 = x_i^1 + \Delta x_i^1$	1	2.5	3.5	4.2	4.9	5.2	9.2	7.4	3.5	9.5	11

Table 4. Illustration of the execution of two moves based on a SRK vector

activity i	1	2	3	4	5	6	7	8	9	10	11
x_i^0	1	2	4	5	7	5	9	8	2	9	11
$\Delta x_i^0 = \lambda^1 \|F_i^1\|(u_i - x_i^0)$		1.03				0.11	0.57		0.96		
$\Delta x_i^0 = \lambda^1 \|F_i^1\|(x_i^0 - l_i)$	0		-0.21	-0.27	-0.62			-0.3		-0.17	0
$x_i^1 = x_i^0 + \Delta x_i^0$	1	3.03	3.79	4.73	6.38	5.11	9.57	7.7	2.96	8.83	11
$x_i^1 = \pi(x_i^1)$	1	2	4	5	7	5	9	8	2	9	11
$\Delta x_i^1 = \lambda^2 \|F_i^2\|(u_i - x_i^1)$	0					0.09			0.63	0.57	
$\Delta x_i^1 = \lambda^2 \|F_i^2\|(x_i^1 - l_i)$	0		-0.37	-0.58	-1.75		-0.32	-0.36			0
$x_i^2 = x_i^1 + \Delta x_i^1$	1	2	3.63	4.42	5.25	5.09	8.68	7.64	2.63	9.57	11
$x_i^2 = \pi(x_i^2)$	1	2	4	5	7	5	9	8	2	9	11

Fig. 3(a). After the second move, from x^1 to x^2, the RK-representation belongs to the schedule depicted in Fig. 3(b). Table 4, however, cannot escape from the schedule of Fig. 3(a) since the second move from $\pi(x^1)$ to $\pi(x^2)$ also results in the same schedule. This is due to the transformation of the RK vector x^1 to the SRK vector $\pi(x^1)$ after a makespan evaluation, which cancels out the gradual shift of the move from x^0 to x^1. In the next sub-section, we describe our mutation approach to overcome this cancel-out problem.

3.1.2 Diversification Using Mutation (MUT)

In order to prevent the population from becoming overly homogeneous, we introduce a basic version of diversification using mutation, by modifying randomly chosen priority values of the vector x to a value uniformly chosen between $l_i = |P_i'| + 1$ and $u_i = n - |S_i'|$. This mutation is imposed right after a force is executed, followed by a makespan evaluation.

In table 1, we use p_{MUT} to denote the percentage of activities that are subject to this mutation per move. The results reveal that mutation is only beneficial for the J30 and J60 instances. For the J30 instances, we modify 10% of the priority values per move, i.e. in each move, three activities receive a new priority value randomly generated from the interval $[l_i, u_i]$. The average deviation decreases from 0.22% to 0.10%. For the J60 instances, mutation is only beneficial to a small extent (from 11.50% to 11.47% deviation) with a mutation rate of only 1%. For the J90 and J120 instances, mutation has no beneficial effect. These results confirm that mutation can help to overcome the problem mentioned in the previous section. Since the use of the SRK notation could possibly cancel out the gradual shifts in the solution space, diversification using mutation will be necessary to escape from a particular schedule. This problem was particularly relevant for the J30 instances and - to a smaller extent - for the J60 instances.

3.1.3 Extended Neighbourhood (NBH)

In the original procedure [4] a parameter λ is generated from a uniform distribution between 0 and 1, i.e. $\lambda \in U(0, 1)$, in order to move from one schedule to another (see eq. 3). We have extended this method by generating more schedules out of a schedule by generating more values for λ. To that purpose, we have divided the interval $[0, 1]$ in equal parts and tested a number of scenarios with 2, 3, 4, 5 and 6 different values for λ and, consequently, up to 6 new schedules per move. Tests have revealed that improved results can be found for J60, J90 and J120 by generating two new schedules with the following parameter values: $\lambda^1 \in U(0.2, 0.6)$ and $\lambda^2 \in U(0.6, 1)$. Afterwards, the algorithm selects the best schedule to enter the population. Note that moves with $\lambda < 0.2$ are excluded since this often leads to the cancel-out effect described in section 3.1.1.

3.1.4 Exert the Force F on a Sub-schedule (SUB)

Based on the calculated forces and resulting attraction or repulsion, points are transformed, i.e. moved in the Euclidian space, resulting in a new population. During each move, forces are exerted on the priority value of each activity. We generalize this concept by allowing forces to act only in a particular subset of the dimensions or activities. We randomly select $p_{min} \in [1, n-1]$ and set $p_{max} = p_{min} + \tau$ with τ chosen randomly within $[1/4.e_n(\sigma(x)), 3/4.e_n(\sigma(x))]$. Then, we update only the RK values between p_{min} and p_{max} (inclusive) according to the forces exerted in these dimensions. Note that due to the SRK

Table 5. Comparative computational results

Algorithm	J30 Avg.Dev.Lb	J30 Rank	J60 Avg.Dev.Lb	J60 Rank	J120 Avg.Dev.Lb	J120 Rank
Valls, Ballestin and Quintanilla [32]	0.06	2	11.10	1	32.54	1
Debels, De Reyck, Leus, Vanhoucke [7]	0.11	5	11.10	1	33.10	2
Valls, Ballestin and Quintanilla [33]	0.20	12	11.27	5	33.24	3
Kochetov and Stolyar [13]	0.04	1	11.17	3	33.36	4
Alcaraz, Maroto and Ruiz [2]	0.06	2	11.19	4	33.91	5
Our procedure	**0.10**	**4**	**11.29**	**6**	**33.94**	**6**
Valls, Ballestin and Quintanilla [33]	0.28	16	12.35	16	34.02	7
Tormos and Lova [29]	0.13	7	11.62	7	34.41	8
Hartmann [9]	0.22	13	11.70	8	35.39	9
Merkle, Middendorf and Schmeck [23]	-		-		35.43	10
Tormos and Lova [28]	0.17	10	11.82	9	35.56	11
Tormos and Lova [27]	0.16	8	11.87	11	35.81	12
Alcaraz and Maroto [1]	0.12	6	11.86	10	36.57	13
Hartmann [8]	0.56	23	13.32	24	36.74	14
Bouleimen and Lecocq [5]	0.23	14	11.90	13	37.68	15
Nonobe and Ibaraki [24]	0.16	8	12.18	15	37.88	16
Coelho and Tavares [6]	0.33	17	12.63	18	38.41	17
Hartmann [8]	0.25	15	11.89	12	38.49	18
Schirmer [25]	0.44	18	12.58	17	38.70	19
Kolisch [15]	0.53	21	13.23	22	38.75	20
Kolisch [15, 17]	1.28	26	13.21	21	38.77	21
Kolisch and Drexl [16]	0.52	20	13.06	20	40.45	22
Coelho and Tavares [6]	0.54	22	13.31	23	40.46	23
Leon and Ramamoorthy [20]	1.59	29	13.49	26	40.69	24
Kolisch [15]	1.29	27	13.53	27	41.84	25
Hartmann [8]	1.12	25	12.74	19	42.25	26
Kolisch [14]	1.00	24	14.30	28	43.05	27
Kolisch [14]	1.48	28	15.17	29	47.61	28
Klein [12]	0.17	10	12.03	14	-	
Baar, Brucker and Knust [3]	0.44	18	13.48	25	-	

representation, the thus updated activities all start within a particular time interval. The other RK components are not left unchanged, but are updated as follows. We subtract the constant value n from all RK values lower than p_{min}, and add the same constant to all values higher than p_{max}. Doing this preserves the priority structure since the activities outside the interval $[p_{min}, p_{max}]$ are unaffected by the forces. Table 1 reveals that this leads to an additional improvement for the J60 (from 11.43% to 11.29%), J90 (from 11.05% to 10.89%) and J120 (from 34.21% to 33.98%) instances.

3.2   Comparison with the State-of-the-Art Heuristics

To be able to compare procedures for the RCPSP, [10] presented a methodology in which all procedures can be tested on the PSPLIB datasets by using the number of generated schedules as a stop condition. Based on the methodology they also report state-of-the-art results. In [18] an update is given of these results. In table 5 we compare our algorithm with these results for the datasets J30, J60 and J120 respectively and for a stop condition of 5,000 schedules. In "Avg.Dev.Lb" we report the average deviation from the optimal solution for J30 or from the critical path based lowerbound for J60 and J120. The procedures are ranked according to their performance for the dataset J120. As this ranking slightly differs from the ranking for J30 and J60, we also provide a rank order in the column "Rank". The table reveals that the EM algorithm performs consistently well over all problem sets. Furthermore, the procedures that can outperform the EM procedure are hybrid heuristics. Consequently, we believe that the promising results might contribute to the further development of electromagnetism, possibly in combination (hybridization) with principles from other meta-heuristics.

4   Conclusions

This paper reports on results for the application of a new meta-heuristic procedure for solving combinatorial optimization problems. The procedure is a population-based method that is developed originally for optimizing unconstrained continuous functions based on an analogy with the electromagnetism theory. We illustrate the effective extension of this electromagnetism meta-heuristic to the well-known RCPSP.

The computational results show that the procedure produces consistently good results, compared to the state-of-the-art heuristics in the literature. Furthermore, all procedures that outperform the EM procedure are hybrid heuristics based on principles borrowed from various meta-heuristic approaches. Hence, we believe that the incorporation of ideas from EM in hybrid frameworks might contribute to the development of better meta-heuristic techniques. In the future we want to improve the performance of the EM heuristic for solving combinatorial problems by adding principles from other meta-heuristic techniques.

References

[1] Alcaraz, J., Maroto, C.: A robust genetic algorithm for resource allocation in project scheduling, Annals of Operations Research, 102, 83-109 (2001).

[2] Alcaraz, J., Maroto, C., Ruiz, R.: Improving the performance of genetic algorithms for the RCOS problems. Proceedings of the Ninth International Workshop on Project Management and Scheduling, 40-43, Nancy (2004)

[3] Baar, T., Brucker, P., Knust, S.: Tabu-search algorithms and lower bounds for the resource-constrained project scheduling problem, Meta-heuristics: Advances and trends in local search paradigms for optimization, 1-8 (1998).

[4] Birbil, S.I., Fang, S.C.: An electromagnetism-like mechanism for global optimization, Journal of Global Optimization 25 263-282 (2003).

[5] Bouleimen, K., Lecocq, H.: A new efficient simulated annealing algorithm for the resource-constrained project scheduling problem and its multiple mode version, European Journal of Operational Research, 149, 268-281 (2003).

[6] Coelho, J., Tavares, L.: Comparative analysis of meta-heuricstics for the resource constrained project scheduling problem, Technical report, Department of Civil Engineering, Instituto Superior Tecnico, Portugal (2003).

[7] Debels, D., De Reyck, B., Leus, R., Vanhoucke, M.: A scatter-search meta-heuristic for the resource-constrained project scheduling problem, European Journal of Operational Research, forthcoming.

[8] Hartmann, S.: A competitive genetic algorithm for the resource-constrained project scheduling, Naval Research Logistics, 45, 733-750 (1998).

[9] Hartmann, S.: A self-adapting genetic algorithm for project scheduling under resource constraints, Naval Research Logistics, 49, 433-448 (2002).

[10] Hartmann, S., Kolisch, R.: Experimental evaluation of state-of-the-art heuristics for the resource-constrained project scheduling problem, European Journal of Operational Research, 127, 394-407 (2000).

[11] Klein, R.: Project scheduling with time-varying resource constraints. International Journal of Production Research, 38 (16): 3937-3952 (2000).

[12] Kochetov, Y., and Stolyar, A.: Evolutionary local search with variable neighbourhood for the resource constrained project scheduling problem, Proceedings of the 3rd International Workshop of Computer Science and Information Technologies (2003).

[13] Kolisch, R.: Project scheduling under resource constraints - Efficient heuristics for several problem classes, Physica (1995).

[14] Kolisch, R.: Serial and parallel resource-constrained project scheduling methods revisited: theory and computation, European Journal of Operational Research, 43, 23-40 (1996).

[15] Kolisch, R.: Efficient priority rules for the resource-constrained project scheduling problem, Journal of Operations Management, 14, 179-192 (1996).

[16] Kolisch, R., Drexl, A.: Adaptive search for solving hard project scheduling problems, Naval Research Logistics, 43, 23-40 (1996).

[17] Kolisch, R., Hartmann, S.; Heuristic algorithms for solving the resource-constrained project scheduling problem: classification and computational analysis. In: Weglarz, J. (Ed.), Project Scheduling - Recent Models, Algorithms and Applications, Kluwer Academic Publishers, Boston, pp. 147-178 (1999).

[18] Kolisch, R., Hartmann, S.: Experimental investigation of Heuristics for resource-constrained project scheduling: an update, working paper, Technical University of Munich (2004).

[19] Kolisch, R., Sprecher, A.: PSPLIB - A project scheduling library, European Journal of Operational Research, 96 205-216 (1997).

[20] Leon V. J., Ramamoorthy, B.: Strength and adaptability of problem-space based neighbourhoods for resource-constrained scheduling, OR Spektrum, 17, 173-182 (1995).

[21] Li, K.Y., Willis, R.J.: An iterative scheduling technique for resource-constrained project scheduling, European Journal of Operational Research, 56, 370-379 (1992).

[22] Merkle, D., Middendorf, M., Schmeck, H.: Ant colony optimization for resource constrained project scheduling, IEEE Transaction on Evolutionary Computation, 6(4), 333-346 (2002).

[23] Nonobe, K., Ibaraki, T.: Formulation and tabu search algorithm for the resource constrained project scheduling problem (RCPSP). In: Ribeiro, C.C., Hansen, P. (Eds.), Essays and Surveys in Meta-heuristics, Kluwer Academic Publishers, Boston, pp. 557-588 (2002).

[24] Schirmer, A.: Case-based reasoning and improved adaptive search for project scheduling, Naval Research Logistics, 47, 201-222 (2000).

[25] Sprecher, A.: Scheduling resource-constrained projects competitively at modest resource requirements, Management Science, 46 710-723 (2000).

[26] Tormos, P., Lova, A.: A competitive heuristic solution technique for resource-constrained project scheduling, Annals of Operations Research, 102, 65-81 (2001).

[27] Tormos, P., Lova, A.: An efficient multi-pass heuristic for project scheduling with constrained resources, International Journal of Production Research, 41, 1071-1086 (2003).

[28] Tormos, P., and Lova, A.: Integrating heuristics for resource constrained project scheduling: One step forward, Technical report, Department of Statistics and Operations Research, Universidad Politecnica de Valencia (2003).

[29] Valls, V., Quintanilla, S., Ballestin, F.: Resource-constrained project scheduling: a critical activity reordering heuristic, European Journal of Operational Research, 149, 282-301 (2003).

[30] Valls, V., Ballestin, F., Quintanilla, S.: A population-based approach to the resource-constrained project scheduling problem, Annals of Operations Research, 131, 305-324 (2004).

[31] Valls, V., Ballestin, F., Quintanilla, S.: A hybrid genetic algorithm for the Resource-constrained project scheduling problem with the peak crossover operator, Eighth International Workshop on Project Management and Scheduling, 368-371 (2002).

[32] Valls, V., Ballestin, F.: Quintanilla, S.: Justification and RCPSP: A technique that pays, European Journal of Operational Research, Forthcoming.

Applications of Racing Algorithms: An Industrial Perspective

Sven Becker[1], Jens Gottlieb[2] and Thomas Stützle[3]

[1] VEGA Informations-Technologien GmbH,
Robert Bosch Str. 7, Darmstadt 64293, Germany
sven.becker@vega.de
[2] SAP AG,
Neurottstr. 16, Walldorf 69190, Germany
jens.gottlieb@sap.com
[3] Darmstadt University of Technology, Computer Science Department,
Hochschulstr. 10, Darmstadt 64289, Germany
stuetzle@informatik.tu-darmstadt.de

Abstract. Stochastic local search (SLS) methods like evolutionary algorithms, ant colony optimisation or iterated local search receive an ever increasing attention for the solution of highly application relevant optimisation problems. Despite their noteworthy successes, several issues still hinder their even wider spread. One central issue is the configuration and parameterisation of SLS methods, which is known to be a time- and personal-intensive process. Recently, several attempts have been made to automate the tuning of SLS algorithms. One of the most promising directions is the usage of the racing methodology, which is a statistical method for selecting promising candidate configurations. We present results of a study on the application of this methodology to the tuning of a complex SLS method for an industrial vehicle scheduling and routing problem, and compare the performance of two racing methods.

1 Introduction

Common to many stochastic local search (SLS) methods [1] like evolutionary algorithms [2] and memetic algorithms [3], ant colony optimisation [4], or iterated local search [5] is their high versatility for the effective solution of complex, real-world optimisation tasks. This versatility is due to the many design and parameter choices the general SLS methods leave to the implementer. For example, memetic algorithms require defining appropriate recombination, mutation, and local search operators, choosing the population size and the selection method, and setting a large number of adjustable parameters like the probabilities for applying recombination or mutation. It is well known that the choice of operators and the parameter settings can have a very strong influence on the final SLS algorithm's performance. So far, the problem of taking the right design choices in the development stages of SLS algorithms have mainly been resolved by the experience of the implementer and the algorithm configuration and parameter tuning has often been done using a trial–and–error approach. Only recently, this problem has

E. Talbi et al. (Eds.): EA 2005, LNCS 3871, pp. 271–283, 2006.

been tackled by (semi-)automatic techniques for deciding on an algorithm's configuration and parameter settings. These techniques include the usage of experimental design techniques [6, 7, 8] or racing algorithms [9].

Racing algorithms have been shown to be especially appealing since they can simultaneously handle design choices (type of recombination, local search etc.) and the optimisation of (discretised) parameter settings, and they do not rely on strong assumptions on the distribution of the underlying data. Racing algorithms start with a set of candidate configurations of the algorithm under development and test them on a sequence of problem instances. After each test run on a problem instance by all surviving candidates, those candidate configurations are eliminated, against which enough (statistical) evidence is given [9]. This procedure is repeated until either a limit on the available computation time is passed or only one candidate configuration remains. In [9, 10], one particular racing method, called *F-races*, was proposed; the F in the name stems from the usage of the (non-parametric) Friedman-test [11, 12].

So far, the usefulness of racing algorithms and, in particular, F-races has been assessed for the design of SLS algorithms for academic combinatorial optimisation problems like the travelling salesman problem [9], the quadratic assignment problem [10], or the university course timetabling problem [13]. F-races have not yet been tested under more realistic, real-life settings. In this article, we evaluate the usefulness of racing algorithms for the optimisation of design aspects and parameter choices for a complex SLS algorithm for the highly-constrained real-life vehicle scheduling and routing problem (VSRP). This application differs from the academic examples, because (i) the involved algorithm has a large number of different operators and parameters, (ii) the benchmark set comprises a very heterogeneous set of instances that differ strongly in the constraints involved, the objective function, as well as their size, and (iii) the benchmark instances require, due to their complexity, rather high computation times of at least several minutes. These considerations forbid an exhaustive evaluation of all possible parameter and operator choices and, hence, the racing algorithms are used to optimise specific aspects of the SLS algorithm. In this real-world environment, we compare two racing algorithms: the F-race approach and a new straightforward variant based on removing a predefined portion of the worst candidate configurations after each iteration.

The paper is structured as follows. Section 2 discusses the underlying ideas of the racing methodology and Section 3 introduces the vehicle scheduling and routing problem and shortly describes the algorithm for its solution from a high-level perspective. We present the benchmark instances and the experimental environment in Section 4. Our experience with the racing methodology is discussed in detail in Section 5 and we end with some concluding remarks in Section 6.

2   Racing Methodology

Racing algorithms were first applied to the model selection problem in memory-based supervised learning [14, 15] and later adapted to the problem of tuning SLS algorithms by Birattari et al. [9]. Racing algorithms can select in a fully automatic way a configuration for an SLS method from a given set of candidate configurations C.

A racing algorithm works by sequentially processing a given set of instances I. Let $inst(i)$ denote the ith instance and let $C(i)$ be the set of candidate configurations at iteration i. Initially $C(1) = C$, and then, at iteration i of the race, all candidate configurations in $C(i)$ are run once on instance $inst(i)$. When all results are available, the candidates in $C(i)$ that are shown to be statistically inferior are eliminated, resulting in a possibly smaller set $C(i + 1)$. This procedure is iterated until either only one candidate remains, a maximum limit on the overall computation time of the racing procedure expires, or the race is stopped interactively because the further progress is very low.

There are several possibilities for the technical implementation of races. The most widely explored possibility is given by the *F-race*, introduced by Birattari et al. [9]. The F-race is based on non-parametric statistical tests using *ranking* and *blocking* (a block corresponds to one instance). In the F-race, after each iteration i the *Friedman-test two-way analysis of variance by ranks* [11, 12] is first used to detect whether there exists sufficient statistical evidence that there is a difference among the outcomes of $C(i)$ (the corresponding null hypothesis H_0 is that all candidate configurations perform the same). If H_0 is not rejected, all candidates in $C(i)$ pass to $C(i + 1)$. If H_0 is rejected, pairwise tests between the best configuration (e.g. the one with the lowest sum of ranks) and all others are done and significantly worse candidates than the best one are discarded. For this second test, the *Wilcoxon matched-pairs signed-ranks test* is adopted [16]. The only parameter for the F-race is the significance level α for the two tests.

As an alternative to the F-race, we consider a method that *deletes the worst ρ percent* of the remaining configurations after each iteration. This approach, which we call *DW-race*, is probably the simplest approach for racing and, when compared to F-races, it has the advantage that its overall computation time is exactly predictable and that further progress of the race is forced until only one candidate configuration remains. However, DW-race may (i) eliminate candidates that are not statistically worse than the current best configuration and, therefore, it is somewhat more error-prone, and (ii) fail to eliminate configurations against which enough statistical evidence has been gathered.

Overall, a race is a three step procedure, which can be outlined as follows.

1. Select a set of problem instances I on which the SLS method should be tuned.
2. Select a set of candidate configurations C for the SLS algorithm.
3. Run the race and select the best performing candidate configuration.

The first two steps are equally important as the run of the race. Regarding the first step, the set of instances should be (i) representative of the final set of instances that will be tackled and (ii) the more instances are available, the better usually is the quality of the finally selected configuration, because the bias towards specific instances is reduced with more instances. The set of candidate configurations can be built based on discrete choices like different possible operators in the SLS algorithm (e.g., local search operators or mutation operators) and, simultaneously, varying parameter choices. However, continuous parameters need to be discretised and for this step pre-knowledge on a reasonable range of parameter settings may be useful. The number of candidate configurations increases typically exponentially with the number of algorithm choices under investigation in the application of a race. If choice j has n_j possible values, a total of $\Pi_{j=1}^{k} n_j$ candidates results, where k is the number of choices to be made (for $n_j = 2, j = 1, \ldots k$, this would result in 2^k candidates). If too many values for the particular choices are allowed because of a fine-grained discretisation, this may result in a very large number of candidates. To avoid these problems, races may also be run with several levels of discretisation or in a hierarchical manner.

3   Vehicle Scheduling and Routing Problems

The *vehicle routing problem* (VRP) is a classical combinatorial optimisation problem that is frequently used to study and develop new algorithmic ideas. It considers the delivery of goods from a depot to a set of customers, and its goal is to assign customers to vehicles and to find routes for the vehicles such that total transportation costs are minimised and certain constraints like loading capacities or time windows of the customers are met [17]. Although the VRP forms the core of many real-world applications, e.g. in transportation management systems, real-world scenarios are typically more complex since they involve multiple objective functions, many constraints and decisions to be made. Here, we sketch the most important features of the *vehicle scheduling and routing problem* (VSRP), for which an optimisation algorithm is offered in SAP's supply chain management solution[1], a commercial software that allows to plan and optimise the whole supply chain, including demand planning, supply network planning, production planning, transportation planning and vehicle scheduling.

The VSRP consists of a set of orders, each representing a transportation requirement from a source to a destination location. An order is described by its quantity (volume, weight, etc.), some characteristics (material, frozen or non-frozen, etc.), material availability date at the source and required delivery date at the destination, and a non-delivery penalty that represents an order's priority. There is a fleet of vehicles, each having a certain cost structure (duration costs, fixed costs, distance costs, quantity costs, stop costs) and driving capabilities (speed, reachability between locations). A vehicle may have a fixed start or end location, and a time availability interval, potentially interrupted by a set of breaks (weekends, legal holidays, etc.). Vehicles may have different loading capacities and limits on travelled distance or number of visited locations, and some may be incompatible with orders of given characteristics (e.g. frozen goods require special vehicles). Loading and unloading at locations may require additional capacitive resources (e.g. docks, workers) that are available only during several time windows, and goods with certain characteristics must not be shipped together on the same vehicle (e.g. food and chemicals). There may be hub locations, where orders can be unloaded by one vehicle and loaded again by another vehicle that brings the goods to their final destination location.

The goal of the VSRP is to minimise total costs while satisfying all constraints. The total costs are the weighted sum of different cost terms per vehicle, as indicated above, and per order (earliness costs, lateness costs, non-delivery costs).

The VSRP generalises the VRP in many aspects, and therefore most algorithmic techniques specifically tailored to the VRP are not directly applicable. SAP has developed an SLS algorithm for the VSRP, built on top of several constructive heuristics and a suite of more than 20 atomic variation operators that focus on specific aspects of a

---

[1] See http://www.sap.com/scm for more details.

candidate solution: the choice of hubs, the assignment of orders to vehicles, the routing per vehicle, and the scheduling of activities.

The SLS algorithm is population-based and heavily relies on local optimisation. As selection pressure is used on the population level and certain random variation steps are performed, this approach is called evolutionary local search. The population is rather small, e.g. of size three up to eight, and each individual in the population has a certain role. These roles represent different search behaviours. One role is called iterated local search (ILS), another depth-first search (DFS), and another random walk (RW). The latter role intends to diversify search by frequent random perturbation steps. DFS and ILS are conceptually similar, but DFS uses a more narrow search when compared to ILS. Each of the three roles orchestrates the available variation operators sequentially, with certain operator probabilities that reflect the role. One key difference in the three roles is the frequency of perturbation moves. Since the key idea of ILS (in general) is inherent in all roles, the overall SLS algorithm can also be seen as a population-based iterated local search.

We omit a formal statement of the VSRP and more details about the employed approach for several reasons: (i) the available space does not allow to give more details, (ii) it is not SAP's goal to disclose too many details of its commercial software, and (iii) our intention in this paper is to study racing algorithms on a real-world problem, for which neither the formal problem description nor the detailed algorithm is needed. Therefore, SAP's optimisation approach for the VSRP is perceived as a black-box algorithm with some parameters that shall be fine-tuned by racing algorithms.

4 Benchmark Instances and Experimental Environment

The VSRP is used by SAP's customers to model and solve their transportation planning scenarios. Each customer's transportation business has special requirements that are mapped into a certain family of VSRP instances which share structural similarities. VSRP instances of different customers may differ significantly. In our experiments, we use a total of 47 real-world instances, taken from several customers. These instances are representative in the sense that they cover many customers' scenarios and all instances differ in one or more aspects from the others.

The instances have different numbers of orders, ranging from 19 to 1101 orders. The numbers of source locations and destination locations vary between 1 and 19, and 1 and 548, respectively. 23 instances involve time windows for loading and unloading activities, and 6 instances require capacitive resources for loading and unloading. 17 instances involve at least one possible hub, whereas the remaining 30 instances do not allow indirect shipment via hubs. This heterogeneous benchmark suite contains instances with different objective functions, constraints, and decision variables.

The goal of our racing experiments is not to fine-tune the algorithm's parameters for a single customer's instance family, but to find robust general parameter settings that work well for all instances, or at least for an easily definable and sufficiently large subset of instances.

The run time limits of single optimisation runs differ significantly between the customers. One the one hand, some customers run the optimiser for a few minutes,

which allows several consecutive optimisation runs being evaluated, possibly manipulated manually, and finally executed by the human transportation planner. On the other
hand, other customers make long optimisation runs over night, which are then processed
by the human planner in the next morning or after the weekend. In our experiments we
chose a run time limit of 10 minutes per single optimisation run, which is acceptable
for most customers and allows reasonable results even for the biggest and most difficult
instances under consideration.

5   Computational Experience

In this section we report our computational experience with the F-race and DW-race.
Given the overall complexity of the SLS algorithm and the high computation times
per instance, we focused on two classes of experiments, (pure) parameter optimisation
and structural optimisation including limited settings of parameters. For both cases, we
study an example that was known to have influence on the performance of the overall
SLS algorithm.

We use a modified racing algorithm that was applied to configure SLS algorithms for
the university course timetabling problem (UCTP) [13]. Basically, one iteration means
running all candidate configurations on *all* instances from the benchmark set under
consideration. The reason for this choice is the heterogeneity of the benchmark set and
to avoid a result biased by the order in which the instances are considered in the race.

The experiments are performed on 6 PCs with a same configuration (2.6 GHz Pentium IV with 512 MB of RAM). Since these PCs are also used for other purposes and
their availability is not known a priori, distributed computing is used to make best use of
available time slots on these machines. Here, distribution means that a central instance,
the master, running on one PC, is dynamically fed with the optimisation runs specified by the racing algorithm. On each PC, a client reports availability of the machine
to the master, and the master assigns each required optimisation run to one available
client. Without this grid-like distributed computing environment, a single PC running
exclusively for this project would have required more than 100 days for the experiment
described in Section 5.1.

5.1   Parameter Optimisation: Frequency of Block-Inserts

The optimiser contains many atomic variation moves, one of which is the block insert
operator that assigns several unscheduled orders with identical characteristics and due
dates to a vehicle. This operator is faster than several consecutive single insert operations for the same orders, but it also causes some solutions being more difficult to obtain.
We are therefore interested in this trade-off and analyse the probability for applying this
operator. In principle, all values in the interval $[0, 1]$ are valid. However, in order to reduce the number of alternatives, we discretise this range, yielding the configuration set
$C = \{0, 0.2, 0.4, 0.6, 0.8, 1\}$ that represents the considered operator probabilities.

Our experience before starting racing experiments was that this operator worked
well on instances involving hubs, but its success on other instances was somehow inconclusive. We considered this knowledge by partitioning the set of all instances I into

Table 1. Results for F-Race, different values of α and instance sets

α	I_H				I_{NH}				I			
	runs	days	it	win	runs	days	it	win	runs	days	it	win
0.1	204	1.4	2	0.2	810	5.6	6	0	2538	17.6	18	0
0.05	204	1.4	2	0.2	900	6.3	6	0	3478	24.2	26	0
0.02	680	4.7	16	0.2	960	6.7	6	0	3807	26.4	22	$\{0, 0.2\}$
0.01	1734	12.0	20	0.2	960	6.7	6	0	4136	28.7	22	$\{0, 0.2\}$
0.001		—			1890	13.1	15	0		—		

I_H and I_{NH}, the subsets of instances involving hubs and not involving hubs, respectively. Among the $|I| = 47$ instances, 17 contain hubs, and 30 none. In order to study the impact of hubs on the outcome of racing, three configuration problems were investigated separately: (C, I_H), (C, I_{NH}), and (C, I).

Results for F-Race. Table 1 shows the results for F-Races with different significance levels for the statistical tests on the three instance sets. For each experiment, the number of single optimisation runs, the amount of CPU time measured in days, the number of iterations in the race, and the winning configuration is given. We performed only one race for $\alpha = 0.001$ since the other races on I_H and I were already very CPU intensive for $\alpha = 0.01$. Races without a clear winner were terminated interactively after observing that no more progress can be expected; the remaining configurations are listed as winners in this case.

Some general trends are obvious and intuitive. Firstly, the lower α, the more optimisation runs and iterations are needed, because the test is less aggressive in detecting differences among the configurations. Secondly, all races on I_H terminated with the same result, indicating that the best probability is 0.2; all races on I_{NH} determined 0 as the best probability. This confirms our past experience. However, if we consider the races on I, the probability 0 is the winner for $\alpha \in \{0.1, 0.05\}$. For lower significance levels, the racing algorithm does not indicate significant differences between the two remaining configurations, which were the winners of the separate races on I_H and I_{NH}, even after many days of computation time.

Figure 1 shows the average ranks (upper part) and p-values of tests (lower part) during two typical F-races. The average rank of a configuration in an iteration is defined as the average over all ranks of this configuration on all instances and optimisation runs performed so far. The p-value gives the probability of wrongly rejecting the null hypothesis if in fact it were true; if the p-value is smaller than the significance level, the null hypothesis is rejected. On the instance set I_{NH}, the configurations 0.8 and 1 are eliminated after the third iteration, and after six iterations the winner is found by eliminating the other configurations except 0. The F-race on I shows why it took so many iterations to detect the winning configuration 0: the two best configurations are very close together regarding their average ranks, and it took very long until the difference was proved to be significant. This also indicates why the F-races on I with lower significance levels had to be stopped interactively. For significance levels $\alpha \in \{0.02, 0.01\}$, no significant difference could be found within the allowed time.

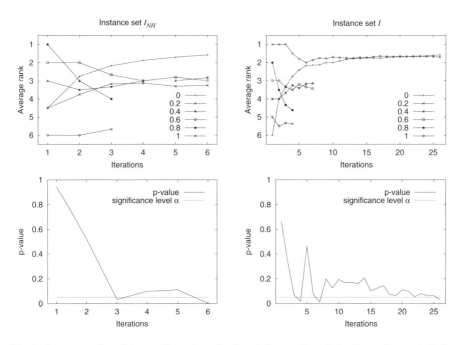

Fig. 1. Average ranks of the configurations (top) and the p-value of the F-test (bottom) during F-Races with $\alpha = 0.05$ on I_{NH} (left) and I (right)

The more heterogeneous set of instances I makes it more difficult to find the best configuration. If rather high values are chosen for α, the probability 0 wins, but for lower significance levels both configurations perform quite well. Perhaps this race is even unfair, because $|I_{NH}| = 30 > 17 = |I_H|$. In order to have a fair instance set, we performed the following experiment five times. 13 instances from I_{NH} are randomly removed from I, and then F-race with $\alpha = 0.05$ is run on the resulting instance set of size 34. In four of the experiments the result was inconclusive, with 0 and 0.2 being among the final candidates when stopping the race after 50 iterations. The average number of optimisation runs was 5310, which requires a CPU time of 36.9 days per experiment. Thus, if both structures (involvement of hub or not) are represented by the same number of instances, the configuration problem becomes even more difficult.

The above discussion shows that a heterogeneous instance set represents a challenge for racing algorithms. The main reason is that a mixture of two different classes of instances, for which different best parameter settings would be obtained, do not lead to statistically significant differences among the configurations, differently from what is observed for the two races on the two separated instance sets. The separation of the instance set into two classes allows to set the investigated parameter by the following simple rule: if an instance involves hubs, apply the insert block operator with probability 0.2, and do not use the operator otherwise.

Results for DW-Race. Analogously to the F-race, we also examined the results of the DW-race in dependence of different delete rates and the three instance sets. With a fixed

Table 2. Results for DW-Race, different delete rates and instance sets

rate	I_H				I_{NH}				I			
	runs	days	it	win	runs	days	it	win	runs	days	it	win
0.15	629	4.4	9	0.2	930	6.5	9	0	1457	10.1	9	0.2
0.1	901	6.3	14	0.2	1410	9.8	14	0	2209	15.3	14	0.2
0.05	1683	11.7	29	0.2	2790	19.4	29	0	4371	30.4	29	0

deletion rate of ρ, one can determine the number of surviving candidates after iteration i of the race as $|C(i+1)| = \lfloor |C| \cdot (1-\rho)^i + 0.5 \rfloor$, if the usual rounding procedure is used. Using this approach, the results of the DW-race are given in Table 2. If the instances are separated into the two classes I_H and I_{NH}, the same configurations as for the F-race were returned, however, taking in most cases more time than the F-race with the standard significance level $\alpha = 0.05$. Surprisingly, for DW-race on set I, the same winner configuration as in the F-race was returned only by the lowest delete rate. The reason for this effect is that the two configurations with operator probabilities 0 and 0.2 result in rather similar ranks and are statistically distinguishable only after 26 iterations in the F-race at the significance level $\alpha = 0.05$. Hence, in this case the DW-race takes the (statistically) wrong decision because of forcing too early convergence of the race.

5.2 Structural Optimisation: Shape of the SLS Method

In a second set of racings, we considered the configuration of the overall structure of the SLS algorithm. As described in Section 3, the SLS algorithm works on a population of individuals, where each individual belongs to one role of ILS, DFS, and RW. For a first race, we considered configurations that differed in the number of individuals for each role and that varied two parameter settings: the number of restart points for the DFS strategy and the perturbation strength for ILS and RW, respectively. To keep the computational effort within reasonable limits, in a first trial only 20 different such configurations were defined and we limited the experiments to the instances in I_{NH}. As before, experiments were run with F-races and DW-races using various settings for α and deletion rate, respectively.

The results of the F-race and the DW-race are given in Table 3. As can be seen, all races return the same winner configuration, which uses only one ILS individual and a perturbation of strength 2. In this experiment, a clear advantage of the F-race appears: with $\alpha = 0.05$, after only 3 iterations one single configuration is declared as the winner. In fact, a more careful examination of the progress of the race shows that only

Table 3. Results for F-race and DW-race in the structural optimisation

α	F-race				rate	DW-race			
	runs	days	it	win		runs	days	it	win
0.1	900	6.25	3	(ILS,2)	0.5	925	6.42	4	(ILS,2)
0.05	900	6.25	3	(ILS,2)	0.25	1 725	11.98	8	(ILS,2)
0.01	1 475	10.24	7	(ILS,2)	0.1	3 350	23.26	14	(ILS,2)

Table 4. Results of F-race and DW-race for additional configurations in the structural optimisation

α	F-race				rate	DW-race			
	runs	days	it	win		runs	days	it	win
0.1	2 250	15.63	21	(ILS,1)	0.5	1 175	8.16	5	(ILS,1)
0.05	2 650	18.40	24	(ILS,1)	0.25	2 275	15.80	10	(ILS,1)
0.02	2 850	19.79	23	{(ILS,2),(ILS,1),(2ILS,2)}	0.1	5 625	39.06	27	(ILS,1)

configurations with at least one ILS individual survive the first iteration; this is consistent with the fact that the winning configuration is the only that consisted purely of one single ILS individual (that is, the configuration without any DFS and RW individuals). Since the differences between the configurations are very strong, the F-race is able to quickly eliminate the poor performing candidate configurations. The DW-race returns the same winning configuration. However, to return this result in the same computation time as the F-race, a very high deletion rate of 0.5 is needed; however, as shown also in the previous example, such a high deletion rate can be quite problematic and lead to statistically unfounded (or even wrong) decisions.

Based on this initial race, others were run, the results of which we summarize next.

Additional configurations. Additional candidate configurations with only ILS individuals (one or two) and a range of parameter settings for the perturbation strength were added to the configurations of the first race; the results of this race are summarised in Table 4. In this new race, for all significance levels, the winning configuration was ILS,1, that is a configuration with only one ILS individual and perturbation strength one – this configuration was not considered in the first race. Interestingly, in the F-race already after the second iteration every algorithm that uses at least one DFS or RW individual was eliminated, leaving only pure ILS configurations that only differed in the perturbation strength and the number of ILS individuals. However, the differences among the various ILS configurations appear to be not too large and, hence, the race still takes quite a few iterations to remove the other candidates. The very rapid elimination of many competing configurations also explains the relatively short computation times for the F-race when taking into account the number of iterations until it was stopped. To reach a similar computation time limit, for the DW-race a rather high deletion rate of 0.25 needs to be used.

Convergence limit. In this race, 30 configurations were examined that, in addition to the population composition of ILS, DFS, and RW individuals, differed mainly in the number of unsuccessful local search moves examined (convergence limit) before the local search is stopped and the probability of making a large step for RW individuals. Similar to the previous two races, here only the three configurations that only used ILS individuals survived the first iteration of the F-race; DW-race was not run because of its inferior performance in the previous races. Among the remaining three configurations, the order of elimination suggested that the smaller the convergence limit is chosen, the higher is the survival probability.

Higher computation times. Here, the same configurations as in the race on the convergence limit were examined, but this time the SLS algorithms were run for 30 minutes instead for 10 minutes. The motivation for this additional race is that the configurations with a stronger diversification through RW individuals or the intensification through DFS individuals may profit from the higher computation times. However, as for the previous race, after the first iteration all candidate configurations that did not make exclusive use of ILS individuals were eliminated. A difference to the previous race was that the influence of the convergence limit was diminished; for example, for the significance level $\alpha = 0.05$ all ILS configurations that differed only in the convergence limit remained in the race.

Overall, for the combined structural and parameter optimisation as done here, the F-race is clearly superior to the DW-race. After only a few iterations, inferior candidate configurations are deleted. Interestingly, the previously chosen configuration of the optimiser was also among those eliminated, indicating that still significantly better performance may be reached by fine-tuning the overall structure of the SLS algorithm.

5.3   General Remarks

In addition to the comparison between the F-Race and DW-Race, several issues were found to be important when running racing algorithms in a real-world environment. Firstly, the rank-based approach for evaluating configurations is essential, especially with such a heterogenous instance suite as ours. This is the case because the instances have quite different ranges of objective function values, their distribution is unknown, and even may have some anomalies. Secondly, the racing method can be examined, analysed and modified interactively. Interactive features are appropriate when, for example, it becomes obvious that the race should be restarted with additional configurations or to stop the race when further progress appears to be very minor; see also [13] for the usage of interactive racings. Thirdly, since in our setting each single optimisation run is rather time-consuming, a true re-start of the race is very costly. Therefore, we used a database that stores the results of already executed optimisation runs per instance, configuration, and seed. New trials are only started if there is no corresponding database entry. This saves much time and even allows to run racings on the same configuration problem, without too much additional CPU costs. Fourthly, by the usage of survival analysis (e.g. by analysing commonalities among the surviving candidates like only using ILS individuals), one may generate profiles of the main components responsible for high quality configurations, which can then allow to refine the algorithm. Finally, the usage of distributed computing, like grid computing or the architecture we used here, is essential to speed-up the experiments. The usage of this type of parallel processing has the advantage that the speed-up of the experiments is essentially linear with the number of computers available. For realistic settings, where individual trials of an SLS algorithm on an instance can take several minutes or even longer, and many configurations are examined, such a parallelisation is essential to keep the overall computation times within manageable limits.

6   Conclusions

We presented results of an experimental study of racing algorithms on real-world vehicle scheduling and routing problems and a commercial SLS algorithm. This applicaton of racing differs from previous studies in several aspects: the high computation time per run, the high complexity of a real-world problem – due to multiple objective functions, many structurally different constraints and decisions to be made – and the heterogeneity of the benchmark suite.

While the computation time of a DW-race is predictable accurately, the CPU time required by a F-race depends on the differences identified in the configurations. If strong differences in performance are observed for the configurations, F-races tend to quickly reduce the set of candidates, as done in the structural optimisation task. A further advantage of F-races is that they are based on sound statistical tests, which may allow to delete significantly worse configurations early in the race and prevents deleting configurations that are not significantly inferior.

One of the most promising results for the usefulness of racing algorithms is that the best configurations identified in an automatic way in this study improved over the previous version of the commercial software, at least for the considered instances, despite the previous efforts to experimentally fine-tune the software. These positive results together with the increasing availability of cheap computation time, for example, through small PC clusters or grid computing, will further increase the applicability of automated techniques for the configuration of algorithms in applications of high industrial relevance.

References

1. Hoos, H.H., Stützle, T.: Stochastic Local Search—Foundations and Applications. Morgan Kaufmann, USA (2004)
2. Mitchell, M.: An Introduction to Genetic Algorithms. MIT Press, Cambridge, MA, USA (1996)
3. Moscato, P., Cotta, C.: A gentle introduction to memetic algorithms. In Glover, F., Kochenberger, G., eds.: Handbook of Metaheuristics. Kluwer Academic Publishers, Norwell, MA, USA (2002) 105–144
4. Dorigo, M., Stützle, T.: Ant Colony Optimization. MIT Press, Cambridge, MA, USA (2004)
5. Lourenço, H.R., Martin, O., Stützle, T.: Iterated local search. In Glover, F., Kochenberger, G., eds.: Handbook of Metaheuristics. Kluwer Academic Publishers, Norwell, MA, USA (2002) 321–353
6. Xu, J., Chiu, S., Glover, F.: Fine-tuning a tabu search algorithm with statistical tests. International Transactions in Operational Research 5(4) (1998) 233–244
7. Coy, S.P., Golden, B.L., Runger, G.C., Wasil, E.A.: Using experimental design to find effective parameter settings for heuristics. Journal of Heuristics 7(1) (2001) 77–97
8. Adenso-Díaz, B., Laguna, M.: Fine-tuning of algorithms using fractional experimental designs and local search. Operations Research (In press)
9. Birattari, M., Stützle, T., Paquete, L., Varrentrapp, K.: A racing algorithm for configuring metaheuristics. In Langdon, W.B., et al., eds.: Proceedings of the Genetic and Evolutionary Computation Conference (GECCO-2002), Morgan Kaufmann, USA (2002) 11–18
10. Birattari, M.: The Problem of Tuning Metaheuristics. PhD thesis, IRIDIA, Université Libre de Bruxelles, Belgium (2004)

11. Siegel, S., Jr., N.J.C., Castellan, N.J.: Nonparametric Statistics for the Behavioral Sciences. second edn. McGraw Hill, NewYork, NJ, USA (2000)
12. Sheskin, D.J.: Handbook of Parametric and Nonparametric Statistical Procedures. second edn. Chapman & Hall / CRC, Boca Raton, Florida, USA (2000)
13. Chiarandini, M., Birattari, M., Socha, K., Rossi-Doria, O.: An effective hybrid approach for the university course timetabling problem. Journal of Scheduling (Submitted)
14. Maron, O., Moore, A.W.: Hoeffding races: Accelerating model selection search for classification and function approximation. In Cowan, J.D., Tesauro, G., Alspector, J., eds.: Advances in Neural Information Processing Systems. Volume 6., Morgan Kaufmann Publishers, Inc. (1994) 59–66
15. Moore, A.W., Lee, M.S.: Efficient algorithms for minimizing cross validation error. In: International Conference on Machine Learning, Morgan Kaufmann Publishers, Inc. (1994) 190–198
16. Conover, W.J.: Practical Nonparametric Statistics. third edn. John Wiley & Sons, New York, NY, USA (1999)
17. Toth, P., Vigo, D., eds.: The Vehicle Routing Problem. Society for Industrial and Applied Mathematics, Philadelphia, PA, USA (2002)

An Immunological Algorithm for Global Numerical Optimization

Vincenzo Cutello, Giuseppe Narzisi, Giuseppe Nicosia, and Mario Pavone

Department of Mathematics and Computer Science,
University of Catania,
V.le A. Doria 6, Catania 95125, Italy
{vctl, narzisi, nicosia, mpavone}@dmi.unict.it

Abstract. Numerical optimization of given objective functions is a crucial task in many real-life problems. The present article introduces an immunological algorithm for continuous global optimization problems, called OPT-IA. Several biologically inspired algorithms have been designed during the last few years and have shown to have very good performance on standard test bed for numerical optimization.

In this paper we assess and evaluate the performance of OPT-IA, FEP, IFEP, DIRECT, CEP, PSO, and EO with respect to their general applicability as numerical optimization algorithms. The experimental protocol has been performed on a suite of 23 widely used benchmarks problems. The experimental results show that OPT-IA is a suitable numerical optimization technique that, in terms of accuracy, generally outperforms the other algorithms analyzed in this comparative study. The OPT-IA is also shown to be able to solve large-scale problems.

Keywords: Artificial Immune Systems, Clonal Selection Algorithms, Immune Algorithm, Aging operator, Global Numerical Optimization.

1 The Immunological Algorithm

Clonal Selection Algorithms (CSAs) [1] are a special class of Immune algorithms (IAs) [1, 2] which are inspired by the Clonal Selection Principle [3] of the human immune system to produce effective methods for search and optimization. In this research paper an immune algorithm inspired by the Clonal Selection Principle, OPT-IA [4, 5, 6, 7, 8], is applied to global numerical optimization.

The OPT-IA algorithm uses a population of candidate solutions, i.e. points of the search space (B cell or B cell receptor according to immunological terminology). At each time step t, we have a population $P_d^{(t)}$ of size d. The initial population of candidate solutions, time $t = 0$, is generated uniformly at random in the relative domains of each function (see table 1) The function *Evaluate(P)* computes the fitness function value of each B cell $x \in P$. The implemented IA uses three immune operators, cloning, hypermutation and aging. The cloning operator, simply, clones each B cell *dup* times producing an intermediate population $P_{N_c}^{(clo)}$ of size $d \times dup = N_c$.

E. Talbi et al. (Eds.): EA 2005, LNCS 3871, pp. 284–295, 2006.

The hypermutation operator acts on the B cell receptor of $P_{N_c}^{(clo)}$. The number of mutations M is determined by *mutation potential*. Our IA uses an *Inversely Proportional Hypermutation* operator, where the number of mutations is inversely proportional to the fitness value, that is, it decreases as the fitness function of the current B cell increases. Two different mutation potential are used, they are defined by the following equations:

$$\alpha = e^{(-\rho*f)}, \qquad \alpha = \left(\frac{1}{\rho}\right) e^{(-f)} \tag{1}$$

where α represents the mutation rate, and f is the fitness function value normalized in $[0,1]$. The number of mutations of a clone with fitness function value f is equal to $\lfloor L * \alpha \rfloor$ where L is the length of the clone receptor, that is $L = l \times n$, with l being the number of bits used to code each variable and n the dimension of the function. The first potential mutation was proposed in [10], while the second potential mutation was introduced in [11]. Figure 1 shows the pseudo-code of the proposed Immune Algorithm.

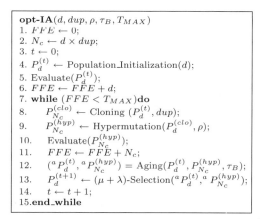

Fig. 1. Pseudo-code of OPT-IA

Aging Operator. By inspecting the pseudo-code of OPT-IA, one can see that an "aging" operator is used. The aging operator eliminates old B cells, in the populations $P_d^{(t)}$ and $P_{N_c}^{(hyp)}$ to avoid premature convergence and to increase diversity in the current population. This operator is the main difference between our algorithm and the other IAs and Evolutionary Algorithms. The parameter τ_B sets the maximum number of generations B cells are allowed to remain in the population. When a B cell is $\tau_B + 1$ old it is erased from the current population, no matter what its fitness value is. During the cloning expansion, a cloned B cell takes the age of its parent. After the hypermutation phase, a cloned B cell which successfully mutates, will be considered to have age equal to 0. In this way, new B cells are given an equal opportunity to effectively explore the given computational landscape. The best B cells which "survived" the aging operator, are selected from the populations $^a P_d^{(t)}$ and $^a P_{N_c}^{(hyp)}$. In this way, we obtain the new population $P_d^{(t+1)}$, of d B cells, for the next generation $t+1$. If $d' < d$ B cells survived, the $(\mu + \lambda)$-*Selection operator* creates $d - d'$ new B cells (*Birth phase*).

The evolution cycle ends if a maximum number of Fitness Function Evaluations (FFE) is reached.

2   Numerical Optimization

Numerical optimization problems are fundamental for every field of engineering and science. The task is that of finding global optima of a generic objective function. However, often, the objective function is difficult to optimize because of numerous local optima. Moreover, this difficulty increases proportionally with the problem dimension.

In this paper we consider the following numerical minimization problem:

$$\min(f(\boldsymbol{x})), \qquad \boldsymbol{B_l} \leq \boldsymbol{x} \leq \boldsymbol{B_u} \qquad (2)$$

where $\boldsymbol{x} = (x_1, x_2, \ldots, x_n)$ is the variable vector in $\mathcal{R}^n$, $f(\boldsymbol{x})$ denotes the objective function to minimize and $\boldsymbol{B_l} = (B_{l_1}, B_{l_2}, \ldots, B_{l_n})$, $\boldsymbol{B_u} = (B_{u_1}, B_{u_2}, \ldots, B_{u_n})$ represent, respectively, the lower and the upper bound of the variables, such that $x_i \in [B_{l_i}, B_{u_i}]$.

We used binary string representation: real values x_i are coded using bitstrings of length $l = 32$. The mapping from the binary string $\boldsymbol{b} =< b_1, b_2, \ldots, b_l >$ into a real number x consists of two steps:

(i) convert the bitstring $\boldsymbol{b} =< b_1, b_2, \ldots, b_l >$ from base 2 to base 10 using the equation: $\sum_{i=1}^{l} b_i * 2^i = x'$;

(ii) finding the corresponding real value: $x = B_{l_i} + \frac{x'(B_{u_i} - B_{l_i})}{2^l - 1}$, where B_{l_i} and B_{u_i} are the lower and upper bounds of the variables.

Test Functions. We selected twentythree functions from three categories [12]. This relative large set is necessary in order to reduce biases in evaluating algorithms. Table 1 lists the 23 functions and their key properties (for a complete description of all the functions and the parameters involved see [12]). These functions can be divided into three categories of different complexities:

-   unimodal functions ($f_1 - f_7$), which are relatively easy to optimize, but the difficulty increases as the problem dimension increases;
-   multimodal functions ($f_8 - f_{13}$), with many local minima, they represent the most difficult class of problems for many optimization algorithms;
-   multimodal functions which contain only few local optima ($f_{14} - f_{23}$).

Some functions possess unique features: f_6 is a discontinuous step function having a single optimum; f_7 is a noisy quartic function involving a uniformly distributed random variable within $[0, 1)$. Optimizing unimodal functions is not a major issue, so in this case the convergence rate is of main interest. However, for multimodal functions the quality of the final results is more important since it reflects the algorithm's ability in *escaping* from local optima.

Table 1. The 23 benchmark functions used in our experimental study; n is the dimension of the function; f_{min} is the minimum value of the function; $S \subseteq \mathcal{R}^n$ are the variable bounds (for a complete description of all the functions and the parameters involved see [12])

Test function	n	S	f_{min}
$f_1(\boldsymbol{x}) = \sum_{i=1}^{n} x_i^2$	30	$[-100, 100]^n$	0
$f_2(\boldsymbol{x}) = \sum_{i=1}^{n} |x_i| + \prod_{i=1}^{n} |x_i|$	30	$[-10, 10]^n$	0
$f_3(\boldsymbol{x}) = \sum_{i=1}^{n} \left(\sum_{j=1}^{i} x_j \right)^2$	30	$[-100, 100]^n$	0
$f_4(\boldsymbol{x}) = \max_i \{ |x_i|, 1 \le i \le n \}$	30	$[-100, 100]^n$	0
$f_5(\boldsymbol{x}) = \sum_{i=1}^{n-1} [100(x_{i+1} - x_i^2)^2 + (x_i - 1)^2]$	30	$[-30, 30]^n$	0
$f_6(\boldsymbol{x}) = \sum_{i=1}^{n} (\lfloor x_i + 0.5 \rfloor)^2$	30	$[-100, 100]^n$	0
$f_7(\boldsymbol{x}) = \sum_{i=1}^{n} i x_i^4 + random[0, 1)$	30	$[-1.28, 1.28]^n$	0
$f_8(\boldsymbol{x}) = \sum_{i=1}^{n} -x_i \sin(\sqrt{|x_i|})$	30	$[-500, 500]^n$	-12569.5
$f_9(\boldsymbol{x}) = \sum_{i=1}^{n} [x_i^2 - 10 \cos(2\pi x_i) + 10]$	30	$[-5.12, 5.12]^n$	0
$f_{10}(\boldsymbol{x}) = -20 \exp\left(-0.2\sqrt{\frac{1}{n}\sum_{i=1}^{n} x_i^2}\right)$ $- \exp\left(\frac{1}{n}\sum_{i=1}^{n} \cos 2\pi x_i\right) + 20 + e$	30	$[-32, 32]^n$	0
$f_{11}(\boldsymbol{x}) = \frac{1}{4000}\sum_{i=1}^{n} x_i^2 - \prod_{i=1}^{n} \cos\left(\frac{x_i}{\sqrt{i}}\right) + 1$	30	$[-600, 600]^n$	0
$f_{12}(\boldsymbol{x}) = \frac{\pi}{n}\{10\sin^2(\pi y_1)$ $+ \sum_{i=1}^{n-1} (y_i - 1)^2[1 + 10\sin^2(\pi y_{i+1})] + (y_n - 1)^2\}$ $+ \sum_{i=1}^{n} u(x_i, 10, 100, 4),$ $y_i = 1 + \frac{1}{4}(x_i + 1)$ $u(x_i, a, k, m) = \begin{cases} k(x_i - a)^m, & \text{if } x_i > a, \\ 0, & \text{if } -a \le x_i \le a, \\ k(-x_i - a)^m, & \text{if } x_i < -a. \end{cases}$	30	$[-50, 50]^n$	0
$f_{13}(\boldsymbol{x}) = 0.1\{\sin^2(3\pi x_1)$ $+ \sum_{i=1}^{n-1} (x_i - 1)^2[1 + \sin^2(3\pi x_{i+1})]$ $+ (x_n - 1)[1 + \sin^2(2\pi x_n)]\} + \sum_{i=1}^{n} u(x_i, 5, 100, 4)$	30	$[-50, 50]^n$	0
$f_{14}(\boldsymbol{x}) = \left[\frac{1}{500} + \sum_{j=1}^{25} \frac{1}{j + \sum_{i=1}^{2}(x_i - a_{ij})^6}\right]^{-1}$	2	$[-65.536, 65.536]^n$	1
$f_{15}(\boldsymbol{x}) = \sum_{i=1}^{11} \left[a_i - \frac{x_i(b_i^2 + b_i x_2)}{b_i^2 + b_i x_3 + x_4} \right]^2$	4	$[-5, 5]^n$	0.0003075
$f_{16}(\boldsymbol{x}) = 4x_1^2 - 2.1x_1^4 + \frac{1}{3}x_1^6 + x_1 x_2 - 4x_2^2 + 4x_2^4$	2	$[-5, 5]^n$	-1.0316285
$f_{17}(\boldsymbol{x}) = \left(x_2 - \frac{5.1}{4\pi^2}x_1^2 + \frac{5}{\pi}x_1 - 6\right)^2$ $+ 10\left(1 - \frac{1}{8\pi}\right)\cos x_1 + 10$	2	$[-5, 10] \times [0, 15]$	0.398
$f_{18}(\boldsymbol{x}) = [1 + (x_1 + x_2 + 1)^2(19 - 14x_1 + 3x_1^2 - 14x_2$ $+ 6x_1 x_2 + 3x_2^2)] \times [30 + (2x_1 - 3x_2)^2(18 - 32x_1$ $+ 12x_1^2 + 48x_2 - 36x_1 x_2 + 27x_2^2)]$	2	$[-2, 2]^n$	3
$f_{19}(\boldsymbol{x}) = -\sum_{i=1}^{4} c_i \exp\left[-\sum_{j=1}^{4} a_{ij}(x_j - p_{ij})^2\right]$	4	$[0, 1]^n$	-3.86
$f_{20}(\boldsymbol{x}) = -\sum_{i=1}^{4} c_i \exp\left[-\sum_{j=1}^{6} a_{ij}(x_j - p_{ij})^2\right]$	6	$[0, 1]^n$	-3.32
$f_{21}(\boldsymbol{x}) = -\sum_{i=1}^{5} \left[(\boldsymbol{x} - a_i)(\boldsymbol{x} - a_i)^T + c_i\right]^{-1}$	4	$[0, 10]^n$	-10.1422
$f_{22}(\boldsymbol{x}) = -\sum_{i=1}^{7} \left[(\boldsymbol{x} - a_i)(\boldsymbol{x} - a_i)^T + c_i\right]^{-1}$	4	$[0, 10]^n$	-10.3909
$f_{23}(\boldsymbol{x}) = -\sum_{i=1}^{10} \left[(\boldsymbol{x} - a_i)(\boldsymbol{x} - a_i)^T + c_i\right]^{-1}$	4	$[0, 10]^n$	-10.53

3   Experimental Results

3.1   Experimental Setup

The performance of the proposed IA is assessed by carrying out optimization on the 23 functions listed in table 1 and comparing results with some well-known algorithms for global optimization. For each test function 50 independent runs are performed. At each generation we compute the mean value of the best fit individuals for all 50 runs in order to plot the evolution curves. Moreover the

Table 2. Parameters used by OPT-IA for each function $(f_1, \ldots, f_{23})$

Function	$\alpha = e^{(-\rho * f)}$				$\alpha = \left(\frac{1}{\rho}\right) e^{(-f)}$			
	d	dup	τ_B	ρ	d	dup	τ_B	ρ
f_1	10	2	5	10	10	2	10	150
f_2	10	2	10	10	10	2	10	150
f_3	20	2	20	10	20	2	10	150
f_4	10	2	10	10	20	2	20	150
f_5	10	2	10	10	20	2	20	150
f_6	20	2	20	10	20	2	50	150
f_7	10	2	10	10	20	2	20	150
f_8	20	2	20	10	20	2	20	150
f_9	20	2	20	10	20	2	5	150
f_{10}	20	2	20	10	10	2	10	150
f_{11}	20	2	20	10	10	2	10	150
f_{12}	20	2	20	10	10	2	10	150
f_{13}	20	2	20	10	20	2	5	150
f_{14}	10	5	5	10	20	2	20	150
f_{15}	20	2	20	10	20	2	20	150
f_{16}	10	2	5	6	10	2	20	100
f_{17}	10	2	15	7	10	2	15	125
f_{18}	10	2	10	8	10	2	15	100
f_{19}	10	2	10	9	10	2	15	100
f_{20}	10	2	10	8	20	2	20	150
f_{21}	10	2	25	6	10	2	10	150
f_{22}	10	2	5	7	10	2	15	125
f_{23}	10	2	5	7	10	2	10	100

standard deviation is used to indicate the consistency of the algorithm. Table 2 summarizes the key parameters setting of OPT-IA for each test function and for each mutation potential used.

There are several algorithms designed for numerical optimization. We start comparing OPT-IA with one of the best evolutionary algorithms for numerical optimization in literature: Fast Evolutionary Programming (FEP) [12] and his improved version IFEP. FEP is based on Conventional Evolutionary Programming (CEP [13]) but uses a new mutation operator based on Cauchy random numbers that helps the algorithm to escape from local optima. The performance of OPT-IA is further compared with some other well-established evolutionary algorithms such as CEP with three different mutation operators (Gaussian Mutation Operator GMO, Cauchy Mutation Operator CMO and Mean Mutation Operator MMO) [13], Particle Swarm Optimization (PSO) [14] and Evolutionary Optimization (EO) [14]. Moreover comparison is made, when possible, with a global search algorithm for bound constrained optimization based on Lipschitz constant estimation, DIRECT [15, 16].

3.2   Comparison with FEP and DIRECT

Unimodal functions $(f_1 - f_7)$. Unimodal functions are not the most challenging test problems. There are more efficient algorithms which are specifically designed to optimize them. The aim in this case is to get a picture of the convergence rate of the algorithms. In table 3 we report the optimization results obtained by OPT-IA with respect to those obtained by FEP and DIRECT. All results

Table 3. Comparison between FEP, OPT-IA and DIRECT on functions $f_1 - f_7$

Fun.	T_{max}	Direct	FEP	opt-IA	
				$\alpha = e^{(-\rho * f)}$	$\alpha = \left(\frac{1}{\rho}\right) e^{(-f)}$
		Min Found	Mean Best Std Dev	Mean Best Std Dev	Mean Best Std Dev
f_1	150,000	n.a.	5.7×10^{-4} 1.3×10^{-4}	$\mathbf{9.23 \times 10^{-12}}$ $\mathbf{2.44 \times 10^{-11}}$	1.7×10^{-8} 3.5×10^{-15}
f_2	200,000	n.a.	8.1×10^{-3} 7.7×10^{-4}	$\mathbf{0.0}$ $\mathbf{0.0}$	7.1×10^{-8} 0.0
f_3	500,000	n.a.	1.6×10^{-2} 1.4×10^{-2}	$\mathbf{0.0}$ $\mathbf{0.0}$	1.9×10^{-10} 2.63×10^{-10}
f_4	500,000	n.a.	0.3 0.5	$\mathbf{1.0 \times 10^{-2}}$ $\mathbf{5.3 \times 10^{-3}}$	4.1×10^{-2} 5.3×10^{-2}
f_5	2×10^6	27.89	5.06 5.87	$\mathbf{3.02}$ $\mathbf{12.2}$	28.4 0.42
f_6	150,000	n.a.	$\mathbf{0.0}$ $\mathbf{0.0}$	0.2 0.44	$\mathbf{0.0}$ $\mathbf{0.0}$
f_7	300,000	8.9×10^{-3}	7.6×10^{-3} 2.6×10^{-3}	$\mathbf{3.0 \times 10^{-3}}$ $\mathbf{1.2 \times 10^{-3}}$	3.9×10^{-3} 1.3×10^{-3}

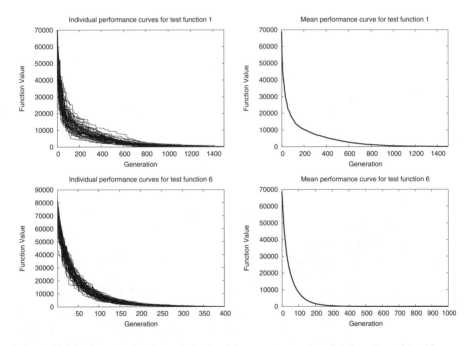

Fig. 2. Evolution curves of OPT-IA algorithm on two unimodal functions (f_1, f_6) over 50 independent runs Individual curves for each run (left plot), mean performance curve (right plot)

have been averaged over 50 independent runs. "Mean Best" indicates the mean best function values found in the last generation, "Std Dev" stands for standard deviation. Second column shows the maximum number of Fitness Function Evaluation allowed (T_{max}); we used the same T_{max} values proposed by X. Yao, Y. Liu

Table 4. Comparison between FEP, OPT-IA and DIRECT on functions $f_8 - f_{13}$

Fun.	T_{max}	Direct	FEP	opt-IA	
				$\alpha = e^{(-\rho * f)}$	$\alpha = \left(\frac{1}{\rho}\right) e^{(-f)}$
		Min Found	Mean Best Std Dev	Mean Best Std Dev	Mean Best Std Dev
f_8	$900,000$	$-4093.0.$	-12554.5 52.6	-12508.38 155.54	$\mathbf{-12568.27}$ $\mathbf{0.23}$
f_9	$500,000$	n.a.	$\mathbf{4.6 \times 10^{-2}}$ $\mathbf{1.2 \times 10^{-2}}$	19.98 7.66	2.66 2.39
f_{10}	$150,000$	n.a.	1.8×10^{-2} 2.1×10^{-3}	18.98 0.35	$\mathbf{1.1 \times 10^{-4}}$ $\mathbf{3.1 \times 10^{-5}}$
f_{11}	$200,000$	n.a.	$\mathbf{1.6 \times 10^{-2}}$ $\mathbf{2.2 \times 10^{-2}}$	7.7×10^{-2} 8.63×10^{-2}	4.55×10^{-2} 4.46×10^{-2}
f_{12}	$150,000$	0.03	$\mathbf{9.2 \times 10^{-6}}$ $\mathbf{3.6 \times 10^{-6}}$	0.137 0.23	3.1×10^{-2} 5.7×10^{-2}
f_{13}	$150,000$	0.96	$\mathbf{1.6 \times 10^{-4}}$ $\mathbf{7.3 \times 10^{-5}}$	1.51 0.10	3.20 0.13

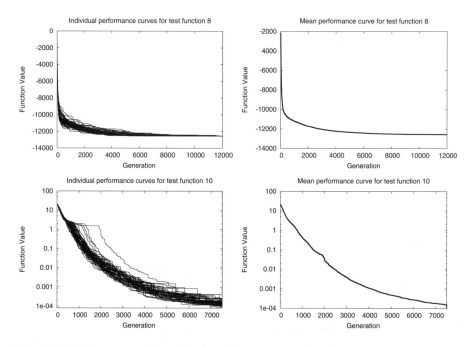

Fig. 3. Evolution curves of OPT-IA algorithm on two multimodal functions with many local minima (f_8, f_{10}) over 50 independent runs. Individual curves for each run (left plot), mean performance curve (right plot).

and G. Lin in [12]. Better results are highlighted in boldface. As it can be seen, OPT-IA is able to always obtain better results than FEP except for function f_3. For most functions, results for DIRECT are not available because the symmetry of the function implies that the optimum is in the center of the variable bounds,

Table 5. Comparison between FEP, OPT-IA and DIRECT on functions $f_{14} - f_{23}$

Fun.	T_{max}	Direct	FEP	opt-IA $\alpha = e^{(-\rho * f)}$	$\alpha = \left(\frac{1}{\rho}\right) e^{(-f)}$
		Min Found	Mean Best Std Dev	Mean Best Std Dev	Mean Best Std Dev
f_{14}	10,000	1.0.	1.22 .56	**1.02** **7.1 × 10^{-2}**	1.21 0.54
f_{15}	400,000	1.2×10^{-3}	**5.0 × 10^{-4}** **3.2 × 10^{-4}**	7.1×10^{-4} 1.3×10^{-4}	7.7×10^{-3} 1.4×10^{-2}
f_{16}	10,000	-1.031	−1.031 **4.9 × 10^{-7}**	**−1.03158** 1.5×10^{-4}	−1.02 1.1×10^{-2}
f_{17}	10,000	0.398	**0.398** **1.5 × 10^{-7}**	**0.398** 2.0×10^{-4}	0.450 0.21
f_{18}	10,000	3.01	3.02 0.11	**3.0** **0.0**	**3.0** **0.0**
f_{19}	10,000	−3.86	**−3.86** **1.4 × 10^{-5}**	−3.72 1.1×10^{-4}	−3.72 1.1×10^{-2}
f_{20}	20,000	−3.30	−3.27 5.9×10^{-2}	−3.31 7.4×10^{-2}	**−3.31** **5.9 × 10^{-3}**
f_{21}	10,000	−6.84	−5.52 1.59	**−9.11** **1.82**	−5.36 2.20
f_{22}	10,000	−7.09	−5.52 2.12	**−9.86** **1.88**	−5.34 2.11
f_{23}	10,000	−7.22	−6.57 3.14	**−9.96** **1.46**	−6.03 2.66

the point from which DIRECT starts the search. Figure 2 shows performance curves of OPT-IA for the two unimodal functions f_1 and f_6.

Multimodal functions with many local minima ($f_8 - f_{13}$). Function $f_8 - f_{13}$ are multimodal function with many local minima. The number of local minima increases exponentially as the function dimension increases. The fitness landscape of these functions is generally very *rugged* and difficult to optimize. Table 4 summarizes the final results obtained bye OPT-IA, FEP and DIRECT. FEP has a better performance on 4 of 6 test problems. In particular, for functions f_9, f_{12} and f_{13} FEP perform significantly better than OPT-IA, except for function f_{11} where results are comparable. On the other hand, OPT-IA performs significantly better then FEP on function f_8 and f_{10}. By comparing results between OPT-IA and DIRECT, we can see that DIRECT is unable to approach the optimum, while on functions f_{12} and f_{13} DIRECT shows a bit better performance.

Multimodal functions with only a few local minima ($f_{14} - f_{23}$). The final results of OPT-IA, FEP and DIRECT on functions $f_{14} - f_{23}$ are summarized in table 5. In this case OPT-IA shows a better performance on 6 of the 10 test functions. Moreover, for for functions f_{16} and f_{17}, OPT-IA is able to obtain the same optimum results of FEP in terms of mean best values found, but FEP shows more consistency in terms of standard deviation. Instead for function f_{15} results are of the same order. Inspecting all the experimental results (reported in the previous tables) it is possible to note an overall better performance of OPT-IA algorithm using a minimum population size ($d \in \{10, 20\}$) while FEP uses a larger population size greater of an order of magnitude.

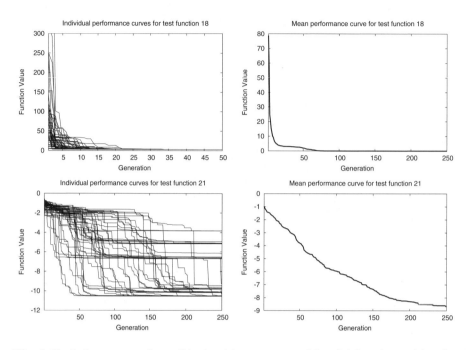

Fig. 4. Evolution curves of OPT-IA algorithm on two multimodal functions with a few local minima (f_{18}, f_{21}) over 50 independent runs. Individual curves for each run (left plot), mean performance curve (right plot).

3.3   Comparison with IFEP

The analyses performed in [12] show that Cauchy mutation performs better when the current search point is far away from the global optimum, while Gaussian mutation is better when search points are in the neighborhood of the global optimum. Based on those observations in [12] was proposed a version of FEP which uses both Cauchy and Gaussian mutations. This improved version was called IFEP. IFEP differs from FEP only in the step of the creation of the offsprings: two new offspring are generated instead of one, the first one using Cauchy mutation and the second one using Gaussian mutation, the better one is chosen. Table 6 shows the comparison between OPT-IA and IFEP on the same functions used in [12]: unimodal functions f_1, f_2, multimodal functions f_{10}, f_{11} with many local minima and multimodal functions f_{21}, f_{22} with only few local minima. OPT-IA has again a better performance on all the functions except for function f_{11} where the results are comparable.

3.4   Comparison with CEP, PSO and EO

Finally, we compare the immune algorithm with some other well-known biologically inspired algorithms: CEP, PSO and EO. Since the optimization results obtained with these algorithms are available in literature only for some of the

Table 6. Comparison between OPT-IA and IFEP on functions $f_1, f_2, f_{10}, f_{11}, f_{21}, f_{22}$ and f_{23}

Function	T_{max}	opt-IA	IFEP
f_1	150,000	1.70×10^{-8}	4.16×10^{-5}
f_2	200,000	7.15×10^{-5}	2.44×10^{-2}
f_{10}	150,000	1.11×10^{-4}	4.83×10^{-3}
f_{11}	200,000	8.36×10^{-2}	4.54×10^{-2}
f_{21}	10,000	-8.29	-6.46
f_{22}	10,000	-9.59	-7.10
f_{23}	10,000	-9.96	-7.80

Table 7. Comparison between OPT-IA and CEP with three different mutation operators (GMO,CMO,MMO) in terms of mean best values found

Function	T_{max}	opt-IA	CEP GMO	CEP CMO	CEP MMO
f_1	150,000	1.70×10^{-8}	3.09×10^{-7}	3.07×10^{-7}	9.81×10^{-7}
f_2	250,000	7.15×10^{-5}	1.99×10^{-3}	5.87×10^{-3}	3.23×10^{-3}
f_3	250,000	260.12	17.60	**5.78**	11.80
f_4	250,000	**0.001**	5.18	0.66	1.88
f_5	250,000	**29**	86.70	114.0	63.8
f_7	250,000	5.85×10^{-03}	12.20	9.42	9.53
f_9	250,000	24.0	120.0	**4.73**	9.52
f_{10}	150,000	1.11×10^{-4}	9.10	1.3×10^{-3}	7.49×10^{-4}
f_{11}	250,000	8.36×10^{-2}	2.52×10^{-7}	2.2×10^{-6}	6.99×10^{-7}

Table 8. Performance Comparison among OPT-IA, PSO, and EO on functions f_1, f_5, f_9, f_{11}

Fun.	T_{max}	opt-IA Mean Best	opt-IA Std Dev	PSO Mean Best	PSO Std Dev	EO Mean Best	EO Std Dev
f_1	250,000	1.70×10^{-8}	3.5×10^{-15}	11.75	1.3208	9.8808	0.9444
f_5	250,000	**29.0**	**0.0**	1911.598	374.2935	1610.39	293.5783
f_9	250,000	**24.0**	**7.69**	47.1354	1.8782	46.4689	2.4545
f_{11}	250,000	8.36×10^{-2}	4.32×10^{-2}	0.4498	0.0566	0.4033	0.0436

23 test functions considered in this work, comparisons will be made accordingly. Results shown in tables 7 and 8 indicate that while OPT-IA maintain a consistent performance, CEP, PSO and EO appear unable to reach the optima for most of the listed functions, although they make a considerable computational effort. An overall better performance of OPT-IA is evident.

4   Conclusions

In this paper we have introduced an immune algorithm based on clonal selection principle, called OPT-IA, for global numerical optimization. The main features of OPT-IA are the following: cloning operator, inversely proportional hypermutation operator, and aging operator. The cloning operator explores the neighborhood of each point of the search space. The inversely proportional hypermutation perturbs each candidate solution inversely proportional to its fitness function value. Finally, the aging operator eliminates the oldest candidate solutions from the current population in order to introduce diversity and to avoid local minima during the evolutionary search process. We tested OPT-IA on 23 well-known benchmark problems. The experimental studies show that the clonal selection algorithm is an effective numerical optimization algorithm in terms of solution quality. The results show that OPT-IA is significant better than the seven tested evolutionary algorithms and one well-known deterministic algorithm (DIRECT).

As future works we plan to perform a deeper statistical analysis of the obtained experimental results and to compare *opt-IA* with the BCA algorithm [17], Evolution Strategies and Differential Evolution [18].

References

[1] Nicosia G.: "Immune Algorithms for Optimization and Protein Structure Prediction", Ph.D. Thesis, *University of Catania*, Italy, December 2004.

[2] De Castro L. N., Timmis J., "Artificial Immune Systems: A New Computational Intelligence Paradigm" London, UK: *Springer-Verlag*, (2002).

[3] Cutello V., Nicosia G.: "The Clonal Selection Principle for in silico and in vitro Computing", *in Recent Developments in Biologically Inspired Computing*, L. N. de Castro and F. J. Von Zuben, Eds., (2004).

[4] Cutello V., Nicosia G., and Pavone M.: "Exploring the capability of immune algorithms: A characterization of hypermutation operators" in *Proc. of the Third Int. Conf. on Artificial Immune Systems (ICARIS'04)*, pp. 263–276 (2004).

[5] Nicosia G., Cutello V., Pavone M.: "An Immune Algorithm with Hyper-Macromutations for the Dill's 2D Hydrophobic-Hydrophilic Model", *Congress on Evolutionary Computation, CEC 2004, IEEE Press*, vol. 1, pp. 1074-1080, (2004).

[6] Cutello V., Morelli G., Nicosia G., and Pavone M.; "Immune Algorithms with Aging operators for the String Folding Problem and the Protein Folding Problem," in *Proc. of the Fifth Europ. Conf. on Comp. in Combinatorial Optimization (EVOCOP'05)*, LNCS, vol. 3448, pp. 80-90 (2005).

[7] Nicosia G., Cutello V., Pavone M.: "A Hybrid Immune Algorithm with Information Gain for the Graph Coloring Problem", *Genetic and Evolutionary Computation Conference*, GECCO 2003, vol. 2723, pp. 171-182.

[8] Nicosia G., Cutello V., Bentley P. J., Timmis J., "Artificial Immune Systems", *Third International Conference*, ICARIS 2004, Catania, Italy, September 13-16, Springer (2004).

[9] Goldberg D.E.: "The Design of Innovation: Lessons from and for Competent Genetic Algorithms", *Kluwer Academic Publisher*, vol 7, pp. Boston, (2002).

[10] De Castro L.N., Von Zuben F.J.: "Learning and optimization using the clonal selection principle". *IEEE Trans. on Evolutionary Computation*, vol 6, no 3, pp. 239-251, (2002).

[11] De Castro L. N., Timmis J.,: "An Artificial Immune Network for Multimodal Function Optimization", *CEC'02, Proceeding of IEEE Congress on Evolutionary Computation*, IEEE Press, (2002).

[12] Yao X., Liu Y. and Lin G.M.: "Evolutionary programming made faster", *IEEE Trans. on Evolutionary Computation*,vol 3, pp. 82-102, (1999).

[13] Chellapilla, K.: "Combining mutation operators in evolutionary programming," *IEEE Trans. Evol. Comput.*, vol. 2, pp. 91-96, (1998).

[14] Angeline, P. J.: "Evolutionary optimization versus particle swarm optimization: Philosophy and performance differences", in *Proc. Evolutionary Programming VII*, V. W. Porto, N. Saravanan, D. Waagen, and A. E. Eiben, Eds., pp. 601-610, (1998).

[15] Jones, D. R., Perttunen, C. D. and Stuckman, B. E.: "Lipschitzian optimization without the lipschitz constant." *J. of Optimization Theory and Application*, vol. 79, pp. 157-181, (1993).

[16] Finkel,   D.   E.:   "DIRECT   Optimization   Algorithm   User   Guide." *Technical   Report,   CRSC   N.C.   State   University,*   March   2003. (ftp://ftp.ncsu.edu/pub/ncsu/crsc/pdf/crsc-tr03-11.pdf).

[17] Timmis J. and Kelsey J.: "Immune Inspired Somatic Contiguous Hypermutation for Function Optimisation", *Genetic and Evolutionary Computation Conference*, GECCO 2003, Springer vol. 2723, pp. 207-218.

[18] Vesterstrom, J. and Thomsen R.: "A Comparative Study of Differential Evolution, Particle Swarm Optimization, and Evolutionary Algorithms on Numerical Benchmark Problems", *Congress on Evolutionary Computation, CEC 2004, IEEE Press*, vol. 2, pp. 1980-1987, (2004).

Algorithms (X, sigma, eta): Quasi-random Mutations for Evolution Strategies

Anne Auger[3], Mohammed Jebalia[1], and Olivier Teytaud[1,2]

[1] Equipe TAO - INRIA Futurs, LRI, Bât. 490, Université Paris-Sud, Orsay Cedex 91405, France
[2] Artelys, 12 rue du 4 septembre, Paris 75002, France
www.artelys.com
[3] CoLab, ETH Zentrum CAB F 84, 8092 Zürich, Switzerland

Abstract. Randomization is an efficient tool for global optimization. We here define a method which keeps :

- the order 0 of evolutionary algorithms (no gradient) ;
- the stochastic aspect of evolutionary algorithms ;
- the efficiency of so-called "low-dispersion" points ;

and which ensures under mild assumptions global convergence with linear convergence rate. We use i) sampling on a ball instead of Gaussian sampling (in a way inspired by trust regions), ii) an original rule for step-size adaptation ; iii) quasi-monte-carlo sampling (low dispersion points) instead of Monte-Carlo sampling. We prove in this framework linear convergence rates i) for global optimization and not only local optimization ; ii) under very mild assumptions on the regularity of the function (existence of derivatives is not required). Though the main scope of this paper is theoretical, numerical experiments are made to backup the mathematical results.

1 Introduction

Evolutionary algorithms (EAs) are zeroth-order stochastic optimization methods somehow inspired by the Darwinian theory of biological evolution: emergence of new species is the result of the interaction between natural selection and blind variations. Among the class of Evolutionary Algorithms, Evolution Strategies (ES) [12, 17] are the most popular algorithms for solving continuous optimization problems, *i.e.* for optimizing real-valued function f defined on a subset of $\mathbb{R}^{dim}$ for some dimension dim. The common feature of EAs is to evolve a set of points of the search space: at each iteration, some points of the search space are randomly sampled, then evaluated (the f value of the points is computed) and last, some of them are selected. Those three steps are repeated until a stopping criterion is met.

Since the invention of ESs in the mid-sixties, researches to improve the performances of ESs focused on the so-called mutation operator [12, 17, 8]. This operator consists in sampling a gaussian random variable with a given step-size σ and a given covariance matrix C. The main issue has been the adaptation of the step-size parameter σ and of the covariance matrix C. The first step in this direction is the well-known one-fifth rule [12] based on the rate of successful mutations. Then Rechenberg [12] and Schwefel [17] proposed to self-adapt the parameters of the mutation operator, by

E. Talbi et al. (Eds.): EA 2005, LNCS 3871, pp. 296–307, 2006.

mutating the step-size as well (this being usually achieved by multiplying the step-size by a log-normal random variable). For this technique, the so-called *mutative step-size adaptation*, a step size is associated to every individual in the population. This step-size undergoes variations and is used to mutate the object parameters of the individual. The individual is selected with its step-size and therefore the step-sizes automatically adapted. Intuitively unadapted step-sizes can not give successively good individuals.

In this paper, we use a similar concept for adapting the scale of the sampling at each generation but use a uniform sampling in a ball instead of the standard Gaussian distribution. The motivation is that with a ball we have a trust region-effect ([7]), *i.e.* the local operator can be trusted in this ball. Note that though this is not classical in the evolutionary computation community, Rudolph [15] already introduced –mainly for theoretical purposes– sampling of the unit ball instead of a Gaussian sampling. We also make use of a deterministic sampling, or *quasi random sampling*, where we moreover minimize the dispersion of the quasi-random points [11, 20]. Quasi-random numbers have already proved to be successful in many areas one of which is the field of Monte Carlo methods allowing to speed up the convergence of those methods [5, 11] but as far as we know low-dispersion points are new for the evolutionary computation community.

On a theoretical point of view, many papers deal with asymptotic properties of evolutionary algorithms [13, 14] or their finite time convergence in discrete cases [3], but convergence rates are only given under strong assumption (unimodal functions and/or very convex functions and/or very smooth functions and/or only local convergence) [15, 4, 6, 17, 2, 16, 1]. In this paper we investigate the convergence of the new algorithm considered and we prove its convergence with order one.

The paper is organized as follows: Section 2 presents our algorithm, Section 3 presents the theoretical results and Section 4 investigates numerically the theoretical results; Section 5 comments the results obtained and concludes.

2   Definitions and Properties

In this section we introduce the algorithm considered in this paper. As for the self-adaptive Evolution Strategies (SA-ES), a step-size is associated to each individual, moreover for reasons that will become clear in the sequel one individual is a triplet (x, σ, η) and not only (x, σ) as for the SA-ES. To create new points, the so-called *descent* operator is applied. It consists in choosing the best point among N neighbors of x (where the scale of the neighborhood is given by σ) and updating σ with η (see below). At each generation, new individuals are also randomly sampled. Finally individuals created from both sides are submitted to selection. After giving some definitions, we formally describe the *descent* operator and the algorithm:

General Definitions. We consider the minimization of a real valued objective function f defined on X a subset of the real space $\mathbb{R}^{dim}$. We assume that the minimum of f is reached on X and denote $f^* = \min_{x \in X} f(x) \in \mathbb{R}$. Therefore $f := X \mapsto [f^*, \infty[$. Let opt denote the set of optima, *i.e.* opt $= \{x \in X / f(x) = f^*\}$. Let $x \in \mathbb{R}^{dim}$ be a vector of $\mathbb{R}^{dim}$ and r a positive real number. We will denote $B(x, r)$ the closed ball of center x and radius r. For a set E embedded in X we will denote $\overline{E}$ the complementary of X in $E \subset X$. $|E|$ will denote the cardinal of E. The Euclidean distance on $\mathbb{R}^{dim}$ will be denoted $d(.,.)$, *i.e.* let $(x, y) \in \mathbb{R}^{dim} \times \mathbb{R}^{dim}$, $d(x, y) = \sqrt{\sum_{i=1}^{dim}(x_i - y_i)^2}$.

Exploitation operator "descent". Let B be a set of N points of the unit ball, $B = \{B_1, \ldots, B_N\}$, we define *descent* as $descent(x, \sigma, \eta) = (x + \sigma B_\star, \eta\sigma, \eta)$ where $\star = argmin_{j \in [[1,N]]} f(x + \sigma B_j)$ (any of the optimal in case of equality).

Algorithm. The algorithm we investigate in the sequel is an evolutionary algorithm where a population P_n, where n is the iteration or generation index, is evolved. Each individual of the population is a triplet $(x, \sigma, \eta) \in \mathbb{R}^{dim} \times \mathbb{R}^+ \times \mathbb{R}^+$.

1. Sampling of N points $B = \{B_1, \ldots, B_N\}$ included in $B(0, 1)$.
2. Sampling of the initial population P_0 of (x, σ, η)
3. For n varying from 0 à $+\infty$
   (a) Creation of P_{n+1}^a, empty population.
   (b) **Descent step:** for each $(x, \sigma, \eta) \in P_n$, add $descent(x, \sigma, \eta)$ in P_{n+1}^a ; the population at the end of this step is P_{n+1}^b. [1]
   (c) **Random sampling step:** Random sampling of new individuals (x, σ, η) (see the Assumption subsection for the details), $P_{n+1}^{b'}$, the new population is $P_{n+1}^c = P_{n+1}^b \cup P_{n+1}^{b'}$
   (d) **Selection step:** Selection of the best $|P_n|$ element of P_{n+1}^c, the population so generated is P_{n+1}.
   (e) Increase N by 1 and regenerate B, if at least one local descent is interrupted.

Local descent: We call *local descent* a sequence of successive points $((x_1, \sigma_1, \eta_1), \ldots, (x_n, \sigma_n, \eta_n))$ generated at Step 3b, *i.e.* For $i > 1$ $(x_i, \sigma_i, \eta_i) = descent(x_{i-1}, \sigma_{i-1}, \eta_{i-1})$. **Interrupted local descent:** We will say that a local descent is interrupted if for some i (x_i, σ_i, η_i) is removed by the selection step.

Dispersion of B: We note $\Delta(B)$ (or Δ for short) the dispersion of B defined as $\Delta(B) = \sup_{x \in B(0,1)} \inf_{y \in B} \| x - y \|$.

3   Results

The convergence of the algorithm previously defined is analyzed in this Section.

3.1   Assumptions

We consider $V = f^{-1}([f^*, f^* + s[)$ for a given s, and assume that V is a neighborhood of $opt = f^{-1}(f^*)$.

Assumption A.   *1. We require that Step 1 and 3e ensure that $0 \in B$, that Δ is non-increasing in N and that $\Delta \to 0$ as $N \to \infty$. For example, we might assume that each new B generated minimizes $\Delta(B)$ under the constraint $0 \in B$.*
   2. We forbid $\eta \geq 1$ or $\eta \leq 0$; in all cases $\eta \in]0, 1[$.
   3. The generation method (Step c) must generate 3-uples (x, σ, η) in an i.i.d manner ; the number of generated 3-tuples is upper bounded by a given constant G, and the density is lower bounded by $c > 0$ and upper bounded by $d < \infty$ on

---

[1] At the end of this step, we have $|P_{n+1}^b| = |P_n|$.

$V \times]0, 2 \sup_{(a,b) \in V \times V} \| a - b \| [\times]0, 1[$, *and* x, σ *and* η *are independent. Moreover, we generate at each Step c at least one point (which can be removed in the selection step).*

4. *We keep, at Step d, the* $|P_n|$ *best elements for the fitness. This selection depends on* x *only (not on* σ *and* η*) : in particular,* $|P_{n+1}| = |P_n|$ *and* $\forall(x, \sigma_x, \eta_x) \in P_{n+1}^d, \forall(y, \sigma_y, \eta_y) \in P_{n+1}^c \setminus P_{n+1}^d \ f(x) \le f(y)$.

5. *We assume that if* $x \in V$*, the following holds :*

$$f^* + \alpha' d(x, opt)^\beta \le f(x) \le f^* + \alpha d(x, opt)^\beta$$

with $\beta > 0$ *and* $0 < \alpha' \le \alpha$.

6. *For* $\epsilon > 0$ *sufficiently small, the probability of generating (by random generation at Step 3c) an optimal point within* ϵ *is lower bounded by* $K\epsilon^C$ *and upper bounded by* $K'\epsilon^C$ *for some* $C, K, K' > 0$ *(consequence of Assumption* **A**.5 *and Assumption* **A**.3*), i.e.* $K\epsilon^C \le P(f(x) \le f^* + \epsilon) \le K'\epsilon^C$.

Comments: Assumption **A**.5 implies that f is Hölder and anti-Hölder [10, 18] for every isolated point of *opt*. See also [19, 9] for works where fractal quantities are related to the analysis carried out.

The fact that the coefficient β is the same on the left-hand and on the right-hand side in Assumption **A**.5 is, for us, the strongest assumption. Assumption **A**.4 can be removed, with some technical modifications of the proof.

3.2   Preliminary Results

We prove that if $\Delta(B)$ is sufficiently small in front of the constants of the problem and of η_n, and if the optimum is inside the initial ball, then linear convergence occurs.

Lemma 1 (Linear descent). *If* $x_n \in V$ *and* $opt \cap B(x_n, \sigma_n) \ne \emptyset$ *and*

$$\eta_n \ge \sqrt[\beta]{(\frac{\alpha}{\alpha'})} \Delta(B)$$

then $d(descent^k(x_n, \sigma_n, \eta_n), opt) \le \eta_n^k \sigma_n$

PROOF: By induction, we show that all $(c_k, r_k, \epsilon_k) = descent^k(x_n, \sigma_n, \eta_n)$ are in V (by definition of $V = f^{-1}([f^*, f^* + s[))$. By induction also, $B(c_k, r_k) \cap opt$ is nonempty (thanks to Assumption **A**.5). As the radius of the ball is upper-bounded by $\sigma_n \eta_n^k$, the result follows.   □

We now prove the following Lemma:

Lemma 2. *Let* $(x_k, \sigma_k, \eta) = descent^k(x, \sigma, \eta)$*, then either* **P 1** *or* **P 2** *(but not both simultaneously) holds:*

P 1. *for* k *sufficiently large,* $f(x_k) \le f^* + \alpha(\sigma\eta^k/(1 - \eta))^\beta$,
P 2. $f(x_k)$ *is lower bounded by a constant* $> f^*$.

Interpretation: Some sequences converge quickly to the optimum and some sequences are lower bounded. There is no sequence converging slowly or sequence whose successive fitness accumulate around the optimum without converging to it.

PROOF: Assume that $f(x_k) \to f^*$. As $\sigma_k = \sigma \eta^k$, for any $p > 0$ we have

$$d(x_k, x_{k+p}) \le \sigma \eta^k (1 + \eta + \ldots + \eta^{p-1})$$

$$= \sigma \eta^k \frac{(1 - \eta^p)}{(1 - \eta)} \le \frac{\sigma \eta^k}{(1 - \eta)}.$$

Then $(x_k)_{k \in \mathbb{N}}$ is a Cauchy sequence which therefore converges. Let x_∞ be its limit, from the previous equation, the following holds

$$d(x_k, x_\infty) \le \sigma \eta^k / (1 - \eta).$$

Only two situations can occur

Either $f(x_k) \to f^*$ and consequently for k sufficiently large, $x_k \in V$. With Assumption **A.5** we have

$$f(x_k) \le f^* + \alpha (\sigma \eta^k / (1 - \eta))^\beta$$

which is the property **P 1.**.

Either $f(x_k)$ does not converge to f^* but as $f(x_k)$ decreases it is lower bounded by a value $> f^*$ which is the property **P 2.**   □

Satisfactory individual: The 3-uple (x, σ, η) is said *satisfactory* if the property **P 1.** defined in Lemma 2 holds.

Lemma 3. *Let $(n_i)_{i \in \mathbb{N}}$ be the subsequence of the index generation $n \in \mathbb{N}$ such that there exists an individual $(x_{(i)}, \sigma_{(i)}, \eta_{(i)})$ in P_n^c generated at Step c **and** selected at Step d.*

In other words, $(n_i)_{i \in \mathbb{N}}$ is the increasing enumeration of the set of n such that some point is generated at epoch n and selected ; $(x_{(i)}, \sigma_{(i)}, \eta_{(i)})$ is the element among these points with the minimum value of $f(.)$.

When $(x_{(i)}, \sigma_{(i)}, \eta_{(i)})$ is not unique, we choose it arbitrarily among possible points minimizing $f(x_{(i)})$.

Assume that there are infinitely many interrupted local descent (which is equivalent to the fact that there are infinitely many i such that n_i is well defined). Then, for a given C, $P((x_{(i)}, \sigma_{(i)}, \eta_{(i)})$ satisfactory and non-interrupted$) \ge C > 0$ infinitely often.

Interpretation: Lemma 3 states that if infinitely many new local descent occur, then infinitely many of these new descents have a lower bounded probability of being uninterrupted. Lemma 3 will be used in the main Theorem to get a contradiction : if infinitely many new descents are started, by Lemma 3 (almost surely) infinitely many of them are non-interrupted, so there are more and more non-interrupted sequences, so, as the population is bounded after a finite time there is no more room for a new descent (see the Theorem for more details).

PROOF:

1. Assume that n_i is well defined for all $i \in \mathbb{N}$. Note that this implies that Δ decreases to 0 (by Assumption A.1).
2. Note w_n the worst fitness among P_n^b. By construction w_n is non-increasing. As it is lower-bounded, it converges.
3. Let us show that it almost surely converges to f^*. The proof is as follows :
    - Assume, in order to get a contradiction, that w_n is lower bounded by some $f^* + \epsilon$ where $\epsilon = 1/2^k$ for some integer $k > 0$.
    - Then with Assumption A. 6, infinitely many new points (generated in steps 2c) are generated with fitness $< f^* + \epsilon$.
    - The number of points in P_n^b with fitness $\geq f^* + \epsilon$ is decreased of one at each generation of points with fitness $< f^* + \epsilon$. As this occurs infinitely often, after a finite time (almost surely), w_n must decrease below $f^* + \epsilon$. This is true for any $\epsilon = 1/2^k$ with probability 1; by countable intersection, it is true with probability 1 for all $\epsilon = 1/2^k$.
    - Therefore w_n decreases to $f^* + \epsilon$.
4. Note that $f(x_{(i)}) \leq w_{n_i}$ (because if $f(x_{(i)}) \geq w_{n_i}$ then by construction, $x_{(i)}$ would not be selected). Therefore, the fitness of $x_{(i)}$ converges to f^*.
5. Let us show that the event

$$\{(x_{(i)}, \sigma_{(i)}, \eta_{(i)}) \text{ satisfactory and } \eta_{(i)} \leq 0.9\}$$

occurs with probability at least $1 - D$ for some $D < 1$ if i is sufficiently large.
    - The event $\{(x_{(i)}, \sigma_{(i)}, \eta_{(i)}) \text{ satisfactory and } \eta_{(i)} \leq 0.9\}$ in particular holds if the assumptions of Lemma 1 and $\eta \leq 0.9$ are verified. This is the case whenever $\sigma \geq d(opt, \overline{V})$ and $0.9 \geq \eta \geq \Delta \sqrt[\beta]{\alpha/\alpha'}$ and if $f(x) < f^* + \alpha' d(opt, \overline{V})^\beta$.
    - The latter inequality holds if i is sufficiently large, as $f(x_{(i)})$ converges to f^*.
    - Other inequalities occur independently with probability lower bounded by a constant > 0, provided that Δ is sufficiently small.
    - The probability of these three inequalities simultaneously is lower-bounded by a positive constant $1 - D$ ($D < 1$), provided that Δ is sufficiently small. Δ goes to 0 (point 1 above) and therefore Δ is sufficiently small if i is sufficiently large.
6. Note E_i' the event that $\sigma \geq d(opt, \overline{V})$ and $0.9 \geq \eta \geq \Delta \sqrt[\beta]{\alpha/\alpha'}$ and $f(x) < f^* + \alpha' d(opt, \overline{V})^\beta$. We have shown above that $P(\neg E_i') \geq 1 - D$.
7. Note E_i the event $\{(x_{(i)}, \sigma_{(i)}, \eta_{(i)}) \text{ verifies } E_i' \text{ and is never interrupted }\}$ in the sense that its successive sons generated in Step b are never eliminated in Step d.
8. By Lemma 2, if E_i' occurs, then the kth iterate of the local descent (from $(x_{(i)}, \sigma_{(i)}, \eta_{(i)})$) has fitness bounded above by $\alpha^C (\sigma_{(i)} \eta_{(i)}^k / (1 - \eta_{(i)}))^\beta$.
9. Therefore, conditionally to E_i', the probability of interruption of the kth iterate is upper bounded by $K' \alpha^C (\sigma_{(i)} \eta_{(i)}^k / (1 - \eta_{(i)}))^{\beta C}$.
10. So $P(\neg E_i | E_i')$ is upper bounded by the $\sum_{k=0}^{\infty} K' \alpha^C (\sigma_{(i)} \eta_{(i)}^k / (1 - \eta_{(i)}))^{\beta C}$.
11. Now, recall that $P(\neg E_i) = P(\neg E_i | E_i') P(E_i') + P(\neg E_i')$ (as E_i implies E_i'), and therefore $P(\neg E_i) \leq P(\neg E_i | E_i') + P(\neg E_i')$.

12. Then, combining points 11, 10 and 6 above, $P(\neg E_i) \leq 1 - D + \sum_{k=0}^{\infty} K' \alpha^C (\sigma_{(i)} \eta_{(i)}^k / (1 - \eta_{(i)}))^{\beta C}$.
13. σ having a density lower-bounded by a constant > 0 in the neighbourhood of 0, $P(E_i)$ is infinitely often larger than a given $W > 0$ (for example, $W = 1 - D/2$).

Hence the expected result : E_i, having probability $\geq W > 0$ for any i (conditionally to the past and current epochs of the algorithm), occurs almost surely infinitely often. $\square$

3.3 Almost Sure Convergence with Order One

We now investigate the global convergence properties of our algorithm. The delicate part is that it is not enough to have the fact that after a finite number of iterations we are close to the optimum and therefore convergence holds. Indeed, there is always a risk that a local descent is interrupted. Therefore we are going to formalize in the proof below the fact that with probability 1, under minimal assumptions, there is a non-interrupted local descent that converges linearly. We emphasize the fact that this proof could be applied for other operators as well. The only requirement is to have enough fast convergence for the local operator. The heart of the proof can be outlined as follows:

– any non-satisfactory local descent will be interrupted (consequence of Lemma 2 and of Assumption **A**.6) by a new local descent; each new local descent has a probability lower bounded by a constant > 0 of being satisfactory ; so, there are infinitely many satisfactory local descent (this is Step 1 of the proof below) as long as none of them is satisfactory and non-interrupted ; so we always have a satisfactory local descent among the future populations ;
– these local descents have a probability of being interrupted which decreases so quickly (by Lemma 3), that after some time they are no more interrupted (this is the Step 2 of the proof) ;
– hence, the convergence is linear (Step 3) and moreover N is bounded (Step 3).

The detailed proof comes after the Theorem:

Theorem 1. *We have almost sure convergence at least linear of the error to the optimal error, i.e. $\inf_{(x,\sigma,\eta)\in P_n} (f(x) - f^*) \leq A/B^n$ for some $A > 0$ and $B > 1$. Moreover, N is almost surely bounded.*

PROOF:
Step 1: Let us show that with probability 1, there exists infinitely many values of n such that there exists (x, σ, η) satisfactory in P_n^d.

Let us make the hypothesis H1 (to get a contradiction), that for any $n > n_0$, there is no (x, σ, η) in P_n^d such that $f(descent^k(x, \sigma, \eta)) \to f^*$ for $k \to \infty$ (independently of any interruption ; we consider the theoretical sequence of $descent^k(.)$ as $k \to \infty$).

Moreover, let us assume (one again in order to get a contradiction), the hypothesis H2: there exists n_1 such that for any $n > n_1$, the Step c of generation of points does not provide any point better than the worst point resulting from Step b.

Then, if $n > n_1$, $P_n^d = \{descent^{n-n_1}(x, \sigma, \eta) | (x, \sigma, \eta) \in P_{n_1}^d\}$; moreover, N, B and Δ become constant. The $f(descent^{n-n_1}(x, \sigma, \eta))$ are lower bounded by a given $f^* + \epsilon$, for a given $\epsilon > 0$. This is proved by the application of:

- H1 (which states that none of the local descents converges) and
- Lemma 2 (which states that if local descents do not converge to f^* then they are lower bounded).

to the finit set of local descents from $P_{n_1}^d$.

Then for each n, at Step 3c, the probability of generating a new point (x_n, σ_n, η_n) better than the local descents is lower bounded by some P_*, where P_* is provided by Assumption **A**.6.

So, such a generation necessarily occurs, with probability 1.

So, we have a contradiction with H2. So, under hypothesis H1, H2 does not hold, infinitely often, a new point (x, σ, η) generated at Step c is added to P_n^d.

We have assumed H1, and proved that H2 does not hold. Let us now look for a contradiction, so that we can prove that H1 does not hold.

N increases for each n such that the followings holds : "a point generated at Step c is integrated to P_n^d". As this occurs infinitely often (as $H2$ is false), $\Delta \to 0$.

Consider the probability $\Pi = P((x, \sigma, \eta) \text{ satisfactory}|s)$ of generating (x, σ, η) satisfactory ;

$$\Pi \geq \underbrace{P(x \in V|s)}_{\Pi_1} \times \underbrace{P(\sigma \geq sup_{z \in V} d(z, opt)|s)}_{\Pi_2} \times \underbrace{P(\eta \geq \sqrt[\beta]{\alpha/\alpha'}\Delta(B)|s)}_{\Pi_3}$$

where $P(E|s)$ is the probability of an event E conditionnally to the fact that the point (x, σ, η) coming from the generation Step c is selected and is the best selected point.

Π_1 is asymptotically lower bounded by a constant > 0 (and indeed converges to 1), Π_2 is lower bounded by a positive constant thanks to Assumption **A**.3, and Π_3 is lower bounded by a positive constant when Δ is sufficiently small, what occurs as $\Delta \to 0$.

The probability of getting a (x, σ, η) satisfactory and non-interrupted is thus lower-bounded for each step n during which a new point is generated at Step c. Consequently this event occurs necessarily for infinitely many values of n, with probability 1.

So with probability 1, we have contradiction with hypothesis H1. So we can claim that there exists infinitely many values of n such that there exists some (x, σ, η) in P_n^d such that $descent^k(x, \sigma, \eta) \to opt$ if $k \to \infty$.

Step 2: Let us show that finitely many points (x, σ, η) generated in (c) are selected in (d).

Note $(x_{(i)}, \sigma_{(i)}, \eta_{(i)})$ the sequence of 3-uples generated at Step c and selected in P_n^d (not removed by the selection step) and satisfactory (ie, are in the first case of Lemma 2) and are the best (from the point of view of the fitness) among the (x, σ, η) generated in Step c and incorporated in $P_{n,d}$.

Let us do, in order to get a contradiction, the hypothesis that this sequence is infinite (which is equivalent to assuming that there are infinitely many 3-uples generated in Step c selected in Step d).

Then, for i large enough $P((x_{(i)}, \sigma_{(i)}, \eta_{(i)})$ verifies Lemma 1 and is not interrupted) is infinitely often lower bounded by a positive constant (Lemma 3).

So, this occurs, almost surely, infinitely often. As the number of non-interrupted local descents is bounded above by the population size, there is contradiction.

Conclusion: By Step 1, we know that with probability 1, infinitely many 3-uples (x, σ, η) satisfactory are in some P_n^d. By Step 2, we know that these 3-uples can only a finite number of times come from random generations (as only a finite number of points can come from Step c and be included to P_n^d). So, finitely many local descents are interrupted (each interruption is the integration in (d) of a point coming from Step c). So after a finite time, no more local descent is interrupted; N **is now constant** (and so, does not go to infinity) and the satisfactory local descent (whose existence is almost sure thanks to Step 1) **goes to the optimum, with linear convergence** thanks to Lemma 2.   □

4   Practical Experiments

We have experimented our method on different simple objective functions $f_{L_p}(x) = \sqrt[p]{\sum x_i^p}$ satisfying the assumptions we made for the convergence. Figure below shows the linear convergence of the method. We observe the changes of convergence rates due to the changes of η associated to the best point in the population and the increases of N leading to a N-points quasi-random sampling. The choice of B for a given value of N has been performed by optimizing the disrepancy of the points. This part of the procedure is time-consuming when N increases. Note that such sets of points in the ball could of course be evaluated off line. Very efficient and fast algorithms exists for quasi-monte-carlo generation in the sense of standard discrepancy, but as

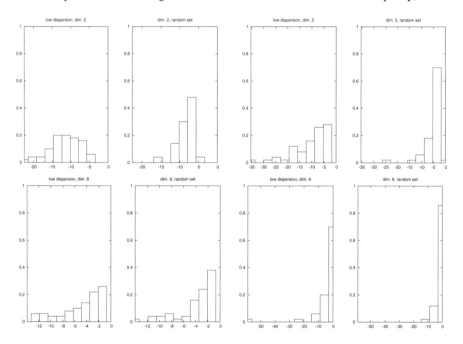

Fig. 1. Histogram of the distribution of $\log(f_{L_5})$ after $500 \times (d/3)^2$ fitness-evaluations for the dimension indicated at the top of the graphs. For each couple of graph, on the left with low-dispersion points resulting from gradient-based optimization on $\Delta(B)$; on the right, with random points.

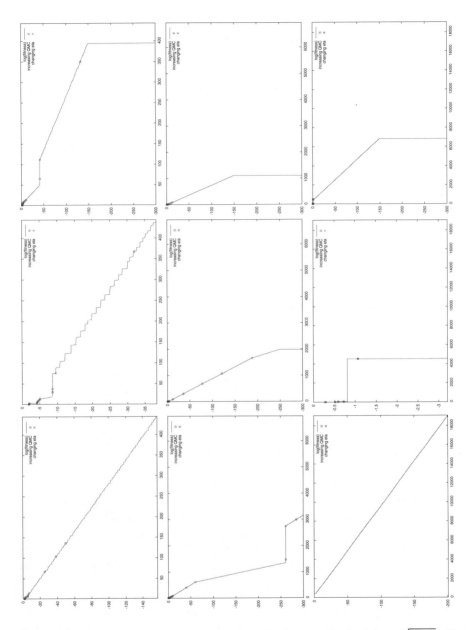

Fig. 2. Fitness value in logarithmic scale vs number of generations for $f_{L_p}(x) = \sqrt[p]{\sum x_i^p}$ with respectively from left to right, $p = 1, 3, 5$. First column $dim = 2$, second column $dim = 5$ and third column $dim = 10$. Due to numerical precisions, $\log(f_{L_p})$ can be equal to $-\infty$. A cross indicates when a new η is chosen. A circle indicates when N is increased by 1. The random generation for x is uniform on $[-1, 1]^d$, η is uniform on $[0, 1]$, 10σ is the absolute value of a standard Gaussian, the population size is 5, the number of random generations at Step 3c is 25 and N is initialized to 1. A cross indicates when a new η is chosen. A circle indicates when N is increased by 1. It may be observed that N quickly stabilizes.

far as we know no equivalent algorithms exist for the optimization of Δ. Interestingly, experiments with random sampling once per increase of N leads to similar results (note that the result about linear convergence remains theoretically true) but the case with one new sampling at each 3c step leads to much worse results. This suggests that quasi-random mutations (at least, stabilizing the random part by keeping the same B until N increases) are not only of theoretical interest (for proving our results of linear convergence on a very large family of fitness functions) but also of practical interest. Note that on the other hand, we need random points for the almost sure convergence and we did not proceed to any quasi-randomization of this random part - in this work globalization remains the work of random.

These results are naive results coming from an Octave implementation. A more optimized implementation, based on EO classes in C++, is in progress. First in dimension 2, for norm L_p with $p = 1$, $p = 3$, $p = 5$; "increasing QMC" denotes epochs at which $N \leftarrow N + 1$.

Figure 1 presents the histogram of the distribution of $\log(f_{L_5})$ after $500 \times (d/3)^2$ fitness-evaluations.

5   Discussions - Conclusions

We have designed a new algorithm using a representation (x, σ, η) instead of (x, σ). This algorithm takes into account different areas of applied mathematics: i) quasi-random points (low-dispersion points, [11]); ii) trust-regions ([7]); iii) adaptive step-size coming from evolution strategies [12, 17]; iv) random diversification of the population for global optimization. A very important remark is that as for classical ES, the algorithm considered here only use the information given by the fitness through the ranking of individuals. Therefore everything is invariant with respect to monotonic transformation of the fitness. In particular all the results holds for $x \mapsto g(f(x))$ where f satisfies the assumptions required for our Theorems and g is a strictly increasing function. This implies notably that convexity is not required for the convergence.

Compared to state-of-the art theoretical results for convergence of adaptive evolution strategies [1], our assumptions are here weaker. Indeed in [1] asymptotic linear convergence is proved for any $x \mapsto g(f(x))$ where g is monotonic and f is the sphere function. The main points here are i) use of (x, σ, η) instead of (x, σ) ; ii) generation of points on a close ball, instead of Gaussian sampling, so that this algorithm can ensure (under some conditions which are asymptotically satisfied with probability 1) that the fact that the optimum lies in $B(x, \sigma)$ is preserved from parents to children ; iii) use of quasi-random sequences ensuring that Δ goes to 0 as $N \rightarrow \infty$.

Experiments confirm the theoretical study but are very preliminary. In fact, we implemented the precise Algorithm, where each generation at Step 3c has to be generated independently with the same distribution at each epoch, whereas intuition suggests that better heuristics for new generations should dramatically reduce the time before reaching linear convergence; such implementations, and the corresponding proofs are yet to be done. Note that even in dimension 10, our very simple implementation, thanks to linear convergence, could reach the limit of the machine precision. These results are not at all results due to multiple attempts and empirical calibration of the parameters;

we simply implemented the algorithm in a naive manner, without any heuristic added; our results are the most immediate consequences of theory above.

Acknowledgments. The authors would like to thank Evelyne Lutton and Jacques Lévy Véhel for pointing out the Hölder property assumption made in this work.

References

[1] A. Auger. Convergence results for $(1,\lambda)$-SA-ES using the theory of φ-irreducible markov chains. *Theoretical Computer Science*, 334(1-3):35–69, 2005.

[2] H.-G. Beyer. *The Theory of Evolution Strategies*. Springer, Heidelberg, 2001.

[3] R. Cerf. An asymptotic theory of genetic algorithms. In J.-M. Alliot, E. Lutton, E. Ronald, M. Schoenauer, and D. Snyers, editors, *Artificial Evolution*, volume 1063 of *LNCS*, pages 37–53. Springer Verlag, 1996.

[4] S. Droste, T. Jansen, , and I. Wegener. On the analysis of the (1+1) evolutionary algorithm. *Theoretical Computer Science*, 276:51–81, 2002.

[5] K. Fang and Y. Wang. *Number-Theoretic Methods in Statistics*. London: Chapman and Hall, 1994.

[6] J. Garnier, L. Kallel, and M. Schoenauer. Rigorous hitting times for binary mutations. *Evolutionary Computation*, 7(2):167–203, 1999.

[7] S. Goldfeld, R. Quandt, and H. Trotter. Maximization by quadratic hill climbing. *Econometrica, vol. 34, no. 3, p. 541*, 1966.

[8] N. Hansen and A. Ostermeier. Completely Derandomized Self-Adaptation in Evolution Strategies. *Evolutionary Computation*, 9(2):159–195, 2001.

[9] Y. Landrin-Schweitzer and E. Lutton. Perturbation theory for eas: towards an estimation of convergence speed. In M. Schoenauer, K. Deb, G. Rudolph, X. Yao, E. Lutton, J. Merelo, and H.-P. Schwefel, editors, *PPSN VI*. Springer Verlag, 2000.

[10] Y. Meyer. *Wavelets, Vibrations and Scaling*. CRM Monograph Series. American Mathematical Society, 1997.

[11] H. Niedereiter. *Random Number Generation and Quasi-Monte Carlo Methods*. Philadelphia: SIAM, 1992.

[12] I. Rechenberg. *Evolutionstrategie: Optimierung Technisher Systeme nach Prinzipien des Biologischen Evolution*. Fromman-Hozlboog Verlag, Stuttgart, 1973.

[13] G. Rudolph. Convergence analysis of canonical genetic algorithm. *IEEE Transactions on Neural Networks*, 5(1):96–101, 1994.

[14] G. Rudolph. Convergence of non-elitist strategies. In Z. Michalewicz, J. D. Schaffer, H.-P. Schwefel, D. B. Fogel, and H. Kitano, editors, *Proceedings of the First IEEE International Conference on Evolutionary Computation*, pages 63–66. IEEE Press, 1994.

[15] G. Rudolph. How mutation and selection solve long path problems in polynomial expected time. *Evolutionary Computation*, 4(2):195–205, Summer 1996.

[16] G. Rudolph. Convergence rates of evolutionary algorithms for a class of convex objective functions. *Control and Cybernetics*, 26(3):375–390, 1997.

[17] H.-P. Schwefel. *Numerical Optimization of Computer Models*. John Wiley & Sons, New-York, 1981. 1995 – 2nd edition.

[18] C. Tricot. *Curves and Fractal Dimension*. Springer Verlag, January 1995. ISBN: 0387940952.

[19] J. L. Vehel and E. Lutton. Holder functions and deception of genetic algorithms. *IEEE transactions on Evolutionary computing*, 2(2), 1998.

[20] S. Yakowitz, P. L'Ecuyer, and F. Vazquez-Abad. Global stochastic optimization with low-dispersion point sets, 2000.

Author Index

Lecture Notes in Computer Science

For information about Vols. 1–3841

please contact your bookseller or Springer

Vol. 3891: J.S. Sichman, L. Antunes (Eds.), Multi-Agent-Based Simulation VI. X, 191 pages. 2006. (Sublibrary LNAI).

Vol. 3890: S.G. Thompson, R. Ghanea-Hercock (Eds.), Defence Applications of Multi-Agent Systems. XII, 141 pages. 2006. (Sublibrary LNAI).

Vol. 3889: J. Rosca, D. Erdogmus, J.C. Príncipe, S. Haykin (Eds.), Independent Component Analysis and Blind Signal Separation. XXI, 980 pages. 2006.

Vol. 3888: D. Draheim, G. Weber (Eds.), Trends in Enterprise Application Architecture. IX, 145 pages. 2006.

Vol. 3887: J.R. Correa, A. Hevia, M. Kiwi (Eds.), LATIN 2006: Theoretical Informatics. XVI, 814 pages. 2006.

Vol. 3886: E.G. Bremer, J. Hakenberg, E.-H.(S.) Han, D. Berrar, W. Dubitzky (Eds.), Knowledge Discovery in Life Science Literature. XIV, 147 pages. 2006. (Sublibrary LNBI).

Vol. 3885: V. Torra, Y. Narukawa, A. Valls, J. Domingo-Ferrer (Eds.), Modeling Decisions for Artificial Intelligence. XII, 374 pages. 2006. (Sublibrary LNAI).

Vol. 3884: B. Durand, W. Thomas (Eds.), STACS 2006. XIV, 714 pages. 2006.

Vol. 3882: M.L. Lee, K.L. Tan, V. Wuwongse (Eds.), Database Systems for Advanced Applications. XIX, 923 pages. 2006.

Vol. 3881: S. Gibet, N. Courty, J.-F. Kamp (Eds.), Gesture in Human-Computer Interaction and Simulation. XIII, 344 pages. 2006. (Sublibrary LNAI).

Vol. 3880: A. Rashid, M. Aksit (Eds.), Transactions on Aspect-Oriented Software Development I. IX, 335 pages. 2006.

Vol. 3879: T. Erlebach, G. Persinao (Eds.), Approximation and Online Algorithms. X, 349 pages. 2006.

Vol. 3878: A. Gelbukh (Ed.), Computational Linguistics and Intelligent Text Processing. XVII, 589 pages. 2006.

Vol. 3877: M. Detyniecki, J.M. Jose, A. Nürnberger, C. J. '. van Rijsbergen (Eds.), Adaptive Multimedia Retrieval: User, Context, and Feedback. XI, 279 pages. 2006.

Vol. 3876: S. Halevi, T. Rabin (Eds.), Theory of Cryptography. XI, 617 pages. 2006.

Vol. 3875: S. Ur, E. Bin, Y. Wolfsthal (Eds.), Hardware and Software, Verification and Testing. X, 265 pages. 2006.

Vol. 3874: R. Missaoui, J. Schmidt (Eds.), Formal Concept Analysis. X, 309 pages. 2006. (Sublibrary LNAI).

Vol. 3873: L. Maicher, J. Park (Eds.), Charting the Topic Maps Research and Applications Landscape. VIII, 281 pages. 2006. (Sublibrary LNAI).

Vol. 3872: H. Bunke, A. L. Spitz (Eds.), Document Analysis Systems VII. XIII, 630 pages. 2006.

Vol. 3871: E.-G. Talbi, P. Liardet, P. Collet, E. Lutton, M. Schoenauer (Eds.), Artificial Evolution. XI, 310 pages. 2006.

Vol. 3870: S. Spaccapietra, P. Atzeni, W.W. Chu, T. Catarci, K.P. Sycara (Eds.), Journal on Data Semantics V. XIII, 237 pages. 2006.

Vol. 3869: S. Renals, S. Bengio (Eds.), Machine Learning for Multimodal Interaction. XIII, 490 pages. 2006.

Vol. 3868: K. Römer, H. Karl, F. Mattern (Eds.), Wireless Sensor Networks. XI, 342 pages. 2006.

Vol. 3866: T. Dimitrakos, F. Martinelli, P.Y.A. Ryan, S. Schneider (Eds.), Formal Aspects in Security and Trust. X, 259 pages. 2006.

Vol. 3865: W. Shen, K.-M. Chao, Z. Lin, J.-P.A. Barthès, A. James (Eds.), Computer Supported Cooperative Work in Design II. XII, 659 pages. 2006.

Vol. 3863: M. Kohlhase (Ed.), Mathematical Knowledge Management. XI, 405 pages. 2006. (Sublibrary LNAI).

Vol. 3862: R.H. Bordini, M. Dastani, J. Dix, A.E.F. Seghrouchni (Eds.), Programming Multi-Agent Systems. XIV, 267 pages. 2006. (Sublibrary LNAI).

Vol. 3861: J. Dix, S.J. Hegner (Eds.), Foundations of Information and Knowledge Systems. X, 331 pages. 2006.

Vol. 3860: D. Pointcheval (Ed.), Topics in Cryptology – CT-RSA 2006. XI, 365 pages. 2006.

Vol. 3858: A. Valdes, D. Zamboni (Eds.), Recent Advances in Intrusion Detection. X, 351 pages. 2006.

Vol. 3857: M.P.C. Fossorier, H. Imai, S. Lin, A. Poli (Eds.), Applied Algebra, Algebraic Algorithms and Error-Correcting Codes. XI, 350 pages. 2006.

Vol. 3855: E. A. Emerson, K.S. Namjoshi (Eds.), Verification, Model Checking, and Abstract Interpretation. XI, 443 pages. 2005.

Vol. 3854: I. Stavrakakis, M. Smirnov (Eds.), Autonomic Communication. XIII, 303 pages. 2006.

Vol. 3853: A.J. Ijspeert, T. Masuzawa, S. Kusumoto (Eds.), Biologically Inspired Approaches to Advanced Information Technology. XIV, 388 pages. 2006.

Vol. 3852: P.J. Narayanan, S.K. Nayar, H.-Y. Shum (Eds.), Computer Vision – ACCV 2006, Part II. XXXI, 977 pages. 2006.

Vol. 3851: P.J. Narayanan, S.K. Nayar, H.-Y. Shum (Eds.), Computer Vision – ACCV 2006, Part I. XXXI, 973 pages. 2006.

Vol. 3850: R. Freund, G. Păun, G. Rozenberg, A. Salomaa (Eds.), Membrane Computing. IX, 371 pages. 2006.

Vol. 3849: I. Bloch, A. Petrosino, A.G.B. Tettamanzi (Eds.), Fuzzy Logic and Applications. XIV, 438 pages. 2006. (Sublibrary LNAI).

Vol. 3848: J.-F. Boulicaut, L. De Raedt, H. Mannila (Eds.), Constraint-Based Mining and Inductive Databases. X, 401 pages. 2006. (Sublibrary LNAI).

Vol. 3847: K.P. Jantke, A. Lunzer, N. Spyratos, Y. Tanaka (Eds.), Federation over the Web. X, 215 pages. 2006. (Sublibrary LNAI).

Vol. 3846: H. J. van den Herik, Y. Björnsson, N.S. Netanyahu (Eds.), Computers and Games. XIV, 333 pages. 2006.

Vol. 3845: J. Farré, I. Litovsky, S. Schmitz (Eds.), Implementation and Application of Automata. XIII, 360 pages. 2006.

Vol. 3844: J.-M. Bruel (Ed.), Satellite Events at the MoDELS 2005 Conference. XIII, 360 pages. 2006.

Vol. 3843: P. Healy, N.S. Nikolov (Eds.), Graph Drawing. XVII, 536 pages. 2006.

Vol. 3842: H.T. Shen, J. Li, M. Li, J. Ni, W. Wang (Eds.), Advanced Web and Network Technologies, and Applications. XXVII, 1057 pages. 2006.